THE CIT

RAILS THROUGH THE CLAY

HISTORY OF THE CITY

RAILS THROUGH THE CLAY

A History of London's Tube Railways

ALAN A. JACKSON & DESMOND F. CROOME

LONDON AND NEW YORK

First published in 1962

This edition published in 2007
Routledge
2 Park Square, Milton Park, Abingdon, Oxon, OX14 4RN
Simultaneously published in the USA and Canada by Routledge
711 Third Avenue, New York, NY 10017

Routledge is an imprint of Taylor & Francis Group, an informa business

Transferred to Digital Printing 2007

First issued in paperback 2013

© 1962 Routledge

All rights reserved. No part of this book may be reprinted or reproduced or utilized in any form or by any electronic, mechanical, or other means, now known or hereafter invented, including photocopying and recording, or in any information storage or retrieval system, without permission in writing from the publishers.

The publishers have made every effort to contact authors and copyright holders of the works reprinted in the *The City* series. This has not been possible in every case, however, and we would welcome correspondence from those individuals or organisations we have been unable to trace.

These reprints are taken from original copies of each book. In many cases the condition of these originals is not perfect. The publisher has gone to great lengths to ensure the quality of these reprints, but wishes to point out that certain characteristics of the original copies will, of necessity, be apparent in reprints thereof.

British Library Cataloguing in Publication Data
A CIP catalogue record for this book
is available from the British Library

Rails Through The Clay

ISBN13: 978-0-415-41817-1 (volume hbk)
ISBN13: 978-0-415-41933-8 (subset)
ISBN13: 978-0-415-41318-3 (set)
ISBN13: 978-0-415-86046-8 (volume pbk)

Routledge Library Editions: The City

RAILS THROUGH THE CLAY

A HISTORY OF
LONDON'S TUBE RAILWAYS

BY
ALAN A. JACKSON
AND
DESMOND F. CROOME
A.M. Inst.T.

London
GEORGE ALLEN & UNWIN LTD
RUSKIN HOUSE MUSEUM STREET

FIRST PUBLISHED IN 1962
Second impression 1964

This book is copyright under the Berne Convention. Apart from any fair dealing for the purposes of private study, research, criticism or review, as permitted under the Copyright Act 1956, no portion may be reproduced by any process without written permission. Enquiries should be addressed to the publisher

© *George Allen & Unwin Ltd, 1962*

PRINTED IN GREAT BRITAIN
in 10 point Plantin type
BY HAZELL WATSON AND VINEY LTD
AYLESBURY, BUCKS

PREFACE

□

THE London Tubes are something different in the realm of railways. Distinguished by a special form of construction, their tunnels, driven through the unobstructed clay, are far deeper than the cut-and-cover subways of their sister lines, the Metropolitan and the District, which venture only as far as basement level, among the mains and drains. This great depth, and the small loading gauge common to all the tubes but one, gives them a special *ambiance*. Far below the streets, they are in a world of their own, self-contained and self-confident. Even when they emerge into the suburban daylight they still retain a distinctive air: the neat, clean stations, tidy car sheds and shrub-adorned banks and cuttings merge pleasantly with the poplar-fringed sports grounds and semi-detached villas of Middlesex and are in character with the orderly minds of the office-workers who make up much of their traffic. This atmosphere of neatness, good planning, quiet precision and confidence is no accident and owes much to the care and competence of two men who were in control for several decades; it was the wisdom and skill of Ashfield and Pick which shaped the original haphazard nucleus to the intricate, interlocked system of today.

In the brief rush hours, when the cars show the extreme possibilities of compressing the human frame this side of asphyxiation, the Londoner regards the tube as a punishment to be endured because he knows it to be the quickest and most reliable means of moving through the car-choked city. Then, as at other times when he travels more comfortably, he takes it as much for granted as the constant supply of pure water to his house. For over sixty years, the tube railways have carried London's millions to work and play—millions of sad and happy journeys in peace and war, of passengers experiencing hate and love, boredom and affection, fear and joyful anticipation. In two world wars, their tunnels became home for many, providing safe refuge from German bombs and rockets, whilst the trains carried thousands of servicemen to and from battle. For all his grumbling, the true Londoner is proud of this unique deep-level railway system that is always at his service, and loves to show it off to strangers as one of the wonders of his great city.

We thought the tube railways deserved a book of their own—there was certainly enough material—and we have recorded their history in detail not hitherto attempted, turning to original sources as much as possible because, so often in the past, books about the London Underground seem merely to have dug over the same old familiar stories and

PREFACE

hand-outs[1]. To make a reference work such as this readable is a sticky task; the perfectionists who demand comprehensive and accurate recording, informed criticism *and* scintillating prose have not been forgotten, and we can only hope that we have succeeded at least some of the time. At any rate, we can perhaps claim that this is something more than a railway history, for, by their very nature, rapid transit railways are an intimate part of the city they serve, closely woven into its life and modern development; in this respect the tubes are no exception, and there is much of our native London in this book.

We gratefully acknowledge willing assistance given by many friends and authorities. Special mention must be made of the following: B. John Prigmore, M.A., M.Sc., D.I.C., A.M.I.E.E., who refined technical references, assisted with rolling-stock information and gave welcome encouragement during the long period in which this work was being prepared; H. V. Borley, who checked many dates and provided others; Wingate H. Bett, F.C.I.I., who wrote much of the Appendix on Fares and Tickets; George Cross, who graciously allowed quotation from his autobiography, *Suffolk Punch*; D. D. F. Adams, who contributed anecdotes and reminiscences of the Edgware line; G. T. Moody, who provided reminiscences; E. Harper Charlton, who undertook research in the USA and obtained photographs; and Edwin Course, Ph.D., and Ralph Taylor, who provided the initial stimulation and encouragement.

The librarians and officials of the following bodies and institutions were unsparing in their efforts to assist our research:

London Transport; Patent Office, London; British Museum; British Transport Commission Historical Record Department; Institute of Transport; Middlesex County Record Office; American Library in London; John Crerar Library, Chicago; Electric Railroaders' Association Inc., New York; Royal Institute of British Architects; Imperial War Museum; Guildhall Library, City of London; Science Museum, London; Feltham Urban District Council; A.C.F. Industries Inc., New York; and many London Public Libraries, especially Acton, Hampstead, Hendon, Westminster and Wood Green.

Among the many publications consulted, we would particularly acknowledge the usefulness of contemporary references in *The Times*, *The Tramway & Railway World* (now *Transport World*), *The Railway News*, *The Railway Times* (both now defunct), *The Railway Gazette*, and *Modern Transport*.

[1] From what is said here we must except two detailed studies of individual lines—*The Central London Railway* (B. G. Wilson and V. Stuart Haram, The Fairseat Press, 1950) and *The City & South London Railway* (T. S. Lascelles, The Oakwood Press, 1955). They are especially valuable for the preliminaries and early histories of these two lines.

PREFACE

Although assistance with information was always willingly given when requested, we must emphasize that London Transport is not responsible in any way for the publication of this book, nor for any fact or opinion stated in it.

ALAN A. JACKSON
DESMOND F. CROOME

London,
    August, 1962

CONTENTS

□

PREFACE   *Page* 7

Hors d'Oeuvre   21

1. EARLY SCHEMES: 1863–1890   25
   Transit in the shallows, 26. Solution in depth, 27. Moving without smoke, 28

2. THE BEGINNINGS OF THE BIG THREE: 1891–1900   33
   City & South London success, 33. Promotions in earnest, 34. Hampstead progress, 36. Baker Street & Waterloo in shady company, 36. District Deep Level, 39. Brompton & Piccadilly Circus, 41. Great Northern & Strand, 42. Difficulties of attracting capital, 44.

3. PIONEER LINES   46
   City & South London, 46. Waterloo & City, 52. Central London, 55.

4. TITAN TURNS EAST: 1900–1901   62
   The Traction King, 62. Dollars for the Hampstead, 64. Ruining the Heath, 66. Bills galore, 67.

5. YERKES TAKES OVER: 1902–1905   70
   Transatlantic pressures, 70. The District in the net, 70. Three more tubes for Yerkes, 71. Financial consolidation, 73. Tube boom, 74. The Ribblesdale Committee–the Yerkes bills, 75. The Ribblesdale Committee–other bills, 76. The Windsor Committee, 77. Final stages, 82. 'A very pretty and clever piece of manoeuvring', 83. Nothing more for north-east London, 85. Pause for enquiry, 85. The ridiculous mouse, 87. Fusion frustrated, 87. Bills through the sieve, 88. Empty pockets, 90. Fires down below, 92. Multiple-unit traction, 93. The Yerkes network takes shape, 94. Construction methods, 96. The odd one out (Great Northern & City), 98.

CONTENTS

6. CONSUMMATION AND DISILLUSION: 1906–1909 — 102

Experiments at Ealing, 102. The Chelsea Monster, 102. Death of the Titan, 105. New men, 105. Cars for the Three, 106. Opening the Three, 107. The Yerkes tubes described, 111. American men and methods, 123. Foreign equipment, 123. A bad press, 124. Vain hopes, 124. Deep water, 125. Stanley arrives, 126. Financial climacteric, 127. Fares co-ordinated, 129. Publicity co-ordinated, 131. Seasons and strips, 133. Improving the services, 134. Sally to the north west, 138.

7. INTEGRATION: 1910–1914 — 140

The London Electric Railway, 140. Charing Cross loop, 141. Oxford Circus, 143. To Liverpool Street and Paddington, 143. Out into the open, 146. The Lots Road deal, 148. The General joins the family, 148. The Metropolitan buys a loser, 150. More additions to the family, 151.

8. THE TUBES IN WARTIME (I): 1914–1918 — 155

Taking cover, 155. Speyer and Stanley attacked, 156. The Common Fund, 157. Extension and Improvement, 158. Matters of power, 162. Overcrowding, 163.

9. GROWING AND GRAFTING: 1919–1926 — 165

A fresh start, 165. Central to Ealing, 167. Edgware's tube, 169. A new life for the veteran, 172. A Surrey struggle, 178. Charing Cross to Morden, 179. New rolling stock, 186. Acton Works, 189. Boosting the power, 190. Reshaping the heart valves, 190. Fares and finances, 192.

10. PICCADILLY PROGRESS: 1919–1933 — 195

Blowing out the cork, 195. LER *versus* LNER, 199. On to Cockfosters, 201. Tracks and tunnels, 203. A new architecture, 204. New escalators, 206. New signalling, 207. New substations, 207. Trawling for traffic, 208. Piccadilly pushes west, 209. A new Hub for the Empire, 217. Central area improvements, 219. New and improved rolling stock, 225. Fighting ice and snow, 231. Service developments, 232. The Monster modernized, 233. Financial reconstruction, 234. New offices, 236. Schemes unfulfilled, 237. Metropolitan to Edgware, 239. Towards full co-ordination, 240.

11. LONDON TRANSPORT: 1933–1939   243

Off with the old, 243. Full co-ordination, 244. A New Works Programme, 245. Troubles on the Morden–Edgware, 248. The Battle of Bushey Heath, 250. Much ado at Morden, 251. And so there were nine, 253. North from Highgate, 254. High Barnet line described, 257. Northern City comes into line, 259. Wasted work on the Northern Heights, 260. A fare row, 263. Bakerloo to Stanmore, 264. Neasden depot, 268. Bakerloo improvements, 269. Uxbridge branch renewed, 270. Central Line modernization, 271. Central Line extensions begun, 273. Improvements in the centre, 275. Better ventilation, 278. The campaign against noise, 278. 1936 stock, 280. 1938 stock, 282. Modernizing the Drain, 286. 'C' for 'Cold Comfort', 291.

12. THE TUBES IN WARTIME (II): 1939–1945   294

Precautions, 294. The best shelters of all, 298. Government deep shelters, 302. Tube tragedies, 303. Wartime changes, 306. Wartime works, 308. Planning for peace, 309.

13. LONDON TRANSPORT: SINCE 1945   313

Return to normal, 313. Changes at the top, 314. Promises fulfilled: The Central Line extensions, 316. Promises unfulfilled: Anti-climax in north London, 327. Blow hot, blow cold for Camberwell, 329. The Victoria line, 330. Retrenchment—and ructions, 332. Station and layout improvements, 333. Signalling improvements, 336. Drico, 337. Winter precautions, 338. Power stations, 339. Rolling stock since 1945, 340. The increasing importance of tube railways, 346.

CONTENTS

14. THE TUBES AND LONDON   348
   The shape of the system, 348. Changing London, 350. Lucrative fields, or perpetual expansion, 351. Changing the traffic flows, 355. Barlow and after, 356. Congestion in the centre, 357. Design and order, 358. The Londoner's tube, 360.

APPENDIX 1. DATES OF OPENING TO PUBLIC TRAFFIC   363

APPENDIX 2. ACCIDENTS AND INTERRUPTIONS OF SERVICE   365

APPENDIX 3. SOME NOTES ON TUBE TICKETS AND TICKET-ISSUING MACHINES   373

APPENDIX 4A. TRAFFIC RESULTS COMPARED WITH ESTIMATES   385

APPENDIX 4B. RESULTS COMPARED WITH ESTIMATES   387

APPENDIX 5. NORTH END STATION AND HAMPSTEAD GARDEN SUBURB   388

INDEX   391

ILLUSTRATIONS

□

1. Waterloo & City Railway train
   Central London Railway train *facing page* 64
2. Central London Railway station architecture
   Great Northern & City Railway car
   Tube railway construction 65
3. Charles Tyson Yerkes 96
4. The Chelsea Monster
   Baker Street & Waterloo Railway car
   'Yerkes' tube station architecture 97
5. Piccadilly Station, 1906
   A blow at the cab trade
   West Kensington tunnel mouths 128
6. Lord Ashfield
   Sir George Gibb;
   'The Underworld' 129
7. Bakerloo tunnel mouths
   Big Ben and Little Len 160
8. Ealing & Shepherd's Bush Railway at East Acton
   Early days at Edgware 161
9. Edgware, rail level
   Hendon Central 192
10. Enterprise at Morden 193
11. Confusion at Finsbury Park
    Order at Manor House 224
12. Sentinel steam wagons at Turnpike Lane
    Sudbury Town station 225
13. Passing at Brent 288
14. A streamlined tube train
    Transition stage at Finchley Central
    Waterloo & City Railway, 1940 train at Bank 289
15. Tube trains on the Metropolitan (Stanmore)
    Steam and electric at Epping 320

ILLUSTRATIONS

16. Central Line in the west (North Acton)
Three types of stock at Ruislip depot   *facing page* 321

In the text

Map of all the London tube railways	*page* 24
City & South London Railway train at Stockwell, 1890	31
Diagram of track layout at Holborn and Strand	109
Diagram of Euston and the Camden Town junctions	173
Map of tube railways between Charing Cross and Kennington	180
Diagram of Kennington junctions and loop	182
Map showing the four uncompleted parts of the 1935–1940 Plan in North London	261

SPECIAL ABBREVIATIONS

□

AEC—Associated Equipment Company
AEG—Allgemeine Elektricitäts-Gesellschaft
ARP—Air Raid Precautions
Bakerloo—Baker Street & Waterloo Railway (after 1910, that section of the LER)
BC&W—Birmingham Railway Carriage & Wagon Company
BET—British Electric Traction Company
Birmingham—Birmingham Railway Carriage & Wagon Company
BPC—Brompton & Piccadilly Circus Railway
BR—British Railways
BTC—British Transport Commission
BTH—British Thomson-Houston Company

cfm—cubic feet per minute
CL—Central Line
CLR—Central London Railway
CLSS—City of London & Southwark Subway
CNESER—City & North East Suburban Electric Railway
CSLR—City & South London Railway
CT—control trailer car
CTC—Centralized Train Control
CWE—City & West End Railway

DDL—District Deep Level Railway
District—Metropolitan District Railway (after 1933, the District Line of London Transport)
DM—driving motor car

E&H—Edgware & Hampstead Railway
ep—electro-pneumatic
E&SB—Ealing & Shepherd's Bush Railway

GCR—Great Central Railway
GEC—General Electric Company (of England)
General—London General Omnibus Company
GER—Great Eastern Railway
Gloucester—Gloucester Railway Carriage & Wagon Company
GNCR—Great Northern & City Railway

SPECIAL ABBREVIATIONS

GNR—Great Northern Railway
GNS—Great Northern & Strand Railway
GWR—Great Western Railway

Hampstead—Charing Cross, Euston & Hampstead Railway (after 1910, that section of the LER)

L&ST—London & Suburban Traction Company
LBSCR—London Brighton & South Coast Railway
LCC—London County Council
LCDR—London, Chatham & Dover Railway
LER—London Electric Railway
LGOC—London General Omnibus Company
LMSR—London Midland & Scottish Railway
LNER—London & North Eastern Railway
LNWR—London & North Western Railway
LPTB—London Passenger Transport Board
LSWR—London & South Western Railway
LT—London Transport (LPTB 1933–1947, LTE since 1948)
LTE—London Transport Executive
LUER—London United Electric Railway
LUT—London United Tramways

M—motor car
MC&W—Metropolitan Carriage Wagon & Finance Company
MDR—Metropolitan District Railway
Met.—Metropolitan Railway
MET—Metropolitan Electric Tramways

NDM—non-driving motor car
NELR—North East London Railway
NLR—North London Railway
Northmet—North Metropolitan Electric Power Supply Company
NWLR—North West London Railway

P&C—Piccadilly & City Railway
PCM—pneumatic camshaft mechanism
PCNELR—Piccadilly, City & North East London Railway
Piccadilly—Great Northern, Piccadilly & Brompton Railway (after 1910, that section of the LER, after 1933 the Piccadilly line of LT)

RA—Royal Assent
REC—Railway Executive Committee

SPECIAL ABBREVIATIONS

SECR—South Eastern & Chatham Railway
SER—South Eastern Railway
SMET—South Metropolitan Electric Tramways & Lighting Company
SR—Southern Railway

T—trailer car
TOT—Train–Omnibus–Tram
tph—trains per hour

UCC—Union Construction & Finance Company (Union Construction Co. till 1929)
UERL—Underground Electric Railways Company of London
UNDM—uncoupling non-driving motor car

W&C—Waterloo & City Railway
W&E—Watford & Edgware Railway
W&S—Wimbledon & Sutton Railway

HORS D'OEUVRE

□

Divine Intervention

'It seems that the beautiful homogenous clay of London has been designed by Nature for the very purpose of having tunnels pierced through it, and it would be a great pity to balk Nature in her design.'
—GRANVILLE CUNINGHAM, General Manager of the Central London Railway, February 25, 1908.

Penny Bizarre

'An advertising company has obtained a concession to operate automatic blinds with advertisements on City & South London Railway car windows. The blinds disappear as the train enters a station.'—*Railway Times*, July 6, 1912.

'Of course, it would not do to encourage this sort of thing, as I am a busy man.'—LORD ASHFIELD after agreeing to be godfather to a baby born on a Bakerloo train on the evening of May 13, 1924. (The child was christened *Thelma Ursula Beatrice Eleanor*.)

'Out of 200 advertisements running alongside the Piccadilly escalators, 48 show women in underwear.'—*Daily Mail*, November 26, 1957, reporting on a *Panorama* television programme.

'Divisional heads of London Transport called today for reports on the "ghost" at Covent Garden Underground station. A West Indian porter at the station had asked for a transfer after claiming that he had seen an apparition of a 6 ft. tall, slim, oval-faced man wearing a light grey suit and white gloves. Other staff at the station have told senior officials of London Transport that they have seen the same apparition.'—*The Star*, November 29, 1955.

Strange and Wonderful Finance

'It would appear to be necessary for someone to come forward and guarantee the guarantor.'—*Railway Times*, January 24, 1903, on a Piccadilly Railway share issue, with 4 per cent interest guaranteed by the Underground Company.

CHAIRMAN: 'We only wanted to understand this strange and wonderful finance of yours.'
HENRY A. VERNET: 'It is really very simple, Sir. I think you will agree now that it has been explained, it is very simple.'—*Minutes of Evidence*, Select Committee on Transport, 1919.

'We may claim to have already suffered twice the term of lean years in Pharaoh's dream. May we now enjoy twice the term of fat years.'
—LORD ASHFIELD, on the Underground Group results for 1928.

Picturesque Tubes

'There was a congestion of traffic in Oxford Street; to avoid the congestion people actually began to travel under the cellars and drains, and the result was a rise of rents in Shepherd's Bush! And you say that isn't picturesque!'—ARNOLD BENNETT, *How to Live on 24 Hours a Day*.

The Inevitability of Straphanging

'... Overcrowding, in the sense of people standing in the cars, must be regarded as a permanent feature of the rush periods of urban traffic operation.'—LORD ASHFIELD, September 1915.

'They're not crammed in. They cram themselves in.'—SIR JOHN ELLIOT, Institute of Transport, December 5, 1955.

Social Distinction

'The *intelligentsia* of Hampstead has the chagrin of witnessing the *canaille* of Highgate lolling at their ease while we have to hang on our straps, usually changing our stance and wishing that we were on a different footing. Is it fair?'—Letter to *The Times*, February 28, 1931.

Not Easy Money

'To put up a fare calls for much courage and obstinacy.'—LORD ASHFIELD, 1915.

'Boring holes in London clay is a species of philanthropy which has seen its best days.'—*Railway Engineer*, 1907.

'It may be a great surprise to you to know that the Underground railways in London have never been, in their whole career, a financial success. In other words, they have failed to earn anything approaching a reasonable return upon the capital invested in them....'—LORD ASHFIELD, 1924.

Fog in the Tubes

'Whilst the thick London fog of January 4th did not upset the Underground as much as other railways, it is worth remarking that late in the evening, visibility in the tube was about 100 yards, one end of a tube station being barely visible from the other.'—*Electric Railway Society Journal*, March–April 1956.

DATES AND MONEY

□

Dates given in the text for the closing of stations or withdrawal of service are the dates of the last day on which service was provided.

The decline in the purchasing power of the £ sterling should not be overlooked when considering money amounts quoted. The 1919 £ was worth only 9s 4d compared with 1914; making the same comparison, the 1923 figure is 11s 6d, 1930, 12s 8d, 1938, 12s 10d, 1948, 6s 7d, 1960, 4s 6d.

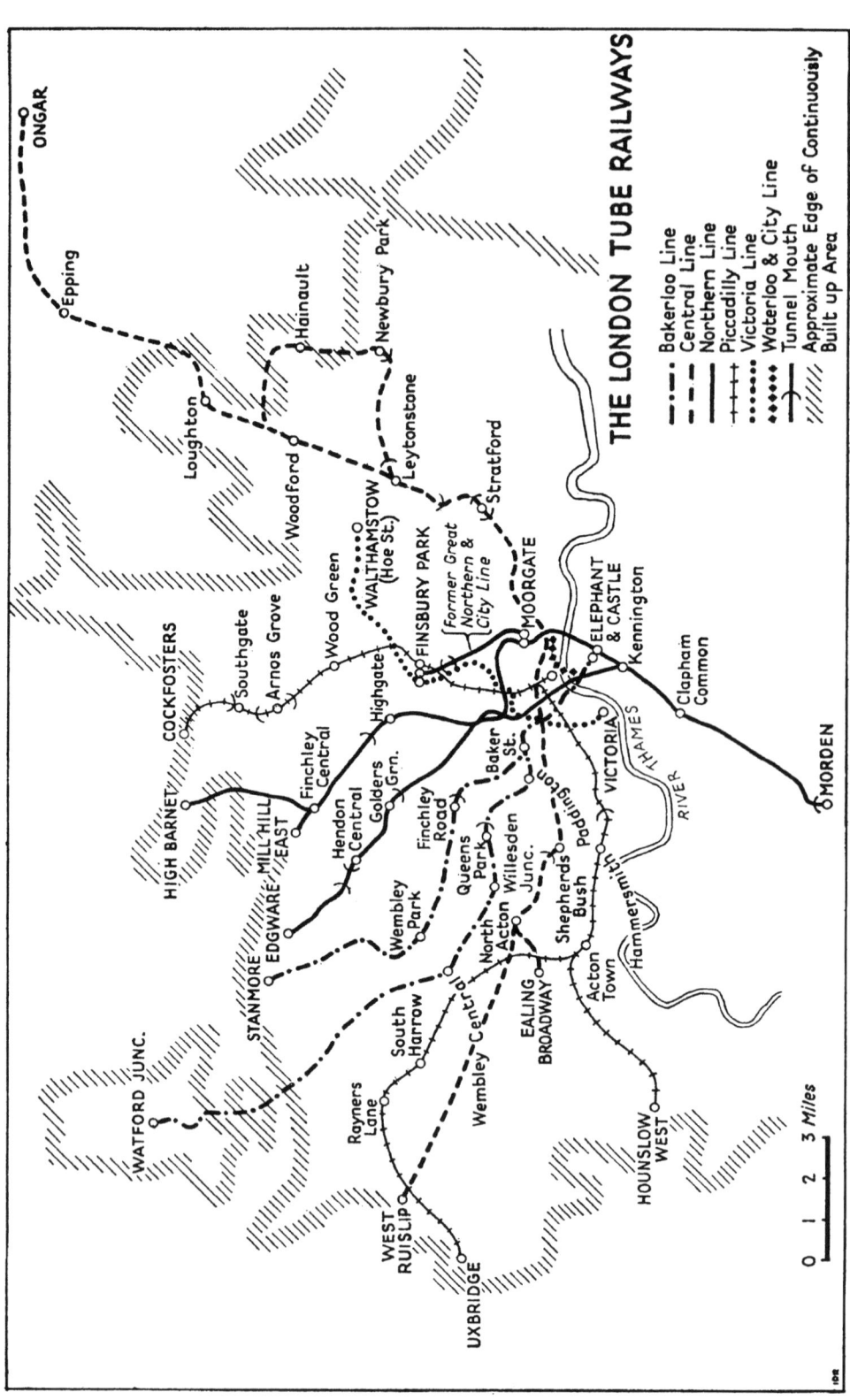

CHAPTER 1

Early Schemes

1863–1890

RAILWAYS are a very enduring feature of the landscape. Meadows may disappear beneath bricks and mortar, cities may be razed, re-planned and rebuilt, highways widened, straightened, diverted, but the railways remain, largely unchanged, to remind us of the hopes, the philosophy, the precepts, of the men of a past age who promoted them, sanctioned them, and built them.

In today's network of London local lines we can see the effects of Victorian economic doctrines, and of other, more permanent English traits—*laissez-faire* economics, with their reliance on the profit motive to ensure the promotion and survival of enterprises best suited to serve the nation's needs, and with small regard for state or civic control; the sanctity of private property; the enthusiastic development of engineering technology, accompanied by the penalties of being pioneers; the tendency to put off the solving of problems until the last possible moment; and the dislike of making present sacrifices for future benefit—all these have had their effect on the London railway scene.

London's first railway problems concerned the provision and location of terminal stations for the main lines. The 1846 Royal Commission on Metropolitan Railway Termini dismissed the idea of a central station for London, and recommended that no railway be built from north of the Thames within the limits of a line following the present Pentonville Road, City Road, Finsbury Square, Bishopsgate, London Bridge, Borough High Street and Road, Lambeth Road, Vauxhall Bridge (and Bridge Road), Grosvenor Place, Park Lane, Edgware Road, then via the present Circle Line alignment back to King's Cross. If in future it should be deemed advisable to admit railways within these limits, this should be done in conformity with a uniform plan and not piecemeal.

TRANSIT IN THE SHALLOWS

With the development of main-line railways and their individual London termini, passenger traffic across the central area grew until it strained the capacity of horse-drawn road transport and the road network. Road travel was slow and congestion rife. Charles Pearson, the City Solicitor, had propounded the idea of underground railways to provide urban rapid transit even before the 1846 Commission sat; his enthusiasm and perseverance overcame the forbidding parliamentary, financial and constructional difficulties and culminated in the public opening on January 10, 1863 of the Metropolitan Railway. This ran between Paddington and Farringdon Street, along the northern border of the 1846 zone. Cut-and-cover, the only feasible method of construction, involved widespread interference with roads and substantial demolition of property.

The opening of the Metropolitan resulted in a spate of schemes for new London railways in the 1864 session, and a Joint Select Committee of both Houses was appointed, with the task of selecting the most worthy proposals. As a result, the lines completing the Inner Circle (recommended by an 1863 Select Committee) were approved by Parliament in July of that year, being authorized partly to the Metropolitan Railway and partly to the Metropolitan District Railway.

Today, it is easy enough to see the demerits of the Circle, with its indirect journeys from north to south and its operating difficulties, but it was a reasonable solution in the circumstances of the time. It appeared adequate for existing and foreseeable demands for rail transport within the central area, and, accepting that the heart of London should be spared the upheavals of cut-and-cover construction, it was in fact an ingenious solution of the problem of serving the main-line termini with minimum construction of new line.

As experience was gained, sub-surface technique improved and the degree of disturbance was lessened, but costs were very heavy. With the growth of London, site values pushed up the compensation payable for land and buildings, whilst the constant and multifarious accretion of sewers, pipes and cables beneath the roads involved mounting expense in diverting these services. The completion of the Inner Circle in 1884 proved very expensive, whilst the ultimate London example of shallow railway construction, the 1902 Whitechapel and Bow Railway, cost roundly £600,000 per mile to build, involving *inter alia*, the reconstruction of half a mile of 5-foot sewer and the diversion of a mile and a half of 36-inch gas main.

Thus the prospect of new cut-and-cover lines for central London had by the middle 1880s become financially and politically impossible, and the further extension of rail rapid transit in London had to await the successful development of tunnelling and traction techniques which would allow railways to be built and operated without disturbing the surface.[1] Fortunately for London, the underlying stratum of blue clay was the ideal working medium for a method of tunnelling which offered an escape from the impasse—tube tunnelling with shields.

SOLUTION IN DEPTH

In 1818 Marc Isambard Brunel took out a patent for a tunnelling method embodying the use of a cellular shield to hold back the material being excavated, in conjunction with a cast-iron tunnel lining (bricks or masonry provided a secondary lining, inside the iron tunnel). Drawings accompanying the patent showed cylindrical tunnels, but when Brunel constructed the Thames Tunnel, from 1825 to 1843, he used a rectangular shield and a brick lining.

The next important step in the development of shield tunnelling was taken by Peter William Barlow (1809–1885), a railway engineer who later undertook the construction of suspension bridges. In 1862, whilst sinking cast-iron cylinders through the blue London clay to form the mid-stream piers of the old Lambeth suspension bridge (demolished 1929), he perceived that such cylinders could also be driven horizontally to form tunnels under rivers; two years later, he patented a circular wrought-iron or steel shield. The accompanying illustration showed iron segments and provision for grouting.[2] Both these features are essential parts of the standard shield tunnelling process. The lining material must be strong enough to take the thrust of the shield-propelling rams immediately it is erected. Bricks and mortar are unsuitable, but pre-cast concrete is permissible and cast iron ideal. Without grouting, the upper soil would settle into the vacant space left by the shield skin, with unhappy effects on existing buildings.[3]

In 1868 parliamentary powers were obtained for the Tower Subway between Great Tower Hill and Vine Street on the Surrey side. Owing to the disastrous experiences of flooding during the construction of the Thames Tunnel, contractors were reluctant to undertake further tunnels beneath the river, but Barlow's former pupil, James Henry Greathead, tendered to build the subway and lift shafts for £9,400.

Greathead was born in Grahamstown, South Africa, in August 1844,

[1] Other countries have, for a variety of reasons, been less averse to subsurface construction. In many cases the streets have been wider, the networks of service pipes less dense, and the subsoil more suited to this type of tunnelling. Furthermore, the climate of opinion has been more favourable towards underground railways, so that there have been active municipal support, less onerous compensation provisions, and a citizenry more prepared to accept temporary disturbance for ultimate benefit.

[2] The process of injecting liquid lime or cement through holes in the segments to fill the space left by the skin of the shield.

[3] In shield-tunnelling techniques recently developed, the segments are forced outwards before the ring is completed, thus obviating the need for grouting.

and came to England in 1859 to complete his education. In 1864 he began a three-year pupilage to Barlow, followed by one year as assistant engineer on the last stages of the Midland Railway extension to London.

Construction of the Tower Subway began in 1869, with a circular shield propelled by screw jacks bearing against the cast-iron segments, giving a single tunnel of 6 feet 7¾ inches internal diameter. This shield incorporated the fundamental principles of driving itself forward into the soil by pushing against the constructed tunnel, and of having the 'tail' of the shield supporting the soil until the lining had been erected. The Subway was officially opened on August 2, 1870, and was equipped with a 2 feet 6 inch gauge railway on which ran a single 12-seat cable car. Steam engines were used to haul the cable to and fro, and to work the lifts.

Here was convincing evidence that the shield system was a practical method of tunnelling through London clay, but the steam lifts and cable railway proved both unreliable and unremunerative, and were removed after a few months. The subway was then used by pedestrians until the opening of Tower Bridge in 1894.

At about the same time as the Tower Subway was being built, Alfred E. Beach was using a shield to construct the Broadway Pneumatic Railway in New York. Opened for public inspection on February 26, 1870, this demonstration line was but 294 feet long, and the tunnels only 8 feet in internal diameter, lined with brick on straight sections and cast-iron segments on the curve. It is notable for the first application of hydraulic rams to a shield.

In 1876 Greathead devised compressed-air tunnelling apparatus and a hydraulic segment lifter for a pedestrian subway through sand and gravel strata between North and South Woolwich. Work began, but the contractor abandoned the contract owing to difficulties elsewhere. Later another contractor attempted to tunnel through the underlying chalk without a shield but was unsuccessful.

During these unpropitious times for further tube railway construction, Greathead worked on many surface railway schemes, including the District's Hammersmith and Richmond Extension Railways. He also devoted his inventive genius to improvements in roller skates, locknuts, hydraulic apparatus and fire hydrants.

MOVING WITHOUT SMOKE

This pause in the development of tube railways was due to the lack of suitable motive power. In the search for a solution to the problem, much energy and money were spent in schemes for pneumatic propulsion. The driving force behind these schemes was the engineer T. W. Rammell, who patented his pneumatic traction system in 1860. He employed an extraordinary centrifugal air pump, formed of 22-feet diameter discs enclosing thirty-two V-shaped pockets, on which *The Engineer* commented, 'A less economical way of employing a steam engine could scarcely be devised.'[1]

His mail-carrying schemes were a moderate success. After a trial in Battersea Park, a 2 feet 6 inch diameter pneumatic tube between Euston and Eversholt Street post office was opened in 1863. A larger tube, 4 feet high and 4 feet 6 inches wide, was opened between Euston and Holborn in 1865, and was later extended to the GPO at St Martins-le-Grand. Although some mail was conveyed, the problem of keeping the tube air-tight proved insuperable and operation ceased in 1876.

Rammell also propounded the use of pneumatic power for passenger-carrying underground railways. In 1864 he organized a successful experimental line in the Crystal Palace grounds. A single saloon coach, surrounded by a ring of bristles to make an air-tight joint, was blown or sucked to and fro in a 600-yard tunnel. This success resulted in the passage in 1865 of the Act incorporating the pneumatic Waterloo and Whitehall Railway, employing a cast-iron tube for the section beneath the Thames. Single 25-seat cars were to be used in a 12-foot tunnel. Some construction did take place, but sufficient finances were lacking, and the company was dissolved in 1882.

Undaunted by these setbacks, Rammell served as engineer for several grandiose pneumatic underground railway schemes, for which unsuccessful applications were made for parliamentary powers. These included the Mid-Metropolitan of 1882 from Shepherd's Bush to Aldgate, Paddington to Bayswater Road and thence to Westminster (coolly passing down Constitution Hill and across St James's Park!) and from Marble Arch to South Kensington; and the 'South Kensington and Knightsbridge and Marble Arch Subways' of 1886 (3 feet 9 inch gauge).

By the early 1880s, cable traction had been developed as a practical system for street tramways, and in 1884 Greathead and others obtained powers for the City of London & Southwark Subway, between the north end of London Bridge and the Elephant & Castle.[2] There were to be two 10 feet 2 inch diameter shield-driven tunnels, with a continuous cable driven from a power station at the Elephant. As with

---

[1] Atmospheric tube railways appear to have been first suggested by George Medhurst in 1812, who then announced 'an improved method of rapid conveyance of goods and passengers on an iron road through a tube of 30 feet area by the power and velocity of air'. On August 19, 1824, a patent specification for an atmospheric tube railway was enrolled in the name of J. Vallance, who demonstrated his tube line at Brighton three years later. Other inventors used the idea of a piston in an atmospheric tube, fastened to a car *above* the tube, on the surface.

[2] Between the Borough and the northern end of London Bridge the 1884 proposals followed a similar alignment to that of an abortive twin 8-feet tunnel scheme for which Barlow had obtained powers in 1870.

street tramways, the driver of each car would have been able to secure his car to the cable or release it at will.

The passage of this City of London & Southwark Subway Act was a great parliamentary triumph for Greathead in the face of severe opposition. He was an excellent parliamentary witness, and was helped by the support of Mr (later Sir) Benjamin Baker, the eminent civil engineer, then engaged on the construction of the Forth Bridge.

Edmund Gabbutt of Liverpool[1] began construction in March 1886, and powers were obtained in the following year for an extension from the Elephant to Stockwell. A shaft was sunk in the Thames behind the Old Swan pier near London Bridge, and tunnels were driven out from the shafts and completed in June 1887. In July 1887, tunnelling started from the site of the Borough station, and all the tunnels were completed in 1890.

Cable traction was still envisaged, *faute de mieux*, but it was by no means an ideal system. It involved the installation and maintenance of a huge number of pulleys and bearings, sheaves and tensioning devices. The main cable, subject to the constant wear of cars gripping and releasing it, wore rapidly, and if it broke the whole system stopped. Maximum speeds were low, and any branches from the basic two running tracks, such as crossovers or turnouts, required their own auxiliary cables.

In 1879, an experiment took place which was destined to revolutionize the development of underground railways. Werner von Siemens demonstrated the possibilities of electric traction powered from a distant source using a diminutive 150-volt coal-mining locomotive to haul three passenger cars on a 900-yard track in the grounds of the Berlin Trade Exhibition. The demonstration was repeated at London's Crystal Palace in 1881, and in the same year the first public service electric railway in the world was opened in Berlin, between Anhalt Cadet School and Lichterfelde Ost. In 1883, Magnus Volk started his seafront electric railway at Brighton, and A. Traill's hydro-electric Giant's Causeway line was also opened in that year, followed by the Bessbrook and Newry tramway in 1885 and the Ryde Pier railway in 1886. Early electric tramways included the Blackpool (conduit) system of 1885 and Frank Julian Sprague's outstandingly successful overhead-trolley installation at Richmond, Virginia, in 1887.

These pioneer systems showed that electric traction was much more than a scientific novelty; it was a practical, reliable system of propulsion, although so far employed only on the lighter type of railway. Railway promoters early appreciated the possibility of applying electricity to underground railways, and the Charing Cross and Waterloo Electric

[1] Walter Scott & Co. of Newcastle were the contractors for the Elephant–Stockwell section, and also completed the northern section after the withdrawal of Gabbutt through illness.

EARLY SCHEMES

Railway was authorized in 1882, but abandoned in 1885. In 1884, unsuccessful applications were made for powers to construct a London Central Electric Railway (from Dudley Street, Soho, to Charing Cross, Piccadilly Circus, and St Martins-le-Grand) and for a Mid-London Electric Railway from Marble Arch to Bank.

Charles Grey Mott,[1] who had accepted the Chairmanship of the Southwark line at an early stage, was not slow to draw the correct inference in relation to his own line, and after full technical investigation, the directors stated in August 1888 that they had decided in principle on electric traction instead of cable. In February 1889 it was announced that Messrs Mather & Platt Ltd of Manchester had contracted to equip the line electrically; they guaranteed that the working costs would not exceed $3\frac{1}{2}$d per train mile.

This decision displayed a commendable degree of courage and technical foresight. Service conditions on the CLSS would be far more arduous than on any existing electric system, and the use of electric traction on a railway wholly in tunnel was a step into the unknown.

In 1890, the company obtained powers to extend the line to Clapham Common and to change its name to the City & South London Railway. Under this name, the line from King William Street to Stockwell was formally opened by the Prince of Wales on November 4, 1890, and to the public on December 18th. It was the first underground electric railway in the world.

This was Greathead's hour of triumph. He had successfully applied the shield-tunnelling system to the construction of a full-length tube

[1] A director of the Great Western Railway from 1868, Mott had been connected, as director, with two early tunnelling works of great magnitude, the Severn Tunnel and the Mersey Railway. His enthusiasm secured the raising of the CLSS's capital, and the commencement of construction. He was an early advocate of unified control for London's underground railways.

One of the first trains on the City and South London Railway at Stockwell Terminus with the cars known as *padded cells*. Drawing made in 1890 for the *Illustrated London News*.

railway, using compressed air in water-bearing strata near Stockwell, devising his own apparatus for compressed-air grouting, and inaugurating the driving of a heading and of piles in front of the shield to break up the soil.

The electric tube railway was born. Greathead had made his name as the exponent of shield tunnelling, and his services were eagerly sought by the promoters of further tube railways. Now that a suitable means of traction was available, there seemed to be no insuperable technical difficulties in the way of extensive underground railway development in London.

CHAPTER 2

The Beginnings of the Big Three
1891–1900

CITY & SOUTH LONDON SUCCESS

JUDGED by criteria other than its financial results, the City & South London Railway was an immediate success. By February 1891 the average daily traffic had reached 15,000 passengers; in the year ended December 31, 1891, 5,161,398 passengers used the new line.

Despite this weight of traffic, the accounts for the half-year to June 30, 1891, showed that net revenue was barely sufficient to meet the debenture interest, and no dividend was paid on the ordinary shares, nor was any ordinary dividend paid in the following half-year. Nevertheless, the *Railway Times*, in reviewing the results for 1891, considered that the line had 'done better than its most sanguine friends had ventured to anticipate'.

Some relatively minor breakdowns and delays occurred during the early days of operation, but the CSLR had proved beyond doubt that electric tube railways were a practical proposition, and company promoters began to take an interest in proposals for new tubes as a possible means of fruitful investment. The promoters could discount the lack of dividends on the CSLR by pointing out that their tubes would benefit by avoiding the pioneer line's mistakes in such matters as tunnel diameter, curves, gradients and layout of termini. They could also argue that, in time, the CSLR's traffic would increase and working expenses decline as experience was gained.

The parliamentary success of another tube railway provided the promoters of new tube lines with a further incentive. An 1890 application for a Central London Railway between Bayswater and King William Street, City, had been passed by the Commons but refused by the Lords. In the next session, the promoters returned with a more ambitious route, between Shepherd's Bush and Cornhill via Oxford Street and Holborn. On this occasion both Houses of Parliament gave the scheme their blessing, and the Central London Railway Bill received the Royal Assent on August 5, 1891, much to the chagrin of the Metropolitan Railway authorities, who alleged that Parliament had promised them that the land within the Circle should remain for ever untouched by local railways.

Further encouragement was given by the steady rise in both the population of Greater London and in the 'travel habit', with the result that the streets in the central area were becoming ever more congested.

The London suburban services of the main-line railways shared the increased traffic in varying degrees. On some lines (such as the London & South Western) traffic was increasing very rapidly, and the main-line directors were not slow to appreciate the advantages of direct tube connections from their termini to points in the heart of London and to termini on the farther side of the central area. (Many of the railways had arranged their own cross-London bus services, either directly operated, or subsidized.) Main-line support for a tube scheme might range from the supply of a witness at the hearing before the parliamentary committee to a full-blooded financial guarantee.

PROMOTIONS IN EARNEST

Thus, when the Bills for new tube railways were deposited for the 1892 session of Parliament, they had few enemies and many friends. There were four main proposals: a Baker Street & Waterloo Railway; a Great Northern & City Railway (Finsbury Park to Moorgate); a Hampstead, St Pancras & Charing Cross Railway; and a Waterloo & City Railway (Waterloo to Bank).

The Hampstead proposal envisaged a line between Hampstead (High Street/Heath Street) and Southampton Street/Strand via Haverstock Hill, Camden Town, Hampstead Road, Tottenham Court Road and Charing Cross Road; at Hampstead Road/Seaton Street a branch diverged to Euston and St Pancras/King's Cross.

Of these schemes, the Hampstead was alone in having no close connection with a main-line railway. Amongst the Baker Street & Waterloo promoters were two prominent LSWR directors, whilst the LSWR connection with the Waterloo & City was so intimate that the latter railway was virtually guaranteed a 3% return on its capital by the former. The Great Northern & City was strongly supported by the Great Northern Railway, and had increased its proposed tunnel diameter to 16 feet at the latter's behest.

The Commons passed a motion on March 1, 1892, that a Joint Committee of Lords and Commons should be appointed to consider the tube schemes currently proposed (the four main schemes plus a proposed Central London extension to Liverpool Street and a CSLR extension to Islington). The House of Lords concurred, and each House appointed five members.

The Committee sat during May 1892. J. H. Greathead was one of

THE BEGINNINGS OF THE BIG THREE

the principal witnesses; he expressed his faith in electric traction, and supported 12-feet diameter tunnels against the costlier 16-feet tunnels which would be needed to take main-line stock. He objected strongly to inter-running between tubes and normal surface railways.

In its report, the Committee considered that additional railways were needed to assist in dispersing the population, and since the current proposals did not appear to prejudice possible future lines, it saw no reason to advise their postponement.

Electricity was endorsed as a suitable form of traction for tube railways, although cable traction also appeared to be useful, especially for steep gradients. The proposed routes appeared fairly satisfactory, as an instalment of a more complete plan.

The Committee favoured a minimum tunnel diameter of 11 feet 6 inches, and recommended that where a tube railway passed under private property the proprietors of the railway should be obliged to acquire a wayleave only, instead of buying the freehold. For lines below the public streets, a free wayleave was recommended subject to the railway's accepting an obligation to run an adequate number of cheap and convenient trains.

Before the Commons Committee on the Baker Street & Waterloo Bill, Greathead (joint engineer with W. R. Galbraith) described the proposed route from a station below Baker Street Metropolitan station, via Regent's Park, Portland Place, Langham Place, Regent Street, Haymarket, Cockspur Street, Charing Cross, Northumberland Avenue, and below the Thames to a terminus under Waterloo LSWR station, with power station and sidings beyond. Intermediate stations were proposed at Oxford Circus, Piccadilly Circus, and at Craven Street, Charing Cross. Objections raised by the Crown estates and the Portland estate precluded a station between Baker Street and Oxford Circus. Greathead anticipated a heavy traffic movement throughout the day, not merely during the peaks, as on the CSLR. Despite opposition from the Metropolitan and South Eastern Railways, Royal Assent was granted on March 28, 1893.

The Commons Committee stage of the Hampstead, St Pancras and Charing Cross Bill was not concluded until March 1893. Owing to the steep gradients, cable traction was proposed, although the Bill also permitted electric traction. This Bill received the Royal Assent, under the name of the Charing Cross, Euston & Hampstead Railway, on August 24, 1893, but the Euston–King's Cross section was not authorized.

Of the other Bills considered by the 1892 Committee, the Great Northern & City, and the Central London extension to Liverpool Street (with a station at the Royal Exchange instead of Cornhill) received the Royal Assent in June 1892; the Waterloo & City, and the CSLR extension to Islington were approved in 1893.

Greathead died on October 21, 1896, before any further tube railways had opened. His claim to fame lay in the successful exploitation of the shield-tunnelling system and posterity has recognized his contribution to civil engineering by linking his name with that of the shield he did so much to develop.

HAMPSTEAD PROGRESS

Subscriptions for 141,600 £10 Hampstead shares (the whole of the authorized share capital) were invited in March 1894, but the amount subscribed was insufficient to justify allotment and was returned. Subsequently it appeared that a syndicate might provide the capital and in 1897 extension of time was obtained. The next Act (1898) abandoned the authorized line south of the Garrick Theatre and substituted a section ending at 23 Craven Street, where there were to be interchange facilities with the SER. A year later, another Act authorized a new section between Euston and Mornington Crescent so that Euston was placed on the through line (the parallel section along Hampstead Road was abandoned). A new branch was to run from Camden Town, 'Mother Redcap', to Kentish Town (Midland) station, with power station and sidings north of Lady Somerset Road, Kentish Town. Shortly after further extension of time was obtained in May 1900, negotiations were concluded which seemed to offer a good prospect of raising the capital, and the story is taken up again in Chapter 4.

BAKER STREET & WATERLOO IN SHADY COMPANY

In 1896, the Baker Street & Waterloo obtained powers for an extension from Baker Street to the authorized Marylebone terminus of the then Manchester, Sheffield and Lincolnshire Railway, but otherwise it remained dormant until taken under the wing of Whitaker Wright's London & Globe Finance Corporation.

Wright was born in Cheshire in 1845, and in 1866 emigrated to the USA where he made a fortune in prospecting for precious metals and as a mining consultant. He later lost heavily in a financial crash, and returned to England in 1889 to set up as a company promoter in the City of London. In 1894 and 1895 he floated new 'holding' concerns, which would invest in mining companies' shares. The public was eager to subscribe, and Wright made handsome personal profits.

After successful mining company flotations in 1896, there was a new flotation in March 1897 to combine the assets of the 1894 and 1895 companies in the 'London & Globe Finance Corporation' with a capital of £2 million. Wright inspired confidence by having titled people as directors. These included Lord Loch, newly returned from a six-year spell as Governor of the Cape and High Commissioner of South Africa. A director of the LNWR, Loch became interested in the investment prospects offered by the Baker Street & Waterloo, and communicated his interest to Wright.

On November 4, 1897, two contracts were signed; the London & Globe, as 'contractors', undertook to build and equip the line for £1,325,000 in Baker Street & Waterloo shares and £441,000 debenture stock; the 'contractors' then made an agreement with the 'sub-contractor', H. H. Bartlett, Tredegar Works, Bow, trading as Perry & Co., to construct the line for £877,000.

In November and December 1897 the old Baker Street & Waterloo board of directors was replaced by a new board identical with that of the London & Globe, and construction was begun.

Following the examples of the CSLR and the Waterloo & City, it was decided to use a staging erected in the Thames as the principal working point for tunnelling, and the first pile was driven on June 20, 1898. The site was upstream from Charing Cross railway bridge, and about 150 feet from the Victoria Embankment. The staging consisted of rows of piles secured with cross timbers, and by 1902 the timber island had grown to 350 feet long and 50 feet wide. Transverse baulks supported the flooring, which carried a variety of engineers' impedimenta, including cranes, stacked segments, workshops, air-compressors and generating plant. Two 18-foot shafts were sunk through the bed of the Thames, and from temporary brick chambers at the base the running tunnels were driven by Greathead shields. Tunnelling began in February 1899 towards Baker Street, and in March 1900 towards Waterloo. By using the river stage, barges could deliver materials and remove spoil, avoiding street cartage.

The 1899 Bill was ambitious. It included a deviation in Lambeth, so that the line would curve to run along the western edge of the LSWR station, ending at the south-east end of Addington Street. There were to be two extensions to the north. One continued from Marylebone to Paddington, thence to a running depot at Blomfield and Randolph Roads, Maida Vale, with a station under the Paddington branch of the Grand Junction Canal, and a subway connection to Paddington (GWR) station. The other was a branch from Regent's Park to Euston.

Sir Benjamin Baker and W. R. Galbraith explained that the Waterloo & City Railway had occupied the original Waterloo siding space and that the LSWR intended to enlarge its terminus, so an alternative depot site had to be found, although there would still be a power station and lift for locomotives at Waterloo. The new running depot was to be below surface level and 'would cause no more inconvenience than cucumber frames'.

The Metropolitan was in furious opposition. It was particularly indignant at support vouchsafed by the GWR, and alleged that this was a breach of the 1865 agreement for the joint operation of the Hammersmith & City Railway, which stated *inter-alia* that neither the Metropolitan nor the GWR should 'promote, directly or indirectly, any extensions from, beyond or connecting with the Hammersmith and City line, which may divert traffic from the Great Western, Metropolitan or Hammersmith and City Railways'. This afforded a glorious opportunity to argue about the meanings of the words in the agreement (particularly 'promote' and 'extensions'), of which the opposing Counsel took full advantage.

The Commons Committee rejected the northern extensions, but the Lambeth deviation and a subway at Trafalgar Square were granted. T. J. Hare, a director of the LNWR and NLR, became a director of the Baker Street & Waterloo in October 1899, and was appointed chairman a year later.

The problem of finding space for a running depot and power station was solved by negotiating the acquisition of the school belonging to the Trustees of the Indigent Blind, at St George's Road, Southwark (the Trustees wanted to sell and move to the country), and a connecting line to the depot was incorporated in the 1900 Bill. Other 1900 proposals were for extensions from Waterloo to the Elephant & Castle (with foot subway connection to the CSLR), and from Marylebone to Bishops Road, Paddington, with a foot subway along Eastbourne Terrace to Paddington station and an intermediate station at Edgware Road.

The Metropolitan returned to the attack, but with less success than in 1899; Royal Assent was signified on August 6, 1900. Lord Loch had meanwhile died on June 20th.

Wright had been thriving by promoting further investment companies, but in 1899 he had lost heavily in Lake View mining shares. The loss was concealed by the inter-company sales of the shares of subsidiaries, but in 1900 further heavy losses in Lake View shares occurred, and desperate attempts were made to secure additional capital by launching new mining companies.

One of the final attempts to prop up the now rapidly collapsing London & Globe edifice was to invite subscriptions for the Baker Street & Waterloo share capital of £2,385,000 (66,000 £10 4% preference shares, and 172,500 £10 ordinary shares) and the subscription list closed on November 14, 1900. With commendable exactness, it was disclosed that the Globe concern had spent £654,705 10s 7d on the railway.

On December 17, 1900, the London & Globe's annual report stated that the recent Baker Street & Waterloo issue was not well subscribed, but mentioned that negotiations were in hand to sell the railway to a syndicate. The *Railway Times* commented, 'The Baker Street & Waterloo baby seems to have outgrown its nurse.'

The most noteworthy feature of the report was the announcement that the company was unable to pay a dividend on its ordinary shares (a 10 per cent dividend had not been unusual in previous years). In fact, the balance sheet had been made to balance by virtue of further complex inter-company share transactions, and gross over-valuation of the shares of subsidiaries. This financial sleight of hand was no substitute for hard cash, and on December 28, 1900, the London & Globe concern and twenty-seven other companies announced their insolvency.

The crash brough ruin to many investors. In March 1903 an order was made for the Official Receiver to prosecute Wright, who had meanwhile fled to the USA, where he was arrested and eventually extradited to England. He was tried at the Law Courts in January 1904 and found guilty, under the Larceny Act 1861, of publishing false balance sheets and accounts. He was sentenced to seven years' penal servitude, but when talking to his legal adviser in a consultation room after being sentenced, he took potassium cyanide and died instantly.

After the London & Globe bankruptcy, construction was continued by monthly payments from the railway direct to Perry & Co., some income being derived from calls on the shares taken by the general public. This hand-to-mouth method of finance could not continue for long, and on May 4, 1901, work ceased on the tunnels driven south from Baker Street, and on the northbound tunnel 107 yards north of Vigo Street (there was continuous tunnel from Waterloo to this point). In July, the board resolved to stop all work forthwith except for the completion of the second river tunnel, Waterloo station surface work and the southbound station tunnel, Oxford Circus station shafts, and the southbound tunnel in Regent Street south of Oxford Circus, which had been driven from Conduit Street to the river.

The second river tunnel was completed in October 1901, and the southbound tunnel between Conduit Street and Oxford Circus by January 1902. Negotiations for take-over by a syndicate were eventually completed in March 1902.

DISTRICT DEEP LEVEL

During the 1890s, the District Railway was beset with troubles. There was the perpetual problem of the smoke-laden atmosphere on the tunnel sections, and the South Kensington–Mansion House section was being worked at its maximum capacity (for steam) of nineteen trains an hour in peak periods. Electrification was the crying need, but the cost was great and in 1897 the company still felt that electric traction had not yet proved itself able to handle traffic as heavy as was sometimes experienced on the District.

Horse-bus competition was an embarrassment, and further competition was promised when the Central London Railway opened and the London United Tramways were electrified and extended. Critics complained of inefficient management. James Staats Forbes, the District chairman until 1904, was also chairman and general manager of the London, Chatham & Dover Railway, as well as being on the boards of several other companies.

No ordinary dividend had been paid for many years. In an effort to improve matters, the District Shareholders' Association announced in October 1896 that it had engaged Sir Benjamin Baker to survey a District Deep Level railway between Earl's Court and Mansion House, and that it was prepared to advance £1 million in 4% second preference stock, this stock to receive no dividend until the existing preference stock had been paid 5%.

In the following month the report was published. It envisaged that the Deep Level would leave the surface tracks halfway between Earl's Court and Gloucester Road, then curve to the right and run below Cromwell Road on a descending gradient of 1 in 42. From Gloucester Road station to Mansion House it would be below the existing District Railway in two normal Greathead-type 12 feet 6 inch tubes. There would be an intermediate station at Charing Cross, 63 feet below the District; the Mansion House terminus would be 71 feet deep. Hydraulic lifts would be provided at these stations. As engine-changing at Earl's Court would cause congestion, it was suggested that the change from steam locomotives to electric should take place in the first 176 yards of the Cromwell Road covered way, the carriages being held by a 'rack-railway brake'.

The estimated cost for works and land came out at £1 million, with a further £453,000 needed for other expenses, equipment, 10 electric locomotives, and 80 'long American type carriages'. Through running was contemplated from all the District's western branches, as the Deep Level would give a much shorter journey time from Earl's Court to Mansion House compared with the surface line.

The District board accepted these proposals for inclusion in their Bill for the 1897 Session. This course of action had several advantages. It helped to placate the shareholders, and if the Bill was passed it would be useful as a 'blocking line' against parallel tube schemes; it might also solve the District's traffic problem without the expense of electrifying the surface lines.

There was little opposition to the Bill, and it received the Royal Assent on August 6, 1897.

The subsequent history of the Deep Level may be dismissed in a few lines. The section between South Kensington and Earl's Court was built as part of the Piccadilly Line by virtue of powers obtained in 1899, 1902 and 1903, but work on the Deep Level proper was confined to the construction in 1903 of a section of station tunnel at South Kensington (level with the westbound Piccadilly Line platform, and forming the rudiments of a flying junction with the Piccadilly). This spare tunnel was used to house a Signal School from 1927 to 1939.

With the advance in the technique of electric traction and automatic signalling, it was clear that the existing District line, when electrified, would have far greater capacity than with steam traction, so that the Deep Level's extra capacity was not needed. The parliamentary powers to build the South Kensington–Mansion House section were relinquished in 1908.

BROMPTON & PICCADILLY CIRCUS

During the horse-bus era, and indeed until the passage of the London Traffic Act 1924, there was no legal restriction on a bus proprietor's choice of route (although the Associations did their best to preclude newcomers from existing routes), and the quantum of service that was provided was a good guide to a route's profitability. This factor was duly noted by the more astute tube promoters, and even today the lines that were chosen on this basis—i.e. the central portions of the Bakerloo, Central, Northern (West End) and Piccadilly lines—carry good off-peak traffic.

By 1897 most of the best bus roads had been covered by existing or authorized tube lines, but one of the few remaining gaps was on the Knightsbridge–Piccadilly–Strand–Fleet Street alignment, although the District gave a parallel facility east of Charing Cross, and had originally sought a route beneath the Strand and Fleet Street.

Piccadilly Circus, with its surrounding theatres and shops, appeared to offer lucrative sources of traffic, and two schemes to fill the gap were presented in the 1897 Session. Both were for deep-level tube railways.

The Brompton & Piccadilly Circus Bill was for a modest line from near South Kensington station, via Brompton Road, Knightsbridge, and Piccadilly to Air Street, just west of Piccadilly Circus. The promoters included C. G. Mott, chairman of the CSLR, and Sir Joseph Dimsdale.

Intermediate stations were proposed at Brompton Road (near Brompton Square), Knightsbridge, Hyde Park Corner, Down Street and Dover Street. The power station was to be at Swan Wharf, Lots Road, Chelsea, and there was to be a branch to a running depot at Yeoman's Row, off Brompton Road.

The other proposal, the City & West End, was more ambitious, with a line from Hammersmith Broadway to Cannon Street, via Kensington, Knightsbridge, Piccadilly Circus, Charing Cross, Fleet Street and Ludgate Circus. The proposed depot and power station were at Hammersmith between Fulham Palace Road and the Thames. The promoters were connected with the Central London Railway.

The City & West End proposal threatened the District with serious competition, whereas the Brompton line would be a useful feeder to the District at South Kensington, and could perhaps be linked with the Deep Level.

The District threw itself wholeheartedly into the struggle against the CWE, alleging that places on the proposed route were already adequately served by itself and by bus services; that what was really needed was a fast route from the western suburbs to the City (which the Deep Level would help to provide); that the CWE could not possibly earn enough to give an adequate return on its capital, and that the scheme was, in any case, premature until the Central London Railway was open.

The Commons Committee passed the Brompton Bill but rejected the City & West End. The Brompton received the Royal Assent on August 6, 1897, with authorized share capital of £600,000 and borrowing powers of £200,000.

In July 1898, the Brompton made an unsuccessful attempt to market its shares, but was taken over by the District Railway later that year. On November 25, 1898, the Brompton board lost all its original directors except Sir Joseph Dimsdale, and received a powerful contingent from the District, including Forbes, who was appointed chairman.

An 1899 Brompton Bill sought to join the Deep Level at South Kensington, and to extend from Air Street to Long Acre/St Martin's Lane (with a Hampstead/Brompton joint station); other clauses were to enable the District to subscribe up to £200,000 of the Brompton's capital, to guarantee interest on its capital and to appoint three directors. Additional capital of £533,000 was included, together with powers to build the South Kensington–Earl's Court section of the Deep Level.

In Parliament the Bill shed the Air Street–Long Acre section, but the other proposals were approved, and Royal Assent was granted on August 9, 1899.

A 1900 Bill for extension of time was withdrawn, but this extension was included in the District Act of the same session. A 1901 Bill, and its fate, will be described in Chapter 4, and the take-over by the parties that acquired the District is recorded in Chapter 5.

GREAT NORTHERN & STRAND

For many years there had been congestion at the Great Northern Railway's approaches to King's Cross, and the directors had done their best to relieve the situation by constructing the Canonbury link line to the North London Railway (opened in December 1874) and by first doubling and then trebling the Maiden Lane and Copenhagen tunnels. Continuing delays to traffic persuaded them to support the 1892 Great Northern & City scheme, but construction of that line did not begin until 1898 because of delays in raising capital; meanwhile, congestion grew steadily worse.

In 1898 Sir Henry Oakley resigned the general managership of the Great Northern to become chairman of the Central London; his place was taken by Charles Steel, who forthwith reported on the suburban traffic problem, and a scheme was prepared for a tube railway between

Wood Green and Holborn. The promoters were independent, but had the Great Northern's 'concurrence and approval', to quote Sir Henry.

Later in 1898 the LCC decided to deposit an 1899 London Improvements Bill for the clearance of the congested area north of St Mary-le-Strand church, by the construction of new streets which were eventually named Kingsway and Aldwych. On hearing this news, the tube proprietors decided to extend their line to the Strand, via the new street, and the Great Northern & Strand Railway Bill was accordingly deposited in the same session.

The Bill provided for an electric tube railway 6 miles 2 furlongs 7 chains long, from Wood Green to Wych Street, Strand (on the island site now occupied by India, Bush and Australia Houses). Between Wood Green and King's Cross the line was to be beneath Great Northern land, but slightly east of the main line. On that section there were to be stations at Wood Green, Hornsey, Harringay, Finsbury Park, Holloway and King's Cross, connected to the corresponding Great Northern stations by lifts or subways. A further station was proposed at Bingfield Street/York Road. South of King's Cross the line cut across the street pattern to a proposed station at Russell Square, then followed the line of Southampton Row and the new street, with stations at Holborn and Strand.

The railway was to be 50 feet below the Great Northern at Wood Green, and 110 feet deep at High Holborn (crossing beneath the Central London). The tubes would be 12 feet in internal diameter, and there was no intention of through-running with the Great Northern as long as the terminus was at Wood Green, although this question would be reconsidered if the line were extended further north. The power station was to be at Holloway, the depot at Wood Green; the new board would include the Great Northern chairman and three other GNR directors.

The chief opposition came from the LCC, whose main contention was that as they were buying the property through which Kingsway would be cut, the GNS should make a financial contribution. They wanted a decision on the Holborn–Strand section of line deferred until their own Bill had been decided.

The new line was approved by both Houses, but clauses were inserted by the Lords giving the LCC powers to approve the elevation and design of station buildings, and to the effect that the GNS was not to take land or easements on the Holborn–Strand section without LCC consent, unless and until Kingsway had been built. The Bill received the Royal Assent on August 1, 1899.

Despite LCC pressure, there was no provision for workmen's fares, and this was a double rebuff to the LCC when one recalls that the 1892 Committee had recommended that tube railways be allowed a free wayleave under the public streets provided that they were put under an obligation to run an adequate number of cheap and convenient trains.

After the Act was passed, the GNS appeared to settle down to a long sleep, from which it was awoken by the siren voices of American capitalists in 1901.

DIFFICULTIES OF ATTRACTING CAPITAL

The major obstacle to the realization of all these schemes was the investing public's reluctance to subscribe enough capital.

There was no doubt that additional lines were needed, nor was there any lack of capital, for large amounts were subscribed to the most wildcat schemes for investment in other spheres, as shown by the career of the London & Globe. Why, then, should investors show such coolness towards tube schemes?

During most of the period under review, the only tube open for traffic was the City & South London. The dividends paid on this company's ordinary stock rose slowly and painfully from nil in 1891 to $2\frac{1}{8}\%$ in 1898, only to fall to $1\frac{7}{8}\%$ in 1899. The financial arrangements of the Waterloo & City (opened in 1898) were peculiar in that the London & South Western Railway undertook to work the line for a maximum of 55% of the gross receipts, on which the payment of a 3% dividend was a first charge. With this generous subvention, the 1894 issue of Waterloo & City shares was attractive to investors, and was over-subscribed. The other promotions lacked the attraction of a virtually guaranteed return, and the CSLR was the only example of a working tube that stood on its own feet. The Central London made several unsuccessful attempts to issue its capital to the public and was finally backed by a mining company, the Exploration Co. Ltd, its stocks being syndicated and issued in blocks. The Electric Traction Co. Ltd, a subsidiary, was formed to buy the land and to build and equip the line in return for £700,000 of the 4% Debenture Stock and £2·54m. in cash.

Other factors tended to discourage investment. The various tube Acts were secured only at the expense of numerous protective clauses, designed to defend the real or alleged interests of public bodies and private land and property owners along the line of route.

Typical clauses for the protection of existing subterranean works such as sewers would include provisions for submitting plans and sections of the appropriate portions of tunnelling to the sewer authority for approval, the inspection of the railway's works by the authority (at the cost of the railway), compensation for damages caused during construction, and restrictions on the alignment and minimum depth of particular sections of tunnel. Such clauses restricted the flexibility with which the line could be laid out (in the absence of these clauses, the engineers were legally free to vary the alignment within the specified limits of deviation) and added to the paper work, thus increasing the engineering and legal costs. The *Railway Times* waxed indignant about the burden of protective clauses, calling many of them 'utterly useless'. In May 1919, Mr (later Sir) William Acworth, a director of the Underground Electric Railways of London Ltd, stated that such clauses had imposed on promoters obligations to spend £5,000 to protect petitioners from possible risks that could be insured against at Lloyds for £1 or £2 per annum.

Furthermore, the London County Council, controlled by its Progressive majority, did its best to secure the inclusion of liberal workmen's fare provisions in every tube Act. During the 1890s (and later) the LCC was enthusiastic about the direct operation of public services such as tramways, markets, water and electricity supply, and it was not beyond the bounds of possibility that the conjunction of a Liberal Government and a Progressive LCC might result in the construction of municipal tubes, or even in the expropriation of existing private-enterprise lines.

Finally, there was the chilling prospect of intensive bus and tram competition at low fares, although as long as these forms of transport were restricted to horse traction, the tube railways could hope to attract medium- and longer-distance traffic by their greater speed.

CHAPTER 3

Pioneer Lines

WE now turn away for a while from Parliamentary Committees and Company Board Rooms to consider the pioneer lines—the three tube railways which were already open in 1900. It was from their mistakes that the later enterprises profited, and in particular their experience in rolling-stock design, ventilation and station layout provided valuable object lessons for those that were to follow.

CITY & SOUTH LONDON

The birth of the City & South London Railway has been described. At the beginning of 1900, it was providing service on its original route between King William Street, City and Stockwell, after nine uneventful and almost troublefree years. The City terminus was 75 feet beneath No. 46 King William Street, close to the Monument; it had an island platform with a line each side, and a scissors crossover at the west end, all in a station tunnel 26 feet wide by 20 feet high. Until December 22, 1895, the terminus had consisted of a single track with a platform on each side, and the altered layout had meant reducing the platform length to three cars. Stockwell terminus, in a tunnel of similar size, was also an island flanked on each side by a single track with a scissors crossover between the platform end and the running tunnels. From the southbound running tunnel just north of Stockwell, there was a 100-yard-long 1 in 3½ incline [1] to the depot and workshops, which were between the Stockwell and Clapham Roads, just south of Spurgeon's Orphanage. Stock to be repaired or overhauled was pulled up the ramp by a chain attached to a winding engine. At the intermediate stations, Borough (at first called Great Dover Street), Elephant & Castle, Kennington New Street and Oval (originally Kennington Oval), each line was in a separate 200-foot brick-lined tunnel 20 feet wide by 16 feet high. Except at the Elephant, where a crossover was provided, the

[1] In 1906, this incline was replaced by a 20-ton Waygood hydraulic lift which could carry one car or one locomotive.

station tunnels for each line were at different depths; at Great Dover Street they were parallel in plan, but at the other stations, platforms were staggered. Greathead wanted the passengers to walk as little as possible, and made the lift landing halfway between the two platform levels. Each station had two 50-passenger hydraulic lifts in a 25-foot diameter shaft, suspended from cables and operated by plant installed by Sir W. G. Armstrong Whitworth & Co. of Newcastle on Tyne. Three 100-h.p. pumping engines at Stockwell pressed the water into the mains at 1,240 lb. per square inch. In the summer of 1897, one of the lifts at Kennington was converted to electric operation by Easton, Anderson & Goolden of Erith. Apart from King William Street, where the station was incorporated in an existing building, the lift mechanisms were enclosed in handsome domes surmounting single-storey brick station buildings designed by T. P. Figgis. Down below, the wooden platforms shimmered in yellow gaslight.

The topography of this tube was rather curious and was related to the original intention to use cable haulage. The northbound or up line was on the lefthand side until a point north of the Elephant, where it crossed to the right, but under the river (both tunnels were west of London Bridge) it ran immediately above the southbound tunnel, resuming the normal arrangement just before reaching the terminus. From King William Street, the southbound line descended to the river bed at 1 in 14 on a severe curve; northbound trains negotiated a reverse curve which fell at 1 in 150 then climbed into the terminus from the river at 1 in 40 and 1 in 70. The worst curves had only a 99-foot radius. Where possible, Greathead arranged that intermediate stations were built on humps to assist acceleration on starting and reduce the need for braking on arrival.

An extension northward to Moorgate Street, built by John Mowlem & Co., was opened to the public on February 25, 1900. This left the old line just north of Borough station and took a new course beneath the river, to the east of London Bridge; the fifty-nine chains from the 'junction' to the King William Street terminus were abandoned. A new station was built at Denman Street (now London Bridge Street) to serve the London Bridge stations of the SER and LBSCR, but the subway to the latter was not ready until December 2, 1901. A second intermediate station, called Bank, was sited at the junction of King William and Lombard Streets, and its booking hall was formed in the crypt of the early eighteenth-century church of St Mary Woolnoth, a privilege which cost the railway company about £170,000 in compensation. Mowlem's skilfully removed the original foundations and left the very tottery fabric of Hawksmoor's church undamaged. From July 30, 1900, a low-level subway gave access to the Central London station, but a high-level subway between the two booking halls was not available until May 3, 1911. All three stations on the extension had up and down platforms in separate tunnels joined by cross passages. The 'wrong running' arrangement was perpetuated as far as a point between Bank and Moorgate Street where the northbound line passed above the southbound, thus righting the position.

The year 1900 also saw an extension southwards as far as Clapham Common, then a select suburb. Constructed by W. Rigby & Co., and opened on June 3rd, this had an intermediate station at Clapham Road (now Clapham North); both stations had single island platforms in 30-foot diameter tunnels.

A further prolongation northwards, to the Angel, Islington, with intermediate stations at Old Street and City Road, also built by Rigby's, was opened on November 17, 1901. Angel had an island platform in a single tunnel, but the other two stations were of the now conventional two-tunnel type.

The final extension of the CSLR was to Euston, constructed by Walter Scott & Middleton, and opened to the public on May 12, 1907, after a formal opening by H. P. Harris, Chairman of the LCC, on the previous day. Euston's platform was an island, 180 feet long by 14 feet wide, in a single 30-foot tunnel with an engine traverser in a 25-feet diameter tunnel about 230 feet west of the far end of the platform. At the east end there was a crossover and siding. A subway was provided to the Hampstead Railway platforms and another to lifts connecting directly with the LNWR terminus. At the east end of the island platform a gallery suspended from the tunnel roof led to lifts communicating with a surface booking hall in Seymour (now Eversholt) Street. Any architectural merit this single-storey building possessed was utterly ruined by a large hoarding on its flat roof listing all the line's stations. The structure, which was elaborately faced in white and green stone, was extinguished by the construction of Euston House in 1934, but it had gone out of public use twenty years earlier, when Euston's tube booking facilities were concentrated in the hall below the main-line station. There were two 239-foot platforms at King's Cross, in separate tunnels, with the lower lift landing in a cross passage beetween them, and a low-level subway to the Piccadilly Railway. The booking hall was below street level, near the Great Northern Hotel, and there were no surface buildings. Between here and the Angel, at Weston Street, a signal box was built in a cross passage between the tunnels, reached by a staircase from the street above. Sited on a hump, like most of the stations, this box was the only example on the tube railways of a mechanically worked cabin placed between stations.

All the extensions had electrically worked lifts, and at the Bank there were five, two for eighty passengers and three for sixty, all with a speed of 180 feet a minute. The architectural style of the stations varied from the elaborate stone cladding of Euston and Bank to the more austere red brick and arch-topped windows of Old Street and Clapham.

The original electrical installation of 1890 was provided by Mather & Platt of Salford Iron Works, whose Dr Edward Hopkinson, of Bessbrook & Newry experience, was appointed consulting engineer to the new line. At the Stockwell power station there were three large 'Edison-Hopkinson' compound dynamos with an output of 450 amps at 500 volts, driven through link belting by 375 (indicated) h.p. marine vertical compound steam engines by John Fowler of Leeds. The 500-volt d.c. traction current was originally distributed to the railway through a simple two-wire system. Soon after the opening, a fourth engine and generator were added, and later a Siemens-Willans generating set with an output of 240 amps at 500 volts was placed between the second and third generators (this was the first direct-coupled set used for traction in Britain). To provide extra power for the extensions, new generating plant was installed in 1900–1901 which gave a total output of 3,250 kW and the original power station was then closed. The distribution was rearranged on the three-wire system with 1,000 volts between the conductor rails of the up and down lines (the running lines earthed midway between). Substations were then provided at Angel and London Bridge, and electric light was available at all stations, together with much improved train lighting. The new power equipment consisted of several generators, the two main ones being 800 kW sets driven by compound Corliss engines by Cole, Marchant & Morley. These were assisted by various high-speed sets.

Signalling was mechanical absolute block, similar to that used on the surface railways of the time, and was operated from ten boxes.[1] C. E. Spagnoletti designed a special lock-and-block apparatus with rail deflection treadles. At the stations there were short semaphore arms, but in the tunnels the signals consisted of oil-lit lanterns with glasses turning at right angles on a horizontal spindle. Gas was later used experimentally for lighting signals, but finally electricity replaced oil. After 1900, vertical sliding spectacles were adopted. Other improvements from 1893 onwards included 'last vehicle' treadles and the provision of extra block sections. Eventually this mechanical signalling was efficiently handling as many as twenty-eight trains an hour.

Trains consisted of three trailers drawn by electric locomotives weighing only 10·35 tons and looking not unlike small steam tram engines. Current was collected from the off-centre conductor rail through flat cast-iron shoes. The first fourteen locomotives were built by Beyer-Peacock and equipped by Mather & Platt. On four 27-inch wheels and only 14 feet long by 6 feet 10 inches wide, they had two 50-h.p. (continuous rating) motors, permanently coupled in series and controlled by a rheostat switch with twenty-six contacts and a revers-

---

[1] Euston, Weston Street, Angel (reversing and sidings only), Old Street (stabling only), Moorgate Street (reversing only), London Bridge, Elephant & Castle (reversing only), Stockwell, Clapham Road, Clapham Common.

ing switch. On all but one machine, the armatures were mounted directly on the axles; the geared engine proved so noisy that it was confined to shunting duties. Brake air was carried in tanks on each side of the cab and was taken up at the start of each trip from a main reservoir at Stockwell which was fed by Westinghouse pumps in the power station. Shortly after the line was opened, two more engines were ordered from Siemens Brothers. These weighed $13\frac{1}{2}$ tons and had curved side tanks. The seventeenth locomotive was built by the railway staff at Stockwell in 1895, and weighed 11 tons 12 cwt.; two years later one of the original fleet was reconstructed for series-parallel working, with a field shunting arrangement. In 1897 and 1898 Crompton & Co., the Electric Construction Co., and Thames Ironworks each delivered one locomotive, and two further machines were constructed at Stockwell in 1899–1900. Ten more came from Crompton's in 1899, followed by twenty more from the same firm in 1900–1901, making a grand total of fifty-two in the fleet. The later machines were stockier in appearance and had electric headlights and brake compressors. Ten of the original engines were modernized from 1904 onwards with two Thomson-Houston 80-h.p. geared motors and tram-type drum controllers with series-parallel control.

Two men were carried on each locomotive in case of emergency, the spare man's duties including coupling and uncoupling. Standby engines stood at Euston, Angel, Moorgate Street, Elephant & Castle, Stockwell and Clapham Common.

Passengers were accommodated in seven-ton cars mounted on two four-wheel bogies and built by the Ashbury Railway Carriage & Iron Co. Ltd. These had wooden bodies, 26 feet long and 6 feet 10 inches wide at waist. Inside, below a curved ceiling 7 feet high, all was grim. Two longitudinal padded benches with high backs, each seating sixteen passengers,[1] were divided in the centre by a body-strengthening semi-partition. Small and narrow quarter lights above the benches provided the only windows. This claustrophobic interior was dimly lit by four 16-candle-power lamps fed from the traction supply, apt to fade to dark red on the climb up to King William Street, and sometimes go out altogether. A notice warned that a fine of £2 would be imposed on anyone who cared to indulge in the near-impossible feat of riding on the roof. At each end of the car, sliding doors led to a gate platform carried on the bogie extensions; entry and exit were made through folding gates operated by a gateman who stood between each pair of cars. These vehicles were soon nicknamed *Padded Cells*, and there were thirty of them when the line opened. Later versions (the fleet had grown to 132 by 1902) had larger windows and were supplied by G. F. Milnes

[1] According to *Lightning*, February 2, 1893, it was possible to squeeze up to allow twenty sitting each side and about the same number could stand, most uncomfortably, in each car.

& Co., Bristol Carriage & Wagon Co., Oldbury Carriage & Wagon Co., and Hurst Nelson & Co. About 1900, the original cars were rebuilt with larger windows to match. A maximum of 170 cars was reached in 1908, the last batch being thirty-three all-steel vehicles built by the Brush Electrical Engineering Co. for the Euston extension. Four-car trains were started on June 3, 1900, and a few years later, five-car trains became standard. (In February 1914 the Underground Group introduced two six-car trains on the line but these had disappeared by 1921.) Motored cars, adapted from existing trailers at Stockwell, were tried experimentally in 1895 with the idea of increasing train length to four cars. Although the trial was pronounced a success, it was realized that the crew lost time changing ends on the congested single platform at King William Street and that extra siding accommodation would be required to store defective trains.

The locomotives, at first reddish-brown, were later painted orange-chrome, and, with their shining brasswork, made a handsome contrast to the dingy varnished wood of the cars.

Running was rather noisy, on track of 60-lb. flat-bottomed rails spiked to 12 feet × 6 inch sleepers laid across the base of the tube without ballast. The conductor rail, of channel steel, $1\frac{1}{4}$ inch square, rested on glass insulators in 24-foot lengths and was offset 1 foot 4 inches to the east of centre just below the level of the running rails. Timber planks were placed between the rails to form an 18-inch-wide walk-way. Litter tended to accumulate in the empty space below the track and small fires were not uncommon. Track gauge was the British standard 4 feet $8\frac{1}{2}$ inches which was adopted for all subsequent passenger-carrying tube railways in London.[1]

Generally the line was about 60 feet below the street but at Stockwell it was only 49 feet. Maximum depths were 75 feet at King William Street, 90 feet at Bank and 109 feet at Angel.

At first, passengers entered the stations through turnstiles, paying a uniform fare of 2d, but after the opening of the Moorgate extension, graduated fares and normal tickets were introduced. The basic headway of 1890 was five minutes; nine years later, headways varied between three and four minutes and the average speed was $13\frac{1}{2}$ m.p.h. In spite of its cramped and inconvenient layout, King William Street was handling 440 trains a day in 1899 and at this time 11 trains and 13 locomotives were used at peak periods. A Sunday afternoon service was worked from April 5, 1891, but Sunday morning trains (from 8 a.m.) did not start until August 1, 1909.

Traffic in the first ten years remained fairly static at around $6\frac{1}{2}$ to 7 million passengers a year. The extensions and the consequent interchange with other lines brought the annual total to nearly 20 million

[1] But London Transport's gauge is now 4 feet $8\frac{3}{8}$ inches on the straight and up to 4 feet $8\frac{3}{4}$ inches on the curves of ten chains and under.

in the early 1900s. Dividends on ordinary stock averaged about 1½% until 1901, when 2% was achieved. A maximum of 3¼% was reached in the following year.

WATERLOO & CITY

London's second tube was the Waterloo & City Railway, opened to traffic on August 8, 1898, after a formal opening by the Duke of Cambridge on July 11th. This line was the realization of the London & South Western Railway's long-held desire to reach the City.[1] Since January 1, 1869, trains on the South Eastern Railway's Charing Cross extension had been available to carry LSWR passengers from a separate station at Waterloo to London Bridge and Cannon Street, but the long walk between the two companies' stations at Waterloo,[2] ticket-inspection delays and the vagaries of SER travel caused continuing dissatisfaction. Some impatient commuters organized their own horse-bus service to the City, and this enterprise grew until it was carrying 2½ million passengers a year in the 1880s. The practicability of tube transport having been demonstrated by the CSLR, the LSWR decided to promote its own line through a nominally separate company. It was to run from the basement of the Waterloo terminus to the heart of the City, 1 mile 46 chains in all, and was authorized in 1893.

John Mowlem constructed the tunnels from a working shaft in the Thames, to a diameter of 12 feet 1¾ inches (12 feet 9 inches on curves and between the river and Waterloo). Spoil was removed by water and dumped on Dagenham Marshes. To make a flush lining and absorb noise, it was decided to fill the tunnel segments with concrete.

The City station[3] was sited just east of the Mansion House and had wooden platforms in 23-foot tunnels connected by cross passages. Its construction demonstrated the first use of cast-iron segments for station tunnels, following a suggestion by Greathead (the CSLR stations had been bricklined and their construction had disturbed the surface). Access to the street, 59 feet above, was through a lengthy subway, sloping at 1 in 7.4, a little bit of pinchpenny meanness which caused much distress to users of the line for over sixty years.[4] At Waterloo, a station was built by Perry & Co. below and at right angles to the arches supporting the main-line terminus; the 100-yard-long platforms were 41 feet beneath the LSWR rails. Here also the unfortunate passengers had

---

[1] An Act of 1846 had authorized an extension to London Bridge, where a terminus would have been shared with the North Kent Railway, but this plan was abandoned in 1849 during the financial crisis. A plan for a shallow subway line to the City was considered some years later but discarded as too expensive.

[2] A single-line connection between the two stations at Waterloo was used by regular trains between July 1865 and December 1867.

[3] Renamed 'Bank', October 28, 1940.

[4] After a time, sets of five steps were built in this subway at 40-foot intervals, and the space between each set of steps was flattened somewhat.

to toil up slopes and stairs without any mechanical aid.[1] Beyond the platforms there was a crossover tunnel under a new single arch followed by car and coal sidings, the extreme ends of which were flanked by retaining walls and open to the sky. Nearby was the power station, equipped with six high-speed Bellis & Morcom engines coupled to Siemens 200-kW bi-polar dynamos which provided 500 volts d.c. to a central conductor rail.[2] Just north of Waterloo, a short branch tube led off the eastbound line to the base of a hoist capable of carrying 30 tons, giving access to the outside world via the sidings on the Windsor side of the main-line station. LSWR standard 87-lb. bullhead rails were used, resting in chairs placed on longitudinal sleepers set in concrete. At stations and crossovers, and in the yard, ordinary ballasted track was provided.

The Waterloo & City was the first tube railway to use motor cars in regular service; its original stock consisted of five four car trains (motor car at each end, two trailers between) and two spare motor cars. These trains were built by Jackson & Sharp of Wilmington, USA, and assembled with Siemens electrical equipment at Eastleigh. Each motor car had two 60-h.p.[3] gearless motors and all four motors in the train were controlled from the leading end through power cables, using a controller handwheel that resembled the operating end of the old-fashioned domestic mangle.[4] Westinghouse compressed-air braking was supplied from car reservoirs which were recharged from a compressor and tank at Waterloo. The seats, some of which were transverse, were in perforated plywood with wooden 'separators', 46 in the motor cars and 56 in the trailers. At one end of the motor cars, the floor was raised over the massive motor bogies with their 33-inch wheels. All cars had roller axle bearings, a feature not adopted by the other tubes until forty years later. The heavy wooden bodies were painted chocolate and salmon and had arch roofs; their general appearance was not displeasing. The trailers had sliding doors, giving access to end platforms, which were protected by folding iron gates; the motor cars had a platform at one end and a sliding door in the side of the car at the other end, which was opened by the platform staff.

In 1899, five motor cars for use singly in the slack hours were ordered

---

[1] From April 9, 1919, LSWR escalators were available between the main-line concourse and the intermediate landing.

[2] In 1915 a power station was opened at Durnsford Road, Wimbledon, to supply the newly electrified LSWR suburban lines. The Waterloo & City was also then supplied from this plant. Three of the old generators were retained at Waterloo as standby equipment. Durnsford Road was closed in 1958 and traction current is now taken from the Central Electricity Generating Board.

[3] One hour rating. All h.p. figures are stated in one-hour rating from this point.

[4] As they handled the full traction current, the controllers were of imposing dimensions (4 feet 3 inches by 4 feet 3 inches by 4 feet).

from Dick Kerr & Co. of Preston. When these arrived, the off-peak headway was shortened from ten to five minutes and the ratio of working expenses to gross receipts fell considerably.[1] These English cars had two 60-h.p. nose-suspended motors with single reduction gear, and 50 seats were obtained by using a half-cab arrangement. Entry and exit were through sliding doors on the car sides. Two more trailers were supplied by Dick Kerr in 1904 and four more were built at Eastleigh by the LSWR in 1922. The extra stock made it possible to run five-car trains in the rush hours. A second man was carried in the driving cab until a 'dead-man' device was fitted (on the single cars in the slack hours the guard would go into the cab with the driver after collecting the fares).

A small Siemens four-wheel shunting locomotive with gearless motors was supplied for use in the sidings at Waterloo, moving the cars and pulling the coal wagons to and from the base of the hoist and the power station. Later, a camel-backed Bo-Bo locomotive with geared motors was added as a standby, and this was transferred to Durnsford Road power station when it opened in 1915.

The line had been planned by W. R. Galbraith, R. F. Church and J. H. Greathead[2] to operate on an economical 'switchback' basis. The eastbound line was declined at 1 in 60 and then 1 in 30 to the river bed, and the westbound line had a long 1 in 88 descent from the City to the river. It was expected that the momentum gained on the descent would assist the train up most of the gradient on the opposite side (1 in 88 eastbound, 1 in 60 westbound). A maximum of 35 m.p.h. was envisaged, with the five-chain curves negotiated at 24 m.p.h. Running time would be four minutes. This pretty plan was ruined by the Board of Trade Inspector who insisted on a maximum of 15 m.p.h. round the five-chain curves, a restriction which increased the running time to 6½ minutes because the drivers could make no use of the full parallel position. The speed restriction was later eased to 20 m.p.h., allowing a minimum journey time of four minutes and giving an average speed of 23 m.p.h. start to stop, higher than that on the other tube railways for many years.

Twopenny single and threepenny return fares were taken at turnstiles at either end (tickets were only issued for return journeys), but after 1900 the fares were collected by conductors who paraded along the train with bell punch merrily ringing, tramcar style. There was a high proportion of through season-ticket traffic from LSWR suburban stations, and in the early 1900s the line was carrying about 5,400,000 passengers a year.

---

[1] From 54·78% at June 30, 1901, to 45·5% at December 31, 1901.

[2] Greathead died during construction and was succeeded by Professor Alexander H. W. Kennedy who designed all the electrical equipment.

Semaphore starting signals at each terminus were worked from signal boxes by mechanical frames and Sykes lock-and-block apparatus. The long block section between the two termini was divided by spectacle-plate electric lamp intermediate signals operated electro-magnetically by plungers in the two boxes. If a train passed a signal at danger, a wiper on the car, touching a short piece of energized rail, would set off a current cut-out and bring it to a halt.

Although the LSWR had owned the rolling stock and operated the line from the opening, the original company remained nominally independent until December 31, 1906.

CENTRAL LONDON

Of the three pioneer tubes, the Central London was by far the most impressive. It had the best route in all London and was a financial success from the start. The penny-saving approach was completely absent; everything was good and solid.

This 5·82-mile line, which cost about £260,000 a mile, ran under the great east–west axis from the Bank of England in the City of London to Shepherd's Bush, where the suburbs met the West End. It passed beneath Poultry, Cheapside, Holborn Viaduct, High Holborn, Oxford Street, Bayswater Road and Holland Park Avenue, establishing a connection between the business and commercial activities of the City and the residential west, serving on the way the main shopping and theatre districts of the capital.

Construction began at Chancery Lane in April 1896, and the engineering was in the hands of Sir John Fowler, Sir Benjamin Baker and Greathead. When Greathead died, his place was taken by Basil Mott, who designed, among other things, the subsurface booking hall at the Bank. The Electric Traction Company, as the main contractor, subcontracted to George Talbot (Bank–Post Office), Walter Scott & Co. (Post Office–Marble Arch) and John Price (the remainder). Some use was made of the Price Rotary and Thomson Mechanical excavators. The Thomson machine consisted of a bucket conveyor on a movable arm which attacked the clay through a large hole in the shield. Price's excavator was still not perfected and will be described in its more successful application to later construction. The Prince of Wales (later Edward VII) ceremonially opened the line on June 27, 1900, and then there followed full service with empty trains for three weeks to give the staff experience and eliminate teething troubles before the public opening took place on July 30th. A uniform fare of twopence applied for single journeys and the *Daily Mail*'s nickname, the *Twopenny Tube*, was almost at once in popular use. In the 1900 revival of Gilbert and Sullivan's *Patience*, 'the very delectable, highly respectable, Twopenny Tube young man' made an appearance, replacing the 'threepenny bus young man' of the 1881 libretto. The nickname could have been applied with equal accuracy to the CSLR in its early days, but that line did not have the fortune to open in the era of the popular Press.

There were eleven intermediate stations, varying in depth between 60 and 92 feet: Post Office, Chancery Lane, British Museum, Tottenham Court Road, Oxford Circus, Bond Street,[1] Marble Arch, Lancaster Gate, Queens Road,[2] Notting Hill Gate and Holland Park. At Bank, the booking hall was below street level in the intersection of eight streets at that point, but all the other stations had surface buildings of a standard design, with elevations of unglazed terracotta, half-baked to obtain a light brown colour. In most cases the flat roof carried several stories of offices and chambers.

The trains were of seven cars, manned front and rear by guards who were assisted by four gatemen on the intervening car platforms. There were 30 electric locomotives, built by the General Electric Co. of America, and each camel-backed machine weighed 44 tons. The steel cabs were mounted on two four-wheel trucks, each equipped with two 170-h.p. gearless motors rigidly and directly fixed to the axles as on the CSLR engines. The livery was crimson lake with gold lettering. Brass rails and fittings were kept brilliantly polished by the two-man crew. Similar locomotives of larger dimensions were produced for the Paris–Orléans Railway and the Buffalo & Lockport Railway at this period, and all three designs were descended from the original machines evolved by the GEC for the Baltimore tunnel electrification of 1895. Strange though it may seem, the CLR also possessed two steam locomotives—Hunslet 0–6–0T built to the tube loading gauge and fitted with condensing apparatus. These oil-fired engines spent most of their time shunting at the depot.

The 168[3] cars each seated forty-eight passengers and were 45 feet 6 inches long over platforms, 8 feet 6 inches wide at waist and 9 feet $4\frac{1}{2}$ inches high overall (ceiling height 7 feet 5 inches). Teak and mahogany bodies, strengthened with steel, rested on two Leeds Forge trucks equipped with 29-inch wooden-centred Mansell wheels. Tare weight was 14 tons. Brush supplied twenty-five cars, Ashbury the remainder. Inside, they put the *Padded Cells* to shame. Glass-shaded bulbs shone brilliantly from the centre of the clerestory, showing off the polished brass hat hooks and basket racks above the large square windows. Seats were resplendent in blue and crimson velvet and mostly longitudinal (there were four sets of double transverse each side of the gangway in the centre). Each seat was properly divided from its neighbour by a heavy leather armrest in massive contrast to the hip-scratching

---

[1] As the lifts were not ready, this station was not opened until September 24, 1900.
[2] Renamed Queensway, September 1, 1946.
[3] Eighteen were not delivered until 1901.

'separators' of the Waterloo & City. Wooden rods, carrying straps for the support of the standing load, were attached to the deck rails. Outside, the cars were purple-brown below waist and white above, with gold lining, CENTRAL LONDON in four-inch gold block letters on the waist panels, and the Company's coat of arms.[1] Access was gained at the ends, through gated platforms covered by the clerestory. Wooden doors, at first unglazed, led from the gate platforms into the car. The end cars of each train were reserved for smokers and were distinctively upholstered in maroon haircloth.

Following Greathead's plan for the CSLR, each CLR station was built on a hump to assist deceleration and acceleration, the gradients being 1 in 60 up and 1 in 30 down. There were crossovers between the eastbound and westbound lines at each terminus and sidings between the running tunnels at Queens Road, Marble Arch and British Museum. Marble Arch also had a locomotive standby spur. Except at Post Office, Chancery Lane and Notting Hill Gate, the 325-foot wooden platforms [2] were in separate and parallel white-tiled tunnels connected by crosspassages. At the other three stations, one platform was sited directly above the other to avoid passing outside the line of the street. There were fifty-four lifts, hydraulically operated at Shepherd's Bush but otherwise Sprague electric. Lift capacity was generally 50 per car, but the four lifts at Shepherd's Bush carried 55 and each car at the Bank accommodated 90.

Beyond Shepherd's Bush station, a single line curved sharply north and climbed to the surface where a back shunt connection led into locomotive and car sheds, which occupied the site of a large house called Woodhouse Park. Here also were the power house, car washing sheds, repair shop and machine shop, and a spur running east to the West London Railway.

There were seventeen signal boxes controlling mechanically worked semaphore starting signals and sliding spectacle inner and outer homes. Most boxes were sited on the ends of the platforms against the tunnel headwall. A brush on the last truck of each train contacted a copper treadle plate which released a lock-and-block mechanism and allowed the signalman to lower the starting signal in rear. In 1912–1913, a fully automatic electro-pneumatic, a.c. track-circuited colour-light signalling system with train stops was installed by the McKenzie, Holland & Westinghouse Power Signal Co. All signal boxes were then closed except seven needed for reversal, shunting and train regulation. Train stops had appeared earlier, with the mechanical signalling, in January 1908, and their adoption, together with the dead-man's handle, enabled the Company to dispense with assistant drivers.

[1] The only tube companies to affect coats of arms were the Central London and the City & South London—see *The Railway Magazine*, Vol. 96, pp. 381–2.
[2] Replaced by stone and iron platforms from 1908 onwards.

The tunnels were 11 feet 8¼ inches[1] in diameter (12 feet 5 inches on curves) and contained 60-foot, 100-lb. bridge-type running rails fastened to longitudinal sleepers, either side of an 85-lb. channel section conductor rail carrying the traction supply at 550 volts d.c. At the Wood Lane power house, sixteen Babcock & Wilcox boilers supplied steam to six 1,300-h.p. reciprocating Corliss engines which drove 25-cycle 850-kW American GEC alternators. Output was at 5,000 volts, three-phase a.c., which was fed to rotary converter substations placed at the base of lift shafts at Notting Hill Gate, Marble Arch and Post Office. Three battery stations supplied lighting current for emergencies or after hours. Another substation was opened in 1903 adjacent to Bond Street booking hall to provide current for emergency and night use, and a further rotary converter was installed as a standby at the power house about the same time. Later, another ground-level substation was built at Post Office, and was intended, like Bond Street, to replace any disabled deep-level station. No provision was apparently made for lighting in the event of a complete failure at the power house of longer duration than the hour the batteries would last. All the original electrical installation, including the locomotive equipment, was by British Thomson-Houston.

Trains ran on a basic headway of five minutes between 5 a.m. and midnight, and the service soon built up to about thirty trains an hour at peak periods. Traffic grew in a most satisfying manner, reaching 45,305,110 in what proved to be the peak year of 1902 and thereafter settling down to just under 45 million a year until the motorbus began to eat into receipts from mid-1906 onwards. At first a 4% dividend was paid on ordinary shares, but from 1907, bus competition reduced this to 3 and then 2%. Undoubtedly the simple fare system and the happy choice of route contributed much to the early success.

But it was not all milk and honey for the Central London. For all its brash modernity and achievement, it rattled and it *smelt*.

'For myself, I could not risk my health by travelling on such a railway as the Central London, with its obnoxious vapour, which I know from meeting it at the stations; I could not risk my health by travelling upon it, and I hear so many complaints too . . . many individuals I know are giving it up. . . .' So spoke Sir Theodore Martin, KCVO, KCB, Parliamentary Agent, in evidence before the Royal Commission on London Traffic (February 19, 1904). A few days later, the Commission heard Granville C. Cuningham, the company's General Manager, admit to a 'peculiar smell' that was difficult to account for. He thought it might be the smell of the earth and added that there had been objec-

---

[1] At first it was intended that the tunnels should be concrete-lined throughout with an internal diameter of 11 feet 6 inches when finished. About 1898 it was decided to confine the lining to the first fifteen rings from each station.

tions from 'delicate people, ladies and others'. Everyone was disarmingly honest about the whole business. The chairman went along to the Commission and told them about a 'huge fan' which had been erected the previous year at Shepherd's Bush.[1] At night they closed the doors of the stations and sucked out the air of the whole line. 'But,' continued Sir Henry Oakley, 'I cannot say we are quite satisfied, there is a certain degree of flatness in the atmosphere.' The LCC made investigations and found no unusual excess of carbon dioxide or microorganisms, but Max Rittenburg, writing in the *Daily Express* in March 1907, attributed the unpleasantness to excessive dryness, finding the average humidity 44·5% against an average street humidity of 75·7%. Another exhaust fan was fitted at British Museum and kept running continuously, but complaints continued until an elaborate pressure system, with 6,000 cubic feet per minute fans injecting filtered and 'ozonized' air was fitted in 1911.[2]

Another snag came first to notice in the autumn of 1900 when occupiers of premises along the route began to complain that the trains were causing vibration. In view of the large amount of tube railway promotion at this time these complaints attracted serious attention, and in January 1901 the Board of Trade appointed a committee to investigate and report. Lord Rayleigh, Sir John Wolfe Barry and Professor J. A. Ewing were joined by A. Mallock as technical adviser. Evidence was given that the trains caused draughtsmen in a Cheapside surveyor's office to draw wavy lines, and a resident of Hyde Park Terrace told how he was awakened by the trains at 5 a.m., when the bed shook, his windows rattled and the vibration generally gave 'a very good imitation of a small earthquake'. Seven other occupiers described their wobbles.

Mr Mallock and pupils of the Professor set themselves up in houses along the route equipped with seismographs, cameras and telephones connected to the nearest signal boxes. The needles waggled obediently at the passing of each train and by May 1901 an interim report showed that the movement was due to the large proportion of locomotive weight unsprung and also to a lack of rigidity in the track.

The CLR took note of the unsprung weight hint and in the summer of 1901 converted three of their thirty locomotives to geared operation, thus reducing the overall weight from 44 tons to 31 and the unsprung weight from 34 to 10¾ tons. At the same time, four of the passenger

---

[1] The engineers had not considered forced ventilation of the line necessary when it was built—it was thought that the piston action of trains in the close-fitting tunnels would produce enough air circulation. It soon became obvious that this was not so, and the first experimental fan had been fitted at Bond Street early in 1902.

[2] Ozone was also provided for the enjoyment of other tube passengers. In 1913–1914 plants were installed at Edgware Road, Euston, Goodge Street and Charing Cross (Embankment).

cars were rebuilt as 20½-ton motor cars[1] equipped for Thomson-Houston multiple-unit operation. One truck of each car was replaced by a power bogie with two 100-h.p. motors. The floor at one end of the car was raised to clear the motor bogie and the driver's cab was built at that end. At first, two motor cars ran with two trailers, but two further trailers were soon added, giving a train formation of M-T-T-T-T-M.

Crouched over their apparatus in the bedrooms of Notting Hill Gate and in City offices, the observers found that the geared locomotives reduced the vibration to less than one-third of that made by the others but the multiple-unit trains produced less than one-fifth. The Committee's report was published in February 1902. It stated that a deeper and stiffer rail would have been preferable to the bridge rails used, but the rail type could not now be changed as the vertical clearances were too tight.[2] Larger clearances and deeper rails were recommended for future tube railways. Uneven rail surfaces were the main source of disturbance, although there would have been no unpleasant vibrations if the vehicles had been mainly spring-borne. Motor cars would so reduce the vibration that serious annoyance would disappear.

The company accepted the report with commendable promptness and ordered 64 new motor cars in May 1902, 24 from the Metropolitan Amalgamated Railway Carriage & Wagon Co. and 40 from the Birmingham Railway Carriage & Wagon Co. These cars, which had BTH non-automatic electro-magnetic control equipment and two 125-h.p. motors, were delivered early in 1903 and were made up with the trailers into seven-car trains. The complete changeover, which was the first general adoption of contactor-type control and multiple-unit operation in Britain, was finished in June 1903. In the same year, the CLR took delivery of the first all-steel railway cars to be operated in Britain, six trailers built by the Birmingham Co. The new motor cars had a half-octagonal roof over the driver's cab, a CLR characteristic which can still be seen on some LT sleet locomotives made up from these cars. The offending locomotives were offered for sale, but no buyers were found for a long time. Two, rebuilt with nose-suspended motors, were kept for shunting in the depot yard, one surviving until about 1943. In 1905 two of the machines were sold for regenerative control experiments by the Metropolitan Railway,[3] several were used on industrial lines and the remainder were broken up.

Both the City & South London and Central London had the unfortunate experience of finding that their rolling stock, on delivery, proved to be too large for the tube tunnels. The Ashbury cars of the

---

[1] Two of these were reconverted to trailers in 1906.
[2] It was noted that the tight clearance increased the air pressure in front of moving trains and thus wasted power.
[3] Painted grey, and broken up at Neasden in 1913.

CSLR were fitted on site with 12-inch diameter wheels instead of 24 inch whilst their roof sticks were straightened by 5 inches, and this work was one of the reasons for the delay in opening the line to the public. When it was found that the CLR locomotives were too high, new springs were fitted and shallower rails laid, which later dropped at the joints. Here, perhaps, lay the root cause of the vibration troubles.

CHAPTER 4

Titan Turns East

1900–1901

THE TRACTION KING

BEFORE carrying the tube story further, we must digress a little in order to take a close look at an American who was to inject new life into the London traffic situation, and to examine the hothouse atmosphere of nineteenth-century American urban capitalism in which this man gained his experience and wealth.

Charles Tyson Yerkes[1] (pronounced to rhyme with 'turkeys') was born in Philadelphia in 1837, into a Quaker banking family. At the age of twenty-two he opened his own brokerage office, and three years later purchased his own banking house. He made his name by selling a City bond issue at par when his predecessors had been able to sell only at a heavy discount, and by the late 1860s he seemed assured of a promising future as a leader of Philadelphia finance.

In October 1871 a disastrous fire burnt out the commercial heart of Chicago, and caused heavy losses for the eastern capitalists who had owned buildings and merchandise there. The ensuing financial panic caught Yerkes with his funds spread very thinly in a scheme to enlarge his Philadelphia tramway holdings, and he was forced into bankruptcy. Worse, he was indicted for technical embezzlement because he could not immediately deliver to the City the money he had received from the sale of municipal bonds. He was sentenced to thirty-three months' imprisonment in the penitentiary, but was pardoned after seven months. After release from prison, Yerkes picked up the threads of his business, but he was ostracized by Philadelphia society and, sensing the westward flow of European gold, he moved to Chicago in the early 1880s.

From the end of the Civil War to the turn of the century, corrup-

---

[1] A good short biography of Yerkes is included in the *Dictionary of American Biography* (Oxford University Press, 1936). Upon his careers in Philadelphia and Chicago were based Theodore Dreiser's novels *The Financier* and *The Titan*, which describe the environment in which Yerkes prospered.

tion was widespread in American federal, state and municipal government. The authorities had a wide variety of blessings to distribute, for which business was prepared to pay its price—public utility franchises, contracts, the fixing of utility rates and charges, favourable legislation. The colloquial term for the monetary bribes was 'boodle', and the recipients, at city level, 'boodle aldermen'.

Tramway operation in Chicago and other rapidly expanding cities was governed by municipal franchises which gave monopoly powers to exploit particular lines for a fixed number of years. Once assured of franchises, the tramway magnates could safely buy up small and inefficient companies, refloat them with a heavy addition of watered stock (which sold well because of the monopoly attractions of the franchise) and make further profits by cabling or electrifying the horse lines, or building new lines, through sudsidiary construction companies which charged exorbitant prices for their services. When the idea of holding companies was developed, the way was clear for endless combinations of holding, construction and operating companies, each new flotation carrying a handsome profit for the financier.

On the Stock Exchanges, the financiers 'made a market' for their stocks, partly by astute market operations and partly by paying dividends which were far higher than earnings justified. These high dividends could ultimately come from only one source—capital, but they lasted long enough for the magnates to sell out at high prices before the crash came.[1]

When Yerkes arrived in Chicago he began in a small way, in a brokerage business, while he was finding the lie of the land. In 1886, with the aid of the Philadelphia traction kings, Widener and Elkins, he gained control of the North Division tramway companies; the West Division companies followed soon afterwards.

Yerkes made substantial physical improvements in the Chicago tramways, cabling or electrifying horse lines, creating a huge suburban network and adapting for tramway purposes two disused tunnels under the Chicago river. Nevertheless, there were many complaints about infrequency, irregularity, breakdowns and overcrowding, although some of them may have been politically inspired. In reply to complaints of overcrowding, Yerkes pronounced his frank and oft-quoted dictum, 'It is the straphanger that pays the dividends'. The complementary quotation is his formula for success in the tramway business—'Buy up old junk, fix it up a little, and unload it upon other fellows'.

Elevated railways introduced a further complication in Chicago. They had been operating successfully in New York for many years, and there was public pressure for such lines, both to rival New York and to feed

[1] An interesting description of the magnates' activities is given in Chapter 5 of *The Age of Big Business* by Burton J. Hendrick, Oxford University Press, 1919.

the 1893 Columbian Exposition. The first elevated line in Chicago, the South Side, fell to Yerkes' rivals, and opened, with steam traction, in 1892. Yerkes was forced to develop further elevated lines to protect his own tramways, but they represented very large investments in structural steel, equipment, engineering fees and labour; it took a long time before traffic built up to remunerative levels.

Most of the franchises of the north and west division tramway companies were due to expire in 1903; if Yerkes could secure 50- or 100-year franchises, his stocks would greatly appreciate and he would make a further large fortune.

A first attempt in 1895 resulted in the Illinois state legislature passing the Humphrey Bills, which renewed the franchises for 100 years without payment to the City, but these were vetoed by Governor Altgeld. Next came the Allen Bills, passed by the state legislature but requiring endorsement by the Chicago City Council before the proposed 50-year franchises became law. By now, popular feeling had been thoroughly aroused against Yerkes, and the Chicago aldermen were badly frightened. On the night the vote was taken in December 1898, mobs with guns and sticks paraded the streets, and a hempen noose was lowered from the gallery of the City Hall. The Allen Bills were rejected, and Yerkes had lost his chance to secure long-term franchises, despite the distribution of roundly $1 million in bribes. In 1897 he had successfully floated the $12 million Union Traction Company (a tramway holding company), which was reported to have yielded him $10 million in cash and $6¾ million in bonds.

In 1899 Yerkes sold his Chicago traction investments to Widener and Elkins for a figure approaching $20 million and moved (with $15 million in cash) to his Fifth Avenue mansion in New York. He had controlled 432 single-track miles of electric tramway, 48 miles of cable tramway, 7 of horse tramway and a prospective 43 miles of elevated railway. The ineluctable result of his financial activities came in April 1903 when the Union Traction Company was declared bankrupt.

After leaving Chicago, Yerkes might well have retired to enjoy his fortune, but his ambition spurred him on to fresh adventures. His Chicago plans had been frustrated, and the social ostracism stemming from his Philadelphia imprisonment had followed him from city to city. He may have wished to prove to the world that he could prosper in a land where there was no 'boodle'.

DOLLARS FOR THE HAMPSTEAD

The first London line with which Yerkes became associated was the Charing Cross, Euston & Hampstead, brought to his attention by R. W. Perks.[1]

Perks had many interests. He was an enthusiastic worker for the

---
[1] 1849–1934. Created a Baronet in 1908. Perks and Yerkes were old friends.

An original train of the Waterloo & City Railway in the yard at Waterloo station (motor car no. 9 leading).—*W. L. Box, courtesy C. E. Box.*

Central London Railway locomotive no. 19 about to leave Wood Lane depot for a trial run, 1900.—*London Transport.*

Central London Railway station architecture—Notting Hill Gate about 1914. The first floor site remains unlet and the large "TUBE" sign, a CLR feature, has not yet been demolished by the new owners.—*London Transport.*

Great Northern & City Railway trailer car 17 in the maker's yard at Preston.—*London Transport.*

Tube construction with a Greathead shield. Miners are cutting the clay in the heading at the rear and the hydraulic jacks can be seen pushing against the last completed ring of segments, ready to propel the shield further forward.—*London Transport.*

Methodist Church, and organized the raising of a million guineas to celebrate the dawn of the twentieth century (part of the money was used to build Westminster Central Hall). As Liberal MP for Louth, Lincs., he sat in Parliament from 1892 to 1910. He had previously qualified as a solicitor, and in 1876 became partner of Sir Henry Fowler (later Lord Wolverhampton) to establish a legal firm which specialized in railway and parliamentary practice. He was legal adviser to the Metropolitan Railway from 1879 to 1892, when he acquired a substantial holding in the District Railway.

The association between Perks and the Hampstead railway appears to have begun at the end of 1897 when Sir Henry Fowler joined the Hampstead board. From then onwards Perks is recorded as 'in attendance' at most meetings, reporting on his negotiations with railway companies, contractors and engineers. He acted as intermediary between Yerkes and the Hampstead company, with the result that on September 28, 1900, the existing directors retired and were replaced by Yerkes (chairman), H. C. Davis (vice-chairman), R. W. Perks and Ernest Halsey.

On Saturday, October 6th, came the official announcement that a Yerkes syndicate had taken over the Hampstead on the previous Monday (October 1st) for £100,000. The syndicate was said to include Widener and Elkins of Philadelphia and Marshall Field (of store fame) of Chicago.

An early result of the take-over was the decision to apply for an extension from Hampstead to Golders Green. There are various versions of the field surveys leading to this decision, but the authors prefer that related by H. H. Dalrymple Hay[1] in 1922. First, Yerkes sent an agent, Lauderbeck, to investigate; the latter arranged for Dalrymple Hay to meet him one morning at the Hotel Cecil, and they took a two-horse cab over the authorized route. On the journey, Dalrymple Hay described the proposed 1 in 24 gradients associated with the original cable-traction proposal, and the depot at Hampstead. Lauderbeck was not at all happy with these proposals, and stated that he could not recommend the proposition to Yerkes unless the line were extended to open country. They accordingly drove on past Jack Straw's Castle and down North End Road. When they came to Golders Green crossroads, Lauderbeck stopped and told Dalrymple Hay that here was the proper site for the terminus, meeting protests about the absence of houses by pointing out that in the USA, railways were built and the people followed.

When plans for the extension to Golders Green had been prepared, Yerkes went over the route with the vice-chairman, Davis. It was a very wet day, but when they alighted near Jack Straw's Castle, the sun

[1] Knighted in January 1933.

came out and illuminated the spires and towers of London below. Yerkes asked, 'Where's London?', and on being shown, turned to his companion with the words, 'Davis, I'll make this railway'.

RUINING THE HEATH

The deposited plans for the 1901 Session of Parliament disclosed an ambitious series of proposals to improve the Hampstead. The extension from the original Hampstead terminus would run to the west of The Grove, then below Whitestone Pond and east of North End Road to terminate just east of the Finchley Road, north of the Golders Green cross-roads. A large tract of land was scheduled at Golders Green for a power station, depot and sidings. The proposed alterations in levels at Hampstead and points south had the effect of making the railway much deeper but considerably flatter. At Haverstock Hill, a 1 in 73·2 gradient replaced 1 in 27, and further north, 1 in 40 replaced 1 in 24. The line would now be 108½ feet deep at Belsize Grove instead of 44 feet, and 153 feet at Hampstead High Street/Church Lane (now Perrins Lane) instead of 53 feet. Other proposals included a deviation to ease the curve between Warren Street and Euston, subways (for a sub-surface booking hall) at Tottenham Court Road, and enlargement of the maximum size for running tunnels from 11 feet 6 inches to 13 feet to avoid vibration trouble.

These proposals were included in the 'No. 1' Bill, but there was also a 'No. 2' Bill which provided for extensions from Kentish Town via the Archway Tavern to Archway Road/Bishops Road (near Highgate GNR station, with which interchange was proposed), and from the Garrick Thea :, Charing Cross Road, via Whitehall and Victoria Street to Victoria station.

The proposal for a Golders Green extension was strenuously opposed by certain Hampstead residents, who believed that the tube would ruin Hampstead Heath. The vehemence of the opposition may surprise the present-day reader, but there had been a long history of struggles to preserve the Heath as an open space.

The Times for December 25, 1900, carried a long account of the proposed extension and its supposed disadvantages, from 'a correspondent'. He assumed, correctly, that the promoters intended to build intermediate stations between Hampstead and Golders Green, but he then proceeded to state, as facts, an extraordinary series of assumptions. 'A great tube laid under the Heath will, of course, act as a drain, and it is quite likely that the grass and gorse and trees on the Heath will suffer from lack of moisture.... Moreover, it seems established that tube trains shake the earth to its surface; the constant jar and quiver will probably have a serious effect on the trees by loosening the roots.'

The local opponents formed an Executive Committee under the chairmanship of Samuel Figgis (who had earlier been instrumental in securing the acquisition of Golders Hill Park for public use) and held a public protest meeting on January 24, 1901.

Meanwhile the Hampstead engineers had issued a statement pouring cold water on some of the wilder rumours. The tunnels would be deep in the London clay, with a minimum depth of 150 feet below the Heath; there was no risk of draining the Heath, and as multiple-unit trains would be used (and not locomotives) there was no risk of vibration. They confirmed the intention to build a station at Jack Straw's Castle, but not a hotel (which had been rumoured). Would it not be wiser for the local residents to concur in the extension, rather than see the authorized power station erected in the centre of Hampstead?

Hampstead Borough Council initially opposed the Bill, but withdrew its opposition upon the company's agreeing to insert certain protective clauses. These included the prohibition of any station within the Borough north of the authorized station at High Street/Heath Street. Figgis and his colleagues were deeply disappointed and pinned their hopes to approaches to the LCC or local MPs. As events turned out, little more was heard of the matter until the 1902 Session, when the whole question was raised again.

BILLS GALORE

Whilst Yerkes had been buying the Hampstead, the Central London was settling down to what promised to be a most successful future. Traffic was good, the flat fare helped the railway to catch the public imagination as the 'Twopenny Tube', and the equipment worked well. Gross revenue increased steadily, although the high prices of fuel and materials caused nearly 60% of the revenue to be absorbed in working expenses. Provided that this ratio could be reduced to nearer the theoretical 50%, there were good prospects of a steady 4% on ordinary capital.

It thus appeared that with modern equipment, London tube railways were capable of yielding substantially better ordinary dividends than the miserable results so far achieved by the CSLR.

Promoters looked at tube schemes with renewed interest, and as the statutory notices appeared for the 1901 session, it became clear that London was to experience a 'Tube Mania'.

The Government appointed a Joint Select Committee to report whether the routes in the current Bills were best suited to present and future traffic trends, and, if not, what modifications were desirable; to decide what special provisions should be made to protect property owners and occupiers from damage and annoyance, what special terms and conditions about construction and working (if any) should be applied to the promoters, and to say whether any of the current Bills should not be allowed to proceed. Five members were appointed from each House, with Lord Windsor as Chairman.

Two Bills early fell by the wayside. Owing to vibration fears, a proposed revival of the 1897 City & West End scheme was not even deposited, and a scheme with the imposing title of the Victoria, City & Southern Electric Railway (Victoria–City–Southwark–Peckham) was withdrawn by the end of January. One other Bill, the City & Brixton, was later withdrawn from the jurisdiction of the Committee because it was mainly for extension of time of a scheme approved in 1898. The remaining eleven schemes considered by the Committee were as follows:

Bill	Main Proposals
Central London	Terminal loops at Shepherd's Bush and Liverpool Street.
Charing Cross, Euston and Hampstead (as described above)	No. 1 Bill—extension to Golders Green. No. 2 Bill—extension to Highgate and Victoria.
Islington and Euston	Extension of the CSLR from Angel to Euston.
West and South London Junction	New line from Paddington to Kennington Church via Marble Arch, Hyde Park Corner, Victoria and Vauxhall.
Kings Road	New line from Victoria (junction with WSLJR) to Eelbrook Common, Fulham via Kings Road).
City and North East Suburban Electric	New line from Cornhill/Gracechurch Street to Waltham Abbey via Hackney Road, Victoria Park, Hackney Marshes and Walthamstow.
North East London	New line from Cannon Street to Tottenham (Page Green) via Bishopsgate, Kingsland Road, Dalston and Stamford Hill, with a branch from Stoke Newington to Walthamstow.
Piccadilly and City	New line from end-on junction with NELR at Cannon Street to junction with Brompton & Piccadilly Circus at Piccadilly Circus via Cannon Street, Carter Lane, Ludgate Hill, Fleet Street and Strand.
Charing Cross, Hammersmith and District	New line from Hammersmith to Charing Cross via Hammersmith Road, Kensington, Knightsbridge, Hyde Park Corner and The Mall.
Brompton and Piccadilly Circus	Extension from Piccadilly Circus to the Angel, Islington, via Shaftesbury Avenue, Theobalds Road, Rosebery Avenue. Another extension from Brompton Road to Chelsea via Fulham Road. (Of these proposals, all except the 6 furlongs to Bloomsbury Square failed to pass Standing Orders.)

The hearings lasted from May 2 to July 2, 1901. Amongst the witnesses was the Chief Inspecting Officer of Railways, Lieut.-Col. H. A. Yorke, RE, who recommended that confluent junctions should be avoided where possible, because of the requirement that two trains should not approach a converging junction simultaneously, in relation to the difficulty of sighting signals in a tube. He disliked long through routes, because of the widespread effects of breakdowns on such lines.

The Committee's report was issued on July 26th. The main recommendations were that the lines to be preferred should feed buses or trams at their termini, but should be capable of extension; lines should run between traffic centres, or from catchment areas to such centres; 'extension of time' applications should be closely examined, and the City Corporation and County Councils should have *locus* to object thereto; terminal loops should be avoided in the City, as they might block other schemes; confluent junctions should be avoided; stations should not discharge passengers into crowded streets; subway connections should be built at tube intersections; a standard workmen's fare clause should be adopted. It also recommended more direct control and supervision of underground railway projects, and that the City Corporation and the County Councils be given powers to construct, or financially support new underground lines.

With regard to the individual proposals, only one scheme was completely rejected, that of the Brompton & Piccadilly Circus to Bloomsbury. The others were referred to the normal committee, subject to the following observations:

Central London	City loops not favoured.
Hampstead	One of the branches should be worked as a shuttle.
Kings Road	Need for interchange facilities at Victoria. A Putney Bridge extension was desirable.
NELR	Extension further north was desirable. The Walthamstow branch was unnecessary if the CNESER was built.
CNESER	The City–Victoria Park routeing was undesirable; a more easterly route was recommended.
Charing Cross, Hammersmith and District, Piccadilly and City	There should be an end-on junction at Piccadilly Circus to give a through route between Hammersmith and the City.
Brompton and Piccadilly Circus	Not approved, but if the above through route were secured, favourable consideration should be given to an Angel extension.

It was now too late for the Bills to proceed through the normal stages of Parliament in the 1901 Session, and leave was given for them to be suspended till 1902. Of the eleven Bills, two (CLR and BPC) were later withdrawn and replaced by more ambitious proposals, three were carried over intact to 1902, and six were carried over with complementary new Bills amending and extending them in various ways.

CHAPTER 5

Yerkes Takes Over

1902–1905

TRANSATLANTIC PRESSURES

THE personal circumstances which induced Yerkes to turn from Chicago to London coincided with a larger movement overseas of American investment funds.

There were several factors that damped enthusiasm for home investment in the USA. In the public utility field, the citizens had grown weary of being fleeced by financiers, and the pressure for reform began to show results. Theodore Roosevelt set out to enforce the existing anti-trust legislation, and the prosecution of the Northern Securities Co. in February 1902 prompted Wall Street to wonder whether any holding company was immune from such an attack. Furthermore the home market for products, raw materials and investments had become temporarily saturated, and the need to invest overseas coincided with the development of imperialist attitudes springing from America's swift and overwhelming victory in her 1898 war with Spain.

THE DISTRICT IN THE NET

In London, the District Railway, under the twin influences of Forbes' blighting chairmanship and competition from the Central London, was sinking ever deeper into the mire. Perks, who had large holdings of District stock, was naturally interested in any possibility of reviving its fortunes, and, according to popular account, wrote to Yerkes to ask him to look into the prospect of electrifying the District. Perks' own version of the matter was more modest. When giving evidence to the Royal Commission on London Traffic in 1904 he said '... we were brought into communication with friends of ours in America who were attracted by the capabilities of the District Railway, and who decided that they would attempt to apply to the District Railway in London ... the same method of successful electric working which they had been instrumental in introducing in New York, Philadelphia, Chicago and elsewhere'.

Yerkes and his associates began by buying £500,000 nominal value of District ordinary stock at the price of 25; a further £450,000 worth of the same stock was bought at 35. Substantial amounts of second preference and debenture stock were also acquired, and by March 1901 Yerkes was in effective control. The change was publicly announced on June 6, 1901, although there had been many earlier hints and rumours. In April the *Tramway and Railway World* had stated that it hoped that the rumours that Yerkes was in control were true. 'Half measures have never characterized Mr Yerkes' undertakings, and he is not accustomed to hesitating at trifles.'

One of the first results of Yerkes' assumption of control was the great electrification controversy with the Metropolitan Railway. The latter supported the Ganz system, using three-phase alternating current, generated at 11–12,000 volts, and stepped down in unmanned transformer stations to three-phase 3,000 volts, passing to the trains by twin overhead wires, the rails forming the third conductor. Yerkes wanted the d.c. system, which had been successful on his Chicago elevated lines, and as the companies could not agree voluntarily, the Board of Trade appointed an arbitration tribunal in September 1901.

The tribunal favoured the d.c. system, because of the severe traffic conditions on the Inner Circle, the reliability of the d.c. and the untried nature of the Ganz. The Ganz company alleged that when Yerkes had visited them at Budapest, they had given him all possible information and assistance, but he had given no hint that arbitration might follow.

Meanwhile provision had been made for financing the District electrification by registering on July 15, 1901, the Metropolitan District Electric Traction Co. Ltd, whose principal objects were to erect and equip the Lots Road power station (originally authorized to the Brompton & Piccadilly Circus company in 1897) and to supply funds for the electrification. The company had a capital of one million pounds in £20 shares, of which Yerkes, as managing director, took a large number. The remaining shares were taken up privately, mainly by Americans. The board of directors had an appropriately Anglo-American flavour (C. T. Yerkes; Walter Abbott of Boston, Mass.; Patrick Calhoun of New York; C. A. Grenfell of Throgmorton Avenue; and Murray Griffith of Austin Friars), and the secretary was J. W. Brown of Baltimore.

THREE MORE TUBES FOR YERKES

Once the District Railway had fallen into the Yerkes net, it was only a matter of time before its adopted child, the Brompton & Piccadilly Circus tube, followed. Formalities were completed during the summer of 1901, and the Brompton board minutes for September 12, 1901, recorded laconically—'Mr Perks reported that the interest of the District Railway in the Brompton & Piccadilly Circus was to be transferred to Mr C. T. Yerkes as his nominee.' Yerkes' nominees thereupon joined the Brompton board, which henceforth took on the semi-puppet nature common to the boards of the subsidiary companies, with all the major policy decisions being made at group level.

Yerkes next turned his attention to the dormant Great Northern & Strand scheme, and when the board of that company met on September 26, 1901, they had before them the heads of a carefully-devised agreement between the GNS, the Brompton company and the Great Northern Railway. The BPC was to take over the GNS powers from Finsbury Park to the Strand, seek powers for a Piccadilly Circus–Holborn connection between the two companies' lines, and, if successful in this application, to work the line from Earl's Court to Finsbury Park as a continuous railway. The combined system was to be known as the Great Northern, Piccadilly & Brompton Railway.

The suburban traffic of the Great Northern Railway was carefully protected. The GNS was to abandon its powers from just south of Finsbury Park to Wood Green, but this would not affect the right of the GNR to build such a line. The BPC was to pay an easement of £2,500 per annum in respect of the section of tube beneath GNR property, and the GNR was to build Finsbury Park (Piccadilly) station and lease it to the tube company at a yearly rental of 4% of the cost.

Then followed three all-embracing paragraphs which effectively bottled up the Piccadilly tube at Finsbury Park. The BPC was not to construct without GNR consent any railway north of Finsbury Park, nor without such consent to acquiesce in the construction of any such railway by any other company. Should any such railway be made north of Finsbury Park, no through trains were to be run without GNR consent. The Brompton company was not to dispose of its railway or its working to any company owning or controlling any railway north of Finsbury Park without GNR consent. The lease of Finsbury Park station was to contain provisos for the re-entry of the GNR to the station if any of these conditions were breached.

The GNR solicitors, doubtless old hands at the game, seemed to have thought of everything, and the Finsbury Park legal cork stayed securely in the Piccadilly bottle until blown out by force of public opinion in the 1920s. The Great Northern & City was similarly tied up at Finsbury Park at about the same time, and both Finsbury Park tube stations remained in main line ownership until nationalized in 1948.

Meanwhile the half-constructed Baker Street & Waterloo, left out in the cold after the London & Globe collapse, was in the market. Negotiations with possible buyers (including Yerkes) dragged on for months. Finally, on March 7, 1902, an agreement was sealed for the Metropolitan District Electric Traction Co. to buy out the London & Globe's interest in the Baker Street & Waterloo, and to act as 'contractor' for the completion of the line. The purchase price was £360,000, plus interest from January 1902. By taking control of the company, Yerkes also assumed obligations to the members of the public holding Baker Street & Waterloo shares, but nevertheless he secured a good bargain, as at least double the purchase price had already been spent on the line, and his group would also benefit from the release of the £107,000 parliamentary deposit when the railway was completed.

A happy early result of the take-over was an increase in salary for Baker Street & Waterloo officials. Even the junior clerk received a 2s 6d addition to his humble 10s a week.

FINANCIAL CONSOLIDATION

By now, Yerkes had bitten off far more than the £1 million Traction company could chew, and in March 1902 he made a request for assistance to the banking and issuing house of Speyer Bros of Lothbury, London. This firm was destined to play a predominant role in the finance of the Yerkes group of undertakings, in conjunction with its associate, Speyer & Company of New York. Edgar Speyer, the senior partner of the London concern, was born in New York in 1862, son of Gustavus Speyer, a Jewish banker from Frankfurt-am-Main, where there was a third associated firm, L. Speyer-Ellisen. In 1884 Edgar became a partner in all three firms, and he came to London in 1887. Speyer Bros specialized in exchange arbitrage and railway finance.

When giving evidence to the Royal Commission on London Traffic in 1904, Edgar Speyer stated that he did not know Yerkes before the latter approached him in 1902, but upon being approached, his firm carefully examined Yerkes' project, found it sound, and agreed to raise £5 million capital for a new company with the resounding title of the Underground Electric Railways Company of London Limited (hereafter referred to as the UERL).

The new company was duly registered on April 9, 1902, with £5 million capital in £10 shares. The objects included the acquisition of the Traction Co., the acquisition and electrification of railways, light railways and tramways, and the generation and supply of electricity. Powers were also taken to act as company promoters, financiers, guarantors and contractors. In practice, the UERL was to act as 'contractor' for the construction of the tubes, receiving shares of the subsidiary companies in lieu of payment, as work progressed. Speyer Bros would market these shares on behalf of the UERL.

The directors of the UERL were C. T. Yerkes (elected chairman at the board of directors' first meeting on June 9, 1902); Walter Abbott of Berkeley Square; Frank Dawes, Old Broad Street, solicitor; the Right Hon. Lord Farrer; Major Ernest St. Clair Pemberton, RE, of Norwich;

C. J. Cater-Scott, of Leadenhall Street, Merchant, and Charles Ainsworth Spofford, Royal Hotel, Blackfriars.

The financial arrangements between the new company, the Traction company, Yerkes and Speyer Bros were extremely complex. The Traction company (in practice the shareholders, as the company was to be wound up) was to receive £500,000 in cash and 100,000 £10 UERL shares, half paid; Yerkes was to be rewarded for his services in connection with the formation of the UERL and Speyer Bros would receive commission for marketing the group's securities. Of the remaining 400,000 £10 shares, it was reported that about half would be subscribed in Britain and half in New York and Boston. Initially, all shares were privately subscribed.

An agreement was signed on June 25, 1902, between the UERL, the Traction company, Speyer Bros, Speyer & Co., and the Old Colony Trust Company of Boston, Mass. This provided, *inter alia*, for the Traction company's rights, assets and liabilities (including control of the four tubes) to be transferred to the UERL by July 8, 1902; that the two Speyer firms and the Trust company should have power to nominate a majority of directors of the UERL for ten years; and that Yerkes should act, if required, as UERL chairman for five years. The agreement was not to be impeached on the grounds that Yerkes, the Speyer firms or the Trust company were the promoters of the new company, or that Yerkes and the nominees of those firms were the first directors of the new company.

TUBE BOOM

While Yerkes' reorganization had been progressing, the Central London had gone from strength to strength. For the half-year to December 31, 1900, it paid a 2½% dividend on its undivided ordinary shares, having carried 14·9 million passengers in the five months since the public opening on July 30th. The following half-year showed a material improvement, with 20·4 million passengers, and the company was able to pay 4% per annum.

This shining example aroused fresh interest in tube railway promotion. Furthermore, apprehensions about vibration troubles had been assuaged by the Rayleigh Committee's interim report that the problem was not insoluble, and the effect of the Joint Select Committee's report of July 1901 was to invite applications for lines giving the Hammersmith–Piccadilly–City connection. As the leaves fell from the trees in Parliament Square in 1901, the pile of petitions in the Private Bill Office grew, until notice had been given for no fewer than thirty-two Bills involving tube railways (including extensions of time for authorized lines, and Bills carried over from 1901).

Three of the Bills failed at the first fence and one did not even start. These were the:

City and Old Kent Road   Not deposited. (This and another abortive 1902 project, the City and Surrey Electric Railway, proposed to use the disused King William Street–Borough section of the CSLR for their terminal section.)
City, Wandsworth and Wimbledon; East London, City and Peckham; Victoria, Kennington and Greenwich Abandoned by January 1902.

The District Railway had a Bill with various minor provisions relating to electrification. This encountered little opposition, and powers were obtained on August 8, 1902, to take additional land at Lots Road; to electrify the Wimbledon, Richmond and Hounslow branches; to extend the time for the construction of the Deep Level, and to constitute it a separate company if the District shareholders so desired.

Nearly all the tube Bills in this session were considered first by the House of Lords. Their Lordships arranged two Select Committees under the chairmanship of Lords Ribblesdale and Windsor. The Ribblesdale Committee dealt mainly with lines running in a north to south direction, proposals generally of a less contentious nature than those handled by the Windsor Committee, which considered the east–west schemes and especially the competing claims for a Hammersmith–Kensington–City connection.

THE RIBBLESDALE COMMITTEE—THE YERKES BILLS

The most important of the Yerkes' group Bills before the Ribblesdale Committee were those affecting the Charing Cross, Euston and Hampstead Railway. The 1901 plans were revived, but the proposed extensions from the Garrick Theatre to Victoria, and from the Archway Tavern to Bishops Road, Highgate, were dropped, leaving extensions from Kentish Town to the Archway Tavern and from Hampstead to Golders Green. The latter proposal was modified by a new Bill (the Charing Cross, Euston and Hampstead No. 3 of 1902) which provided for a deviation between Hampstead town and a point near the south end of The Avenue. At the southern end, the Craven Street terminus was to be replaced by a line down Villiers Street to Charing Cross (MDR) station. There was also a proposed station at Adelaide Street, with a subway to Charing Cross (SECR).

There was an independent Bill for an Edgware & Hampstead Railway, from Edgware (near the GNR station) to an end-on junction with the Charing Cross, Euston & Hampstead, at Holly Hill, Hampstead, via Hendon (The Burroughs), Brent Bridge and Golders Green. After swinging across North End Road near the present Manor House Hospital, the proposed line cut straight across the Heath. The promoters were J. H. Riley, Arthur Stirling and Captain W. K. Trotter.

The prospect of two tube railways beneath Hampstead Heath served to stir the redoubtable Mr Figgis and his associates into a frenzy of memoranda, letters to editors, deputations, and petitions, but the Hampstead Borough Council obtained protective clauses and actively supported the Bills, whilst the Ribblesdale Committee refused to give the Protection Society a *locus standi* to appear before it.

In evidence, Yerkes said that his company would be prepared to take over the powers of the Edgware & Hampstead and build it as a through line. Thereupon an agreement was reached for this railway to be curtailed at Golders Green where it would join the Hampstead, and for its powers to be transferred to Yerkes. (The Edgware & Hampstead board of directors was not, however, replaced by Yerkes nominees until January 1903.)

The Bill passed the Lords and came up for second reading in the Commons on July 16th, to which the House agreed without dividing. Two days later, W. J. Bull, MP [1] for Hammersmith, successfully moved an instruction to the Commons Select Committee to enquire whether the railway would seriously injure the Heath by 'tapping the wells, draining the soil, destroying the verdure and interfering with this public place of resort'. The Committee did as instructed, but was satisfied that the proposed railways were so deep beneath the Heath that all risk of injury to common land would be obviated.

Royal Assent was given on November 18, 1902, and the Hampstead Bills were consolidated into one Act. An interesting provision in Sec. 48 of the Edgware & Hampstead Act envisaged the line being worked as a light railway.

The other Yerkes Bill to come before the Ribblesdale Committee was the Great Northern & Strand, framed in accordance with the agreement of November 1901. At Finsbury Park, a deviation line 4 furlongs 7·4 chains long was to bring the line over to the west side of the GNR station. The 1899 line was abandoned between the southern end of the deviation and Wood Green. An extension was sought from the Strand terminus to Temple station via Norfolk Street, for interchange with the MDR, but this was killed by strong opposition from the Duke of Norfolk, other landowners, and the LCC. Other provisions included amalgamation with the Brompton company, and powers for the GNR to build Finsbury Park (Piccadilly) station. This Bill received Royal Assent on August 8, 1902.

THE RIBBLESDALE COMMITTEE—OTHER BILLS

The fate of the other Bills considered by the Ribblesdale Committee was as follows:

[1] Knighted in 1905 and created a Baronet in 1922; father of Mr Anthony Bull, Member of the London Transport Executive.

YERKES TAKES OVER

Bill	Main Provisions	Result
Great Northern and City	Extension from Moorgate to Lothbury	Royal Assent 8.8.02.
Islington and Euston	Extension of the CSLR from the 'Angel' to Euston	Refused (lack of evidence).
City and Crystal Palace	New line. Queen Street/Cannon Street to Penge West via Old Kent Road, Peckham, Dulwich and Kirkdale	Refused (financial grounds).
Baker Street and Waterloo	Extensions of time	Royal Assent 18.11.02.
North West London (authorized 1899 between Marble Arch and Cricklewood)	Extensions of time. Stations for interchange with CLR (Marble Arch), Baker Street and Waterloo (Edgware Road), LNWR (Kilburn), Hampstead Junction Railway (Brondesbury)	Royal Assent 18.11.02.

THE WINDSOR COMMITTEE

Before describing the main struggle in front of the Windsor Committee, mention may be made of two minor schemes brought forward from 1901. The original Kings Road Railway Bill of 1901 (from Victoria to Eelbrook Common, Fulham) became the 'No. 1' Bill, whilst the recommended extension from Fulham to Putney was embodied in a 'No. 2' Bill. The Committee rejected this scheme, and the associated West and South London Junction Railway (Paddington–Kennington) was thereupon withdrawn.

The 1901 Committee's recommendation for a through Hammersmith–City tube railway, via Kensington and Piccadilly, was so enthusiastically taken up that proposals to provide this valuable connection came from all who promoted Bills in 1901 *and* from the London United Tramways. The various projects were:

Brompton & Piccadilly Circus New lines between (i) Piccadilly Circus and Holborn (running connection with Great Northern & Strand, facing north), (ii) Piccadilly Circus and Charing Cross (running connection with District Deep Level, facing east), (iii) South Kensington and Parsons Green (connection with the District Railway's Putney Bridge line) via Fulham Road. Powers to make an agreement with the District to take over the District Deep Level between South Kensington and Earl's Court. Take-over of the GNS, and change of name to 'Great Northern, Piccadilly and Brompton Railway'.

(The original District Deep Level came to the surface between Gloucester Road and Earl's Court, and the witnesses giving evidence on this Bill at the end of April 1902 referred to Earl's Court as the western terminus, although they agreed that it was physically possible to run through to Hammersmith. However, an agreement dated April 17, 1902, between the District and Piccadilly Railways laid down that the former should provide access between the Deep Level and its surface tracks 'not further west than 20 chains west of West Kensington station'. In the Metropolitan District Railway (Various Powers) Act of 1903 there was a deviation of the Deep Level at Cromwell Road, and provision for a deep level station at Earl's Court, and in the MDR Act of 1904 powers were obtained to widen the District between West Kensington and Hammersmith so as to provide separate tracks for the Piccadilly trains. The District retained running powers over the Deep Level.)

Central London New line joining existing line at each end to form a circle, starting at Shepherd's Bush via Goldhawk Road, The Grove, Hammersmith Broadway, Hammersmith Road, Kensington High Street, Kensington Road, Kensington Gore, Knightsbridge, Piccadilly, Coventry Street, King William Street, Strand, Fleet Street, Ludgate Circus, New Bridge Street, Upper Thames Street, Queen Victoria Street, Bank, Cornhill, then a loop via Leadenhall Street, St Mary Axe, Houndsditch, Liverpool Street and Austin Friars Church, to join the existing line at Old Broad Street. It will be noted that the proposal was not for a figure 8 but for a long, flat loop, with the sides nipped together at Bank.

The route between Hammersmith and Hyde Park Corner via Kensington High Street was followed by three other schemes and will be referred to as 'the Kensington route'.

Charing Cross, Hammersmith and District The 1901 Bill was brought forward together with a modifying 'No. 2' Bill, and what was now proposed was a main line from Castelnau (where the depot and power station would be sited) via Hammersmith and the Kensington route to Hyde Park Corner, then via Constitution Hill to the Mall at the Duke of York's column. A loop line to be worked by a shuttle service was proposed from Hyde Park Corner via Piccadilly and Leicester Square to the Strand at Agar Street. The main line was to continue beyond the Duke of York's column to the Strand at Adam Street, joining the loop line at Agar Street. This section would be used only for rolling-stock movements, unless the proposal for an end-on junction with the Piccadilly & City at Adam Street was implemented.

London United Electric Railways This was a new Bill, promoted by the proprietors of the London United Tramways, who were then on the crest of the financial wave.

The proposed tube had large terminal loops at each end. The western loop tapped both LUT tram terminals. It began at Holland Road/Kensington High Street, proceeding along Holland Road to Shepherd's Bush, then turning south via Shepherd's Bush and Brook Green Roads to Hammersmith where it joined a branch coming from Castelnau (depot) and Rainville Road, Fulham (power station). It then continued along the Hammersmith Road to join the beginning of the loop by a trailing junction (in the open) at Holland Road. The main line then continued by the Kensington route to Hyde Park Corner, and via Constitution Hill and the Mall to the Duke of York's column, where the eastern loop began. This was via Charing Cross, Duncannon Street, Victoria Embankment/Northumberland Avenue and Horse Guards Parade, encircling Charing Cross (SECR) station *en route*. A separate north–south line ran from Clapham Junction to Marble Arch via Lavender Hill, the present Queenstown Road, Chelsea Bridge, Sloane Street, and across Hyde Park, crossing the main line at Knightsbridge, but without physical connection.

Piccadilly & City Following the 1901 recommendation, this line was now extended from Piccadilly Circus, via the Kensington route, to Hammersmith. There was a revised alignment between Piccadilly Circus and Charing Cross. East of Fleet Street/Salisbury Court the 1901 routeing was given up in favour of the associated and connected:

North East London Railway, which proposed a new line between Ludgate Circus and Cannon Street/College Hill via Upper Thames Street, in order to keep well clear of St Paul's Cathedral. Adopting another of the 1901 recommendations, there was an extension from Tottenham to Southgate, swinging north-west from Tottenham High Road to surface at The Avenue, and to cut across the new Tottenham LCC estate. The remainder of the route was via Chequers Green and Palmers Green, with the terminus at Chase Side, Southgate. The Walthamstow branch was dropped in favour of the associated:

City & North East Suburban Electric Railway This was another 1901 Bill, now modified at the Committee's suggestion so that the exit from the City was via Whitechapel Road and the present Cambridge Heath Road. The terminals were Mansion House (MDR) and Waltham Abbey, via Victoria Park, Leyton, Walthamstow, and Chingford.

The Piccadilly & City, North East London and City & North East Suburban schemes were backed by John Pierpont Morgan, the multimillionaire leader of American finance. Morgan had reorganized and retained control over many of America's trunk railroads, organized the General Electric Co. of New York, the International Harvester Co., the huge US Steel Corporation and, early in 1902, had sent shivers down British spines by proposing to amalgamate British and American Atlantic shipping companies into a new American-controlled company, the International Mercantile Marine. He knew London well and visited Europe for about three months each year, spending much of his time in his London home at Princes Gate, Kensington, keeping in touch with the English banking house of J. S. Morgan & Co., which had been founded by his father.

The Windsor Committee began its labours on April 16, 1902, and the Charing Cross, Hammersmith and District was the first casualty, its rejection being announced on May 13th.

On April 18th, Balfour Browne opened the case for the Piccadilly & City Railway. He disclosed that an agreement had been made with the London United promoters for the two lines to be linked at Hyde Park Corner (thus abandoning the London United's Charing Cross loop) and for the Hammersmith–Charing Cross section to be promoted jointly. Sir Douglas Fox gave engineering evidence, stating that the London United's Kensington route would be adopted, instead of the Piccadilly & City's, because the former had the Shepherd's Bush loop and the Marble Arch–Clapham Junction line, and was routed beneath Hyde Park (instead of the Kensington Road). Thus the Hammersmith–Hyde Park Corner section was covered by the technically independent LUER Bill. The spur to Castelnau was withdrawn, but the Fulham power station was retained. The tunnels would be 13 feet or 13 feet 6 inch internal diameter.

Clinton Dawkins, partner of Pierpont Morgan in J. S. Morgan & Co., said that his firm could supply the whole capital for the Hammersmith–Southgate line, but in practice would provide half the capital of the LUER Hammersmith–Hyde Park Corner line, two-thirds of that of the CNESER, and the whole of that of the P&C and the NELR. The four Bills envisaged £16 million of capital for 38 miles of line.

Between Mansion House and Monument, both the NELR and the CNESER would have been beneath the District & Metropolitan Joint line, with the CNESER continuing to follow that line closely as far as Whitechapel. The CNESER had no running connection with the other three lines, but there was to be an interchange station at Mansion House.

At the beginning of May, the heads of agreements with the London United were formally scheduled to the Bills, various redundant sections of line withdrawn, others altered, and revised estimates of costs submitted. Later, the P&C and NELR were formally amalgamated to form the Piccadilly, City & North East London Railway, whilst the whole scheme was known as the London Suburban Railway.

The existing north London railways were in strong opposition. J. F. S. Gooday, the Great Eastern's general manager, thought that the traffic estimates were 'enormously exaggerated', whilst both Gooday and James Holden expounded the virtues of the latter's 0–10–0 tank locomotive, the 'Decapod'. This was virtually a political locomotive built especially to prove that steam performance could equal electric, and that tube railways were therefore unnecessary on GER territory – but alas, it smashed up the track beneath it. Great Northern and North London witnesses also averred that the London Suburban traffic estimates were far too high and the fares 'ridiculously inadequate' (the proposed fare between Cannon Street and Southgate was only 2d!).

Lord Robert Cecil spoke on behalf of the Southgate cricket ground and the Committee rejected the Southgate–Palmers Green section. The only other casualty to the main line was the Shepherd's Bush loop, which was refused on May 16th, but despite further attacks on the traffic estimates by opposition Counsel, the Committee finally found the preambles proved for the Hammersmith–Palmers Green and Marble Arch–Clapham Junction lines. A very useful tube seemed about to be born.

The City & North East Suburban was not so fortunate. The Committee rejected the City–Whitechapel section (to which the line had been diverted at the behest of the 1901 Committee) on the grounds of property disturbance, and the remainder of the line was therefore withdrawn.

The case for the Central London Railway Bill opened on May 5th. Learned Counsel stressed the advantages of the proposed circular route —increased line capacity, alternative routes from Bank, Liverpool Street, Shepherd's Bush or Hammersmith, the simplicity of a universal 2d fare. The CLR was the first to show that a tube could be a financial success and it could raise capital on favourable terms.

There was a strong team of supporting witnesses, but the Central London scheme failed to gain the approval of the Committee, which announced on May 16th that it was not prepared to proceed with the Shepherd's Bush–Hammersmith–Bank lines. Thereupon the City loop was withdrawn, and the Bill finally approved was confined to minor provisions, including land to enlarge Marble Arch station.

As usual with Private Bill committees, no reasons were vouchsafed for this refusal, but probably the Committee disliked the prospect of there being two east–west lines under the same ownership, whereas the London Suburban scheme would encourage competition, and would serve the less remunerative northern and eastern suburbs as well as the richer central area.

Finally, we come to the Brompton & Piccadilly Circus case, which was spread over various days from April 28th to June 5th. Curiously, Counsel blighted the fortunes of his own Fulham branch by describing it as 'not an essential part of the scheme' (both this proposal and the Charing Cross connection may have been included merely to draw the enemy's fire while the important Piccadilly–Holborn link slipped through). He was able to make the useful point that contracts had been let for the authorized section of the BPC, and construction had begun (a shaft had been started at Knightsbridge, and some work done at South Kensington). Tenders had also been invited to construct the GNS, whose Holborn–Strand section would be worked as a shuttle.

After R. W. Perks had given evidence, the witness box was honoured by the presence of the great C. T. Yerkes, who gave an almost philanthropic explanation for his incursion into London transit affairs: 'I have got to a time when I am not compelled to go into this business, but seeing the way things are in London, I made up my mind that this would be my last effort.'

He was asked about allegations that he proposed to buy equipment in the USA, and stated in reply that 98% of the £400,000-worth of contracts made so far had been placed in England. Then followed the promises which later provided ammunition for his critics:

Q. 'Is it the fact that what is not with firms in England is because you could not get at the moment what you wanted?'
A. 'That is so, that is the only reason.'
Q. 'And with that exception, every contract that has been made has been made with English firms?'
A. 'And always will be.'

James R. Chapman, Yerkes' engineer, also gave evidence. The proposed services were given as: 2-minute Earl's Court–Finsbury Park; 4-minute Piccadilly Circus–Mansion House; 4-minute Parsons Green–Brompton Road.

The vital Piccadilly–Holborn connection was approved in principle on May 26th, having encountered little opposition. The Parsons Green proposal met with very strong objection from the authorities at the various Fulham Road hospitals, and was rejected.

Yerkes' attempt to achieve the recommended Hammersmith–Piccadilly–City connection by linking the Brompton & Piccadilly Circus at Piccadilly Circus with the District Deep Level at Charing Cross involved a 3½-chain curve on a 1 in 44 down gradient at Charing Cross and was refused on engineering grounds.

The Windsor Committee sat for the last time on June 5th, and the BPC and London Suburban Bills passed to the House of Commons.

FINAL STAGES

The preamble of the BPC Bill was found proved by the Commons Committee on July 30th. The other Commons stages were completed in October, and the Bill received the Royal Assent on November 18, 1902, with the consequent absorption of the Great Northern & Strand and change of name to 'Great Northern, Piccadilly and Brompton Railway'.

The London United Electric Bill was opposed at the Commons second reading on July 16th, but passed by 250 votes to 69. The Commons was most anxious to ensure that the whole of the line should be built, including the less remunerative portion in North London, and for both this Bill, and the PCNELR, it instructed the Select Committee to 'take security from the undertakers for the completion of the whole scheme of railways comprised in the Bill, either by making the rights of the undertakers conditional upon the due performance of the whole undertaking, or otherwise, as the Committee may think fit.'

The Commons Select Committee (the McIver Committee) began to hear the London United Bill, but then adjourned until after the summer recess.

'A VERY PRETTY AND CLEVER PIECE OF MANOEUVRING'

On October 18th, just before the Commons Committee was due to resume its hearing, the *Railway Times* correctly predicted that the fight was 'likely to be very interesting'. When the hearing began three days later, Committee, Counsel and witnesses were assembled ready to play their allotted parts in the battle for the Hammersmith–Palmers Green line when Sir Edward Clarke (Counsel for the LUER) made a sensational announcement. The London United and Morgan groups had been unable to agree on the joint management of the proposed line, and after the Bills had left the Lords, certain persons interested in the Brompton & Piccadilly Circus had acquired a 'commanding interest' in the London United Tramways. The 'certain persons' were, of course, Speyer Bros—the transaction had been completed in twenty-four hours and kept secret. As Speyers were also interested in the Brompton & Piccadilly Circus Bill, which had been given its third reading on the previous day, and which contained the expensive section of line from Knightsbridge to Hyde Park Corner, they wished to abandon the corresponding section of the London United Bill. Sir Edward said that arrangements could easily be made with the Brompton company to provide a through route. He pointed out that the Commons Instruction could be interpreted either as securing a guarantee for the whole scheme (which could not now be done) or merely for the railways in the LUER Bill. The Committee decided on the former interpretation, and Sir Edward then formally withdrew the LUER Bill.

Balfour Browne, for the Piccadilly, City & North East London, now made desperate efforts to save his Bill, offering a pledge to promote a Hammersmith–Hyde Park Corner Bill in the following session. The Chairman decided that in view of the Commons Instruction, he could not find the preamble proved. Balfour Browne said that the Bill could not be rejected without being heard. The hearing was adjourned for two days to seek an escape from the *impasse*, but the problem of complying with the Instruction in the changed circumstances proved insuperable and the PCNELR Bill was formally rejected.

This turn of events attracted widespread interest, with general sympathy for the Morgan group and hostility towards Yerkes.

G. T. Ashton, MP for Luton, presented motions to recommit the LUER and PCNELR Bills to the Select Committee. The LUER motion was ruled out of order, but there was a $1\frac{1}{2}$-hour debate on the PCNELR motion on the evening of October 29th. Ashton thought that the Morgan group had been 'very hardly used'. They had not the slightest intimation that their allies and partners were going to take action against them. He doubted whether, for a long time, 'such a very dirty transaction was ever done by parties coming before Parliament'. He stressed the enormous amount of parliamentary time spent on the Bills, and the £50–60,000 promotion costs.

Perks emphasized that the Yerkes group would not waste money building duplicate railways, but he promised a Bill for a second Hammersmith–Kensington–City route in the next session.

Sir Lewis McIver, Chairman of the Commons Select Committee on the Bills, contended that the scheme had been wrecked by a 'Stock Exchange ramp'—it was a 'game in which it was proposed to make the London roads pawns on the chequer board of Wall Street'. He spoke of 'the very unsavoury details of the transaction . . . a scandal'.

As the parliamentary machinery did not allow a recommittal, Ashton withdrew his motion. The Morgan group did not again appear on the London underground scene, and no later scheme for the Hammersmith–Kensington–City–Kingsland Road–Tottenham axes came so near the light of day as the London Suburban.

Each party to the affair issued statements justifying its own actions. The Morgan group recalled that their own Bill had beeen withdrawn west of Hyde Park Corner in favour of the London United, but they were to find half the capital for this section. For the whole line, they were to provide £7,690,000 capital and the London United £3,690,000, but the latter had demanded an equal share in the control. Subsequently the LUT had given no formal notice of ending the agreement and had not given the Morgan group an opportunity to buy out the LUER.

The London United story was that they had been trying since April to reach a formal agreement with the Morgan group, but without success. They felt justified in asking for an equal share in control in view of the Instruction, but Morgans would not even discuss their suggestions. By the beginning of August they were so displeased with Morgan methods that they had decided to withdraw their Bill when Parliament reassembled.

Speyer Bros strongly denied the suggestion that they had engineered the take-over of the LUT to bring about the defeat of the London Suburban Bills. They contended that, because of the disagreements, the PCNELR Bill was already dead before they were approached—in fact, the LUT shares had previously been offered to the promoters of a competing tube. They had bought the London United Tramways because they believed it to be a 'splendid property', fitting in well with the Underground group's lines.

In the contemporary Press the general comment on the débâcle was that such occurrences were only to be expected when American interests and methods were introduced, but that Yerkes had cleverly outmanoeuvred the Morgan group. In the words of D. N. Dunlop of *The Railway Magazine*, it was a 'very pretty and clever piece of manoeuvring on the part of the Yerkes combination, made possible by dissension in the enemy's camp'.

NOTHING MORE FOR NORTH EAST LONDON

In the following years, various independent promoters came forward with Bills for part-tube, part-open railways in north east London. A City & North East Suburban Electric Bill of 1903 proposed lines from Mansion House to Palmers Green (Chequers Green) and Waltham Abbey. It was refused on the grounds that the financial support was not guaranteed. Several half-hearted attempts were made to promote a Hammersmith, City and North East London Bill, as shown in the list of Bills given later in this chapter.

In 1905 a North East London Bill for a line from Monument to Waltham Abbey (with connections to the GER) via Shoreditch, Hackney, Leyton, Walthamstow and Chingford received widespread support from the local authorities and local Members of Parliament, and promises of financial help from substantial sources in the City. The Bill was passed despite keen Great Eastern opposition, but the capital could not be raised. After several extensions of time had been granted, a 1910 Bill to allow the Metropolitan Railway to work the line failed, and the scheme came to an end, much to the relief of the Great Eastern, who received compensation for their legal expenses.

PAUSE FOR ENQUIRY

For some time doubts had been growing about the effectiveness of the Parliamentary Private Bill procedure, as a means of securing the best system of urban railways for London. The 1901 Committtee had recommended more direct control and supervision of underground railway projects, and in July 1902 the London County Council urged the need for a statutory authority to deal with all proposals affecting London's transport. The whole London transport and traffic scene was a muddled patchwork of independent operators and authorities, characterized by an absence of planning and co-ordination.

The wrecking of the London Suburban scheme brought matters to a head, and the LCC urged an immediate and complete enquiry into the whole subject. Early in November 1902 the Prime Minister announced that a Royal Commission would probably be the best solution, and the Royal Commission on London Traffic was appointed on February 9, 1903, under the chairmanship of Sir David Miller Barbour. The thirteen-man Commission included Earl Cawdor (Chairman of the Great Western), Viscount Cobham (a Railway Commissioner), Sir John Wolfe Barry (the eminent engineer, with wide experience of underground railways) and George Stegmann Gibb, general manager of the North Eastern Railway.

The terms of reference included the best means of developing and interconnecting railways and tramways; development of other forms of mechanical locomotion; organization and regulation of vehicular and pedestrian traffic; and the desirability of establishing a tribunal to which railway or tramway schemes could be referred.

The Commission was the most thorough investigation ever made into London's transport and traffic problems, and its eight large volumes of recommendations and evidence (including maps and photographs), admirably indexed by Mr (later Sir) Lynden Macassey, remain a most valuable source of information for the transport student.

The LCC, by the very nature of its Progressive political platform, was bound to be opposed to the capitalistic activities of the Underground group, although the vehemence of its opposition gave rise to periodic expressions of pained surprise by UERL spokesmen. Part of the Council's criticism was based on the assumption (derived from a leasing clause in the 1902 Brompton & Piccadilly Circus Act, and similar proposals in 1903 Bills) that the ultimate intention was to lease all the subsidiary companies to the UERL for a fixed annual sum, with the UERL being nominally responsible for operation.

The Commission's report was published on July 17, 1905. The principal recommendation was the appointment of a permanent, full-time, three- to five-man London Traffic Board, with the duty of examining all Bills affecting transport in Greater London, securing amendments to make them harmonize with a general scheme, and reporting to Parliament on such Bills. Other duties of the proposed Board included the submission of an annual report on London Traffic, the preparation of schemes of locomotion for any district (if requested by the Government, or a local authority with the support of the Government), the settlement of disputes about through-running or by-laws, and the facilitation of amalgamations and joint working.

The Commission made numerous useful observations and recommendations on London's railways. Private enterprise was commended; in undeveloped districts where railway extensions were envisaged, the companies should be allowed to buy land so that they could later share in the benefits of increased land values. Local authorities should also be empowered to assist railway companies to extend into new territory by loans or a guarantee of receipts. Attention was drawn to the adverse results of using urban railways for suburban traffic. Amalgamation was approved, provided that the public interest was safeguarded. Speyer and Yerkes had promised to lay an amalgamation scheme before Parliament at the appropriate time. The Commission was satisfied that the formation of the UERL had been of material assistance in raising capital for works which would be of great public benefit.

Particular recommendations for new railway facilities included the extension of the North West London Railway from Marble Arch to Victoria, better connections to north east London, and better interchange between lines running from east to west and those from north to south.

THE RIDICULOUS MOUSE

The implementation of the main proposal for a Traffic Board was bedevilled by the view of the Progressive party on the LCC that the Council should itself be the authority in London traffic matters. At the time of the report, the Government was Conservative and the Council Progressive, but in 1906 a Liberal Government was elected, and in 1907 a Municipal Reform County Council. The railway Press scented 'intrigues' between the Progressives and the Government. Various Government statements promised that a Traffic Board Bill was imminent, but it never materialized, and in 1909 the *Railway Gazette* concluded that the proposal had 'faded into the *ewigkeit*'.

As a conciliatory gesture the Government set up the London Traffic Branch of the Board of Trade in August 1907, with the main duty of preparing an annual report to Parliament on London traffic matters. Officially the Traffic Branch was of a temporary nature, and was without prejudice to future legislation. In practice the branch continued to produce informative annual reports until 1915, when it was disbanded. In 1908 the *Railway Gazette* described it as 'the ridiculous mouse which the Royal Commission has brought forth', but it did good work in its own circumscribed field. Meanwhile London had to rely on the transport operators themselves for measures of co-ordination.

FUSION FRUSTRATED

When the Brompton & Piccadilly Circus Bill of 1902 came before the Commons Select Committee, it had a clause allowing the line to be leased to the UERL for a rent or guaranteed dividend. Sir Lewis McIver, Chairman of the Committee, said that he would like to see the agreement scheduled to the Bill, his object being to protect the investors and to ensure that the line was built. Counsel for the promoters thought that the object could be obtained by making the lease compulsory. This was accepted, and the Act accordingly contained a clause providing that the line should be leased to the UERL at a rent or guaranteed dividend at a minimum rate of 4% on the debenture and share capital. The LCC later complained that it was not given sufficient opportunity to examine this clause before it became law.

The 1903 Great Northern, Piccadilly & Brompton Bills contained clauses for the name of that railway to be changed to 'Underground Consolidated Electric Railways', whilst the complementary Baker Street & Waterloo, and Hampstead Bills contained clauses for the transfer of these lines to the Piccadilly company, and also for perpetual leasing to the UERL. In view of the appointment of the Royal Commission, the promoters withdrew the amalgamation clauses, and the leasing clause in the Hampstead Bill was refused by the Commons committee. Thereupon the promoters withdrew the corresponding clause in the Baker Street & Waterloo Bill, and the three tubes remained legally independent until 1910. In January 1907 the London newspapers carried reports of further proposals to amalgamate the three lines. In fact the UERL board of directors had decided on such amalgamation in November 1906, but the proposal was withdrawn at the request of 'American banking interests' (in the words of the board Minute).

BILLS THROUGH THE SIEVE

Whilst the Royal Commission on London Traffic was sitting, Parliament was naturally reluctant to consider any large tube schemes. Six Bills in the 1903 Session were postponed and later withdrawn (as detailed below), and in August 1903 an official warning was given that major schemes for new London railways could not be considered until the Royal Commission had reported. This warning was repeated in November 1904, although several promoters ignored it, presumably hoping that the Commission would report early enough for their Bills to complete their Parliamentary stages in 1905.

This embargo on new legislation allowed Yerkes and the Metropolitan Railway to press on with electrification and tube construction whilst the rival promoters were marking time. Nevertheless, some quite important schemes were considered (and some approved), and the guiding principle seemed to be that any proposal savouring of the 1902 Piccadilly–Hammersmith schemes should be deferred. The following table indicates the fate of the most important proposals of this period:

Bill	Major Proposals	Result
	1903 Session	
Baker Street and Waterloo	Transfer to Piccadilly Railway, lease to UERL. New station at Lambeth, Christ Church (now Lambeth North)	New station approved. Transfer and lease withdrawn, as described above. Royal Assent, 11.8.03.
Central London	The 1902 Circle proposal revived	Postponed and withdrawn.
Charing Cross, Euston and Hampstead	Transfer to Piccadilly Railway, lease to UERL, take-over of Edgware and Hampstead. New stations at North End, Hampstead and Charing Cross (SECR). Line to depot at Kentish Town abandoned.	New stations approved, lease refused, transfer and take-over of E. & H. withdrawn. RA 21.7.03.
City and North East Suburban Electric Railway	New line—City to Waltham Abbey and Palmers Green	Considered but refused.

YERKES TAKES OVER

Bill	Major Proposals	Result
City and South London	Extension from Islington to Euston; take-over of City and Brixton, with new line below Thames giving better access to City	Approved (despite heavy Metropolitan Railway opposition). RA 11.8.03.
Clapham Junction and Marble Arch	New lines between these points. (Two separate Bills, one Yerkes, one independent—the independent Bill had a junction at Marble Arch with the NWLR)	Postponed and withdrawn.
Great Northern, Piccadilly and Brompton (New lines and extensions)	Extensions from Knightsbridge via High Street, Kensington, to Hammersmith and Shepherd's Bush, and from Piccadilly Circus via Strand to Mansion House, for junction with District Deep Level	Postponed and withdrawn.
Hammersmith, City and N.E. London	New line—Hammersmith–City–Palmers Green and Walthamstow	Late Bill, not accepted by Standing Orders Committee.
Metropolitan District (Various Powers)	Confirmation of 1902 agreement with Piccadilly company for latter to take over powers to build South Kensington–Earl's Court section of District Deep Level, also for Piccadilly line to join District at West Kensington; deviation of Deep Level at Cromwell Curve; additional tracks Turnham Green–Studland Road	Approved, except last-mentioned proposal, which was withdrawn after agreement with the LSWR. RA 21.7.03.
Metropolitan District (Works)	Extensions of District Deep Level to Hammersmith and to beyond Whitechapel	Postponed and withdrawn.
North West London	Extension from Marble Arch to Victoria	Postponed and withdrawn
Watford and Edgware	New line between Watford High Street and the Edgware and Hampstead Railway at Edgware	Approved. RA 11.8.03.

1904 Session

Bill	Major Proposals	Result
Baker Street and Waterloo	Station tunnels at Lambeth (Christ Church), Regent's Park and Edgware Road	Approved. RA 22.7.04.
Charing Cross, Euston and Hampstead	New station at Mornington Crescent. Additional subsoil for Tottenham Court Road station. Additional capital	Additional capital withdrawn (after LCC opposition). Stations approved. RA 22.7.04.
Metropolitan District	Extensions of time for District Deep Level. Agreements with UERL tubes. Additional lands for West Kensington–Hammersmith widening to accommodate Piccadilly trains	Approved. RA 1.8.04.

Bill	Major Proposals	Result
North and South Woolwich	New line	Withdrawn. LCC-inspired conditions were too onerous.

1905 Session

Bill	Major Proposals	Result
Central London (New lines)	A further revival of the 1902 Circle proposal	Postponed and withdrawn.
Charing Cross, Euston and Hampstead	Further subsoil for station at Charing Cross (SECR)	Approved. RA 4.8.05.
Great Northern, Piccadilly and Brompton No. 1	Extension from north of Strand to Waterloo (two tunnels to Surrey Street/Howard Street, then single tunnel, to be operated as shuttle). Improved junctions at Holborn between main line and branch	Only the junctions and the twin-tunnel section were approved. RA 4.8.05.
Great Northern, Piccadilly and Brompton No. 2	New lines—Knightsbridge via High Street, Kensington, and Hammersmith to Chiswick, with branch via Holland Road and Shepherd's Bush to Acton Vale, also branch from Leicester Square to Aldgate	Postponed and withdrawn.
Great Western (New Railways)	New line—Ealing via North Acton to Shepherd's Bush	Approved.
Hammersmith, City and North East London	New line—Hammersmith–City–Palmers Green and Walthamstow	Failed Standing Orders.[1]
North East London	New line—Monument to Waltham Abbey	Approved. RA 11.8.05.

[1] There was also a 1906 Hammersmith, City and N.E. London Bill—equally unsuccessful.

In October 1905 the UERL made an agreement with the Central London that neither company would promote a Bill for an east–west line in the 1906 session. By the end of 1906 the financial wind had begun to blow distinctly chill, and the main preoccupation was to earn sufficient revenue from the existing lines in the face of surface competition rather than seek powers for grandiose new schemes.

EMPTY POCKETS

In January 1903 the Underground group made its first attempt to fulfil its role of selling ordinary shares to the public, when applications were invited for £2 million worth of Great Northern, Piccadilly and Brompton ordinary shares, at par. Two selling points were that the shares were given a prior charge on net earnings and were also guaranteed a 4% return by the UERL.

Only about £821,000 of the Piccadilly ordinary shares were issued. Edgar Speyer later attributed this lack of interest to the disturbed state of the world's money markets.

None the less, it was essential for the group to have further funds to continue with tube construction and District electrification, and in the absence of a market for ordinary shares it was obliged to take the first step in the process of creating a heavy burden of fixed interest securities which was to bedevil the fortunes of the UERL for many long years.

This took the form of offering UERL shareholders the 'privilege', in May 1903, of subscribing for £5 million of Profit Sharing Notes (part of a total issue of £7 million) carrying interest at 5% free of British income tax, and due for redemption in June 1908. The issue price was 96. A Trust Deed was created whereby certain securities of subsidiary companies were deposited with a trustee as security. The 'profit-sharing' description derived from a provision that if the deposited ordinary shares were sold above the price of 95, half the profit should accrue to the note-holders. Any unsubscribed notes, plus the remaining £2 million, were to be taken up by a Speyer/Old Colony Trust Company syndicate. In practice about £3,400,000 of the notes was issued in the USA and about £3,600,000 in the UK. The notes were all sold by February 1905 and the syndicate dissolved.

In May 1903 the Speyer companies and the Trust Company exercised their right to appoint a majority of directors to the UERL board, and the following additional directors were elected:

James A. Blair, of Blair & Co., New York.
T. Jefferson Coolidge (Junior), Chairman of the Old Colony Trust Co.
James H. Hyde, First Vice-President of the Equitable Life Assurance Society of the United States.
Leonor F. Loree, President of the Baltimore and Ohio Railroad.
H. H. McCurdy, of the Mutual Life Insurance Co., New York.
Edgar Speyer, of Speyer Bros, London.
James Speyer, of Speyer and Co., New York.
Jonkheer Henry Teixeira de Mattos, of Teixeira de Mattos Bros, Amsterdam (who raised some capital for the UERL in Amsterdam).

By the end of 1904 the need for funds again became pressing, and British investors were invited to subscribe to a series of debenture issues. Lots Road power station, nearing completion, was too valuable an asset to be left unpledged, so a trust deed was filed to secure £850,000 worth of debentures on it, and £700,000 of 4½% debentures was issued. Next came a series of debenture issues on the three tubes. First, £500,000 of 4% Baker Street & Waterloo debentures, offered in February 1905 at the price of 96, and quickly snapped up. For the other companies' debentures, issued in March 1905, the price had gone up to 97—£1,200,000 of Piccadilly and £800,000 of Hampstead were issued.

In each prospectus announcing a tube debenture issue, there appeared an estimate of traffic and earnings by Stephen Sellon, the 'tramway expert'. These estimates, which are reviewed in detail in the following chapter and in Appendix 4, turned out to be extremely optimistic.

FIRES DOWN BELOW

One factor which increased the cost of the Yerkes lines, and thus brought forward the time when further funds were required, was the need to take stringent fire precautions—far more stringent than on the older tubes.

During the 1890s, the science of electric traction was still in its infancy, and the insulation of train power cables and conductor rails was far from perfect, so that train crews on the CLR and CSLR were sometimes obliged to put out small fires in the tunnels and on the trains. The CSLR had a fairly serious locomotive fire in 1902 and another in 1906, although neither caused casualties. A nasty substation fire occurred on the CLR at Notting Hill Gate in April 1905. These incidents were treated with usual British *insouciance,* but two disasters outside London demonstrated the extent of the potential danger.

The first occurred on the Liverpool Overhead Railway, in the tunnel approach to Dingle, on December 23, 1901. After stopping with a defective motor, a driver made repeated attempts to reset the circuit-breaker, producing arcs beneath the car which eventually ignited its woodwork. A gale blowing into the tunnel carried the flames along the train so that the fire consumed the train, a signalbox and a wooden platform, even attacking a spare train standing 155 yards from the first. Four members of the staff and two passengers perished. The Board of Trade Inspecting Officer recommended that wood should be removed from the proximity of electrical machinery on rolling stock, that the insulation for main current conductors should be incombustible and smokeless, and that platforms and station buildings on underground electric lines should be of incombustible materials.

A second disaster occurred on what is now line 2 of the Paris Métro, during the evening peak period of August 10, 1903. Fire from a motor short-circuit, occurring in a train just north of Couronnes station, spread along the softwood bodywork of the cars. The driver made a reckless attempt to reach the terminus at Ménilmontant, but there was an explosion which set the train well alight and fused the tunnel lights. The crowd in a following train panicked in the smoke and darkness, and eighty-four died from crushing or suffocation.

The Board of Trade promptly drew up a set of draft fire precautions for underground electric railways, and invited the comments of interested railways. The official list of requirements was issued in May 1904, and included the following items:

Hardwood sleepers, not creosoted; the four-foot way to include a ballast or concrete pathway for detrained passengers; tunnel lights to be on an independent circuit; stairs, passages and exits to be well lit, with 25% of the lights from an independent source; exits to be separate, and as nearly as possible in the middle of the platforms; no woodwork to be used for platforms and as little as possible for other underground structures; cars to be of metal construction with a minimum of woodwork, for which hardwood was preferred; seats, panels and fittings to be non-combustible; no power cables to run along the train; no motor cars to be in the middle of the train; means of exit to be available at both ends of the train; for insulation, rubber and all inflammable materials to be avoided; the outer covers of cables to be non-flammable and non-smoking; oil lamps to be carried on trains; driver-station telephone communication to be provided.

In experiments, the Central London allowed the electrical compartment of a motor car to burn out without its affecting the adjoining passenger compartment, and Yerkes had made various attempts in 1903 to ignite cars and fireproofed wood with electric arcs, but the fires went out as soon as the arcs were turned off. On the Yerkes tubes, fire precautions were adopted wholeheartedly, the most notable innovations being the standardization of concrete platforms and steel cars.

MULTIPLE-UNIT TRACTION

The Liverpool Overhead and Waterloo & City Railways had used motor cars, instead of locomotives, from the beginning, but they employed 'direct control' whereby the whole of the traction current passed through the controller at the leading end of the train. This system was perhaps acceptable for a two-car train, but for longer trains with more motors it called for the use of large, heavy and complex controllers, and power cables running along the train, with their attendant dangers. The multiple-unit system, invented by the American, Frank J. Sprague, was a vast improvement.

With this system the electric motors in the various cars of a train are regulated from a single controller, each car drawing its own traction current direct from the conductor rail, but controlled by a low-voltage train line. It was first demonstrated on the South Side Railway of Chicago, where a five-car train was tried in November 1897. By July 1898 the conversion from steam traction had been completed, with 120 motor cars.

The Sprague system was soon adopted for the Brooklyn Elevated, the Boston Elevated, and the Paris (Invalides)–Versailles line of the Western Railway of France, and it quickly came to be used as standard on urban and suburban electric railways. The many benefits include the flexibility of train length, the possibility of uncoupling at junctions and sending a part train to each destination, easy reversing at termini, high acceleration, and simple operation by the driver, with reasonably constant traction characteristics irrespective of train length.

In a discussion at the Institution of Civil Engineers in November 1899, opinion was about equally divided for or against the system. Mr Hudleston, for the Central London, thought it had attractions for elevated railways where the lack of locomotives would allow lighter girders, but he could see no future for it on tubes, where there was no room for the motors beneath the cars.

In America the system was developed commercially by two leading electrical manufacturers into the Sprague-General Electric electro-magnetic and the Sprague-Westinghouse electro-pneumatic, the corresponding British-installed systems being the Sprague-Thomson-Houston and the British Westinghouse. The Thomson-Houston system was adopted on all the Yerkes lines.

THE YERKES NETWORK TAKES SHAPE

Now that the Baker Street & Waterloo was financed by the UERL, construction could be resumed at full speed. London Road depot was started, and by August 1903 the northbound tunnel was complete from Waterloo to Baker Street; the southbound tunnel was completed by November 1903, when workmen started to build the station platforms. By February 1905, lift installation was in hand, the rolling stock was ordered and track laying was under way. October 1905 saw the sub-stations approaching completion, the tiling of the station tunnels nearly complete, the permanent way complete, the conductor rails being bonded, and the depot finished and fully equipped.

On the Piccadilly and Hampstead lines, tunnelling generally had to await the compulsory acquisition of station sites, demolition of existing property and the sinking of shafts. The beginning of Piccadilly line construction at Knightsbridge and South Kensington in April 1902 was followed by the Great Northern's commencement of the low-level works at Finsbury Park in September. By April 1903, twelve out of twenty station sites were in possession. There was a segment dump and shaft at the Cromwell Road triangle. All station sites on the Piccadilly main line were acquired by February 1904, and the boring of the running tunnels was being pressed forward. In October 1904, 80% of the running tunnels had been driven and the District Railway also reported satisfactory progress with the West Kensington–Hammersmith widening.

The construction of the Strand branch, and of all the tunnels at Holborn, was delayed by the provision in the 1899 Great Northern and Strand Act that tunnelling of this portion should not begin without the consent of the LCC unless and until the new road (Kingsway) was completed. An agreement with the LCC was made in June 1905, but the Holborn tunnels lagged behind those on the rest of the line, on which the tunnels were completed by October 1905, when the permanent way was being laid. There was a large labour force at work on the running and maintenance depot at Lillie Bridge (West Kensington) which was finished by July 1906. By the latter month all main-line tunnels were completed, all permanent way and conductor rails complete except for a portion at Covent Garden; lifts, stairways, ventilation and station buildings were well in hand, and the substations complete and equipped except that at Hyde Park Corner.

On the Hampstead line, work began in July 1902 by the demolition of houses at the foot of Haverstock Hill, and by April 1903 the contractors were in possession of the sites at Belsize Park, Chalk Farm, Camden Town, Euston and Leicester Square. Tunnelling began in September 1903; by October 1904, 75% of the running tunnels had been driven and land acquired for Golders Green depot.

The tunnelling contractor's task was almost complete by April 1906, and the laying of the permanent way had begun. The latter was completed by December 1906, when the erection of station buildings was actively progressing. Golders Green depot was finished by March 1907, and work on the remaining surface station buildings was hurried on.

At Charing Cross the original intention had been to build the station north of the Strand, on the site of the Post Office, but powers were obtained in 1903 for a station beneath the eastern part of the SECR station forecourt, with no interference to the surface. Further powers were obtained in 1905 to use the subsoil of the western part of the forecourt, and any temporary occupation of the surface was to be as short as possible. The South Eastern authorities naturally wanted the least possible interference with the cab traffic to and from their station, and plans were in hand to build the underground station by driving a shaft upwards from the station tunnels. However, on December 5, 1905, the arched roof of Charing Cross SECR station collapsed, and the station was closed to traffic for over three months.

The Hampstead company promptly made an agreement with the South Eastern authorities to open up the forecourt for six weeks. During this period one shaft was excavated to its full depth of 73 feet (including a chamber at the bottom for the winding gear, for which there was no room at the top), and the walls of the booking hall erected 12 feet below the surface. A steel girder roof was constructed over the hole, and the forecourt repaved. Further excavation enabled the booking-hall walls to be taken down to 32 feet to rest on blue clay, and a staircase shaft and a second lift shaft were also sunk, the work being completed in September 1906. This scheme for Charing Cross caused the temporary abandonment of the authorized section of line along Villiers Street to Charing Cross (MDR) station; interchange passengers to the

SECR gained direct access to their station, but those changing to the District were faced with the walk down Villiers Street.

CONSTRUCTIONAL METHODS

Most of the tunnels of the three Yerkes tubes were built by driving standard Greathead shields through the London clay. On the Baker Street & Waterloo, the shields for the running tunnels were driven by seven or eight hydraulic rams, and the larger station shields by twenty-two rams. A temporary narrow-gauge railway with electric locomotives drawing current from an overhead trolley wire was used to haul raw materials in and take spoil out. In most cases the lift shafts at the station sites were first used as working shafts from which the running tunnels were driven, but the Baker Street & Waterloo had the advantage of its river shafts.

On the section of the Baker Street & Waterloo beneath the Thames, a pocket of water-bearing gravel was encountered, so that compressed air at between 21 and 35 lb. per square inch had to be used. Several devices were used to protect the miners and to minimize the loss of compressed air. A hooded shield was used, formed by additional steel plates extending forward beyond the normal front surface, at the top of the shield, to form a hood. A staggered pair of half-diaphragms across the tunnel (the upper one in front) formed a 'fountain-trap' to mitigate the effect of a sudden inrush of water, whilst Dalrymple Hay's clay-pocket system was used to retain the compressed air. This involved the hand excavation of a ring of holes, 12 inches square and about 23 inches long, in front of the cutting edge of the shield, and filling them with 'puddled' or 'tempered' clay. When the shield moved forward, its edge was wholly in clay and this greatly assisted in retaining the compressed air (the main face of the excavation being timbered). Nevertheless, some compressed air did escape, and if a large amount was released at once it could form a waterspout of 2 feet 6 inches to 3 feet on the surface of the Thames. In a discussion on the construction of the Rotherhithe Tunnel in 1908, W. R. Galbraith, the civil engineer, related how one of these waterspouts upset a boat in a race from Charing Cross to Putney—the company had to pay damages.

Some water-bearing sand was encountered at Euston on the Hampstead line, but air pressure of 15 to 25 lb. per square inch served to keep the water at bay.

The Price rotary excavator was successfully used on the Piccadilly and Hampstead lines. This was invented by John Price of the contractors Price & Reeves, and of $32\frac{1}{2}$ miles of single tunnel on the Piccadilly and Hampstead, $11\frac{1}{2}$ miles were bored with the improved Price rotaries.

The rotary excavator is based on a Greathead shield equipped with a central shaft on which the rotating part is mounted. Six radial arms

Charles Tyson Yerkes (1837–1905), the American railway and tramway magnate whose enterprise established much of London's original tube railway network. This photograph was taken in 1904.—*Library of the Congress of the United States of America.*

The Chelsea Monster—Lots Road power station seen from the River Thames in October 1916.—*London Transport.*

In its American birthplace—a Baker Street & Waterloo Rly. motor car posed for the official American Car & Foundry Co. photograph before shipment to England. The typical American line name was specially painted for the photograph.—*A.C.F. Industries Inc.*

L. W. Green's station design for the Yerkes tubes. Euston (Hampstead line). This entrance was closed in 1914 and is now a substation.—*London Transport.*

extend from the shaft, and to these arms knives are fixed at different radii, so that when the shaft and arms are revolved, the knives cut concentric grooves in the face of the clay. In front of the central shaft is a toothed triangular plate, and each arm carries a bucket which scoops up the spoil and empties it into a chute at the top of the shield. The six arms are rotated by an electric motor and a series of gears, the final small gear-wheel driving a large ring gear with internal teeth.

At first some difficulty was experienced in keeping the excavators in alignment, and in the Rotherhithe discussion mentioned above, Dalrymple Hay stated that on one section 25% of the tunnel rings had to be readjusted. Later, when the technique of steering the excavators was mastered, a section beneath Piccadilly was driven with a maximum error of half an inch. These machines allowed the work to be speeded up considerably. On the Baker Street & Waterloo, using Greathead shields, average progress was 44 rings (about 73 feet 4 inches) per week per shield, but an average speed of 96 rings (160 feet) was quoted for a rotary on the Hampstead, and a maximum of 109 rings (181 feet 8 inches) on the Strand branch of the Piccadilly. At the very deep section beneath Hampstead Heath, the working face slipped owing to the great 'superincumbent pressure', and crippled the rotary, which was replaced by a Greathead shield. Generally rotaries were a great success; they were also used on the contemporary CSLR extension from the Angel to Euston.

Compared with the above-mentioned tunnelling speeds, construction speed of the 6 feet 7¾ inch diameter Tower Subway of 1869 seems slow at 37 feet 4 inches per week, but screw-jacks (not hydraulic rams) drove that shield forward. The relative stagnation of tunnelling technique is illustrated by the average speed of driving the tunnels for the Central Line eastern extension in 1938/9—45 rings per week (75 feet) for ordinary shields, 100 rings (166 feet 8 inches) for rotaries. All the foregoing figures are for tunnelling in free air; in compressed air the speed is reduced to two-thirds or half of normal. The drum digger shield, tried on the experimental Victoria line tunnels in 1960, drove 934 feet of running tunnel in two weeks.

Two Hurst-Nelson battery locomotives with BTH equipment were delivered in August 1905 and used for removing spoil and bringing in segments, sleepers, rails and other material on the Piccadilly and Hampstead lines. Afterwards one was allocated to each of these lines, and used at night for hauling permanent-way trains or trains of tank cars for emptying drainage sumps. They were both written off in 1918.

On all the Yerkes tubes, the UERL were technically the 'contractors', and the public works contractors who did the main constructional work, the 'sub-contractors'. On the Baker Street & Waterloo, the tunnelling between Edgware Road and Waterloo was done by Perry & Co., whilst the London Road depot and the Elephant and Paddington extensions were built by Mowlems. The Piccadilly tunnelling was divided between Walker Price & Reeves and Walter Scott & Middleton, with Bott & Stennett building the open-air section. Price & Reeves were responsible for the Hampstead tunnelling.

THE ODD ONE OUT

Two years before the first of the Yerkes tubes was ready, there opened in London a unique short tube railway, the only deep-level line in Britain to be constructed to main-line loading gauge.

This was the Great Northern & City Railway, conceived by some Great Northern Railway directors as a tube railway extension of that line into the City of London, to avoid the bottleneck through King's Cross and the Widened Lines. The Act, passed on June 28, 1892, authorized a line from Finsbury Park (junctions with up and down GNR Canonbury lines) passing underground at Drayton Park, and proceeding in twin tunnels to Moorgate (Finsbury Pavement), where the terminus was to be adjacent to the Metropolitan Railway station. The tunnels were to be 16 feet in diameter and the stations 20 feet, to accommodate GNR stock, and the tube railway was given running powers into Finsbury Park station. On January 31, 1894, an agreement was signed between the GNR and the GNCR, in which the GNR undertook to construct the connecting lines, and guarantee at least £20,000 worth of through traffic annually, with a minimum of 50 and a maximum of 150 trains through to Moorgate daily. (Subsequently the GNR agreed in principle to certain alterations of these conditions, but no formal agreement was signed.) If the tube company were unable to raise sufficient capital to complete the line by October 5, 1898, or make substantial progress with construction by June 28, 1900, the agreement would lapse. Great difficulty was experienced in raising money and it was not possible to start work on the line until 1898.

Disputes arose with the GNR over the terms of the 1894 agreement, and by October 1900 the two companies were hardly on speaking terms; the GNR had announced that it wished to cancel its through-traffic agreement and leave the tube company merely with its 1892 running powers into Finsbury Park station. This was confirmed in a fresh agreement of May 21, 1901, and the idea of GNCR electric locomotives hauling GNR suburban trains over the tube line had to be abandoned. Late in 1901 the tube company decided to seek running powers for its own electric trains over the GNR suburban lines to Edgware, High Barnet, Alexandra Palace and Enfield. Meanwhile, it was engaged in a legal battle with the GNR over the running powers into Finsbury Park station, maintaining that these included the use of the crossover on the High Barnet line north of the station. This was indeed an important point, for there was no other means of transferring the luckless GNCR trains from the down side of the station to the up side or vice versa! The hostile attitude of the GNR, which had originally fathered the scheme, probably stemmed from the fact that it no longer had any nominees on the board and had lost all interest in the idea. The main-line company now decided that the tube should be safely bottled up underground and the GNR (No. 2) Act of July 22, 1902, settled the matter. By this Act, the previous agreements were cancelled and the GNCR was to terminate in a new *tube* station below the Finsbury Park main-line platforms, the terminus and its approaches to be built by the GNR and leased to the tube company for 999 years. The surface junctions and running powers of the 1892 Act were wiped out, and the GNCR was restrained from supporting or promoting any tube railway extension north of Finsbury Park. Thus the 'main line tube' lost its original purpose and became just another urban tube railway, albeit with some unusual features.

The tube company's own Act of 1902 authorized a 19-chain southward extension to the Bank of England (Lothbury), lands for a generating station in Shoreditch, and an additional station at Highbury; it also enlarged the maximum permitted diameter of the station tunnels to 23 feet. (The Lothbury extension was started, but the powers lapsed in 1910 with only a few yards built).

An unusual method of construction was employed. The tunnel was first built with normal iron segments with lower key pieces in concrete. The key pieces were then demolished and the segments in the lower half of the tunnel removed for use again elsewhere; as the segments were taken out, they were replaced with three rings of blue vitrified bricks. It was claimed that this type of tunnel, upper half iron, lower half brick, made running quieter and saved the company £30,000. There was certainly a need for economies to offset the high cost of the 16-foot tunnel. (The large size also made possible the use of hydraulic segment erectors in both station and running tunnels.) The line was built by S. Pearson & Son Ltd, who invested large sums in the railway and worked it for three years after opening, taking 60% of gross receipts in the first year, 50% in the other two years, and in return paying a 3% dividend on the ordinary stock.

The 3·42 miles from Finsbury Park to Moorgate Street were opened on February 14, 1904, with intermediate stations at Drayton Park, Essex Road and Old Street (joint with the CSLR). At Drayton Park, the glass-veranda-covered island platform was in an open cutting immediately below and to the west of the GNR's Canonbury lines; north of this station, the twin tunnels of the GNR-owned section led to the underground terminus at Finsbury Park. A further station, at Highbury, between Drayton Park and Essex Road, opened on June 28th. All stations were lit by gas and had 420-foot platforms (450 feet at the termini) in parallel station tunnels reached by lifts (electrically operated at Moorgate, Old Street and Essex Road, hydraulically at Highbury and Finsbury Park). Essex Road became 'Canonbury and Essex Road' from July 20, 1922, but the original name was restored on July 11, 1948; 'and Islington' was added to 'Highbury' from July 20, 1922.

Another feature that made this line unique was the track arrangement. The 85-lb. flat-bottomed running rails were laid on longitudinal timbers, and the 80-lb. positive and negative channel section conductor rails were on opposite sides, outside the running rails, 10 inches from them and 2 inches above them, thus leaving a clear space in the four-foot with nothing to foul the impedimenta of main-line stock or electric locomotives. Direct current at 575 volts was supplied direct from the generating station at Poole Street, Shoreditch, on the banks of the Regent's Canal. There were four Musgrave vertical cross-compound 1,250-h.p. condensing engines, each with standard BTH compound-wound generators directly coupled. Separate generators provided power for the tunnel lighting, which was continuously illuminated.

Although the original intention to haul GNR main-line stock with electric locomotives was scotched by the changed attitude of the GNR and the subsequent abandonment of the running junctions, the large tunnels provided the opportunity for tube rolling stock of a spaciousness not seen before or since. The initial stock consisted of 26 motor cars with 54 seats and 32 trailers with 56 seats. These big teak cars had clerestory roofs and end gate platforms, and were built by the Brush Electrical Engineering Company and Dick, Kerr & Company. The equipment, by the British Thomson-Houston Company, included two 125-h.p. motors to each motor car (one per bogie) and an elementary form of electro-magnetic contactor control. This non-automatic control was the first complete multiple-unit installation on a new tube railway. The lattice gates on the end platforms were controlled by gatemen, but there were also single sliding doors in the centre of each side, only used at termini, where they were locked and unlocked by special porters. Nearly all the rattan-covered seats were transverse and the interiors were somewhat lavishly encrusted in mahogany. (At a later date they were embellished with clocks and cigarette machines.)

The cars were hauled from the factories on their own trucks, and delivered to the depot at Drayton Park, where there was a steep single-line exchange connection with the GNR goods yard. The four-road corrugated-iron depot and adjacent workshop were in a confined space to the west of the station.

In 1905, thirteen steel trailer cars and five steel motor-car bodies were ordered from Brush. These were of similar design to the original stock, whose grained teak was imitated on their aluminium panels. In 1907 a trailer was converted to a motor car. Train formation at peak periods was M-T-T-M-T-T-M (the Board of Trade was more indulgent towards full-size railways and allowed motor cars in the centre of trains). About 1904 an eight-wheel shunting locomotive was built by the GNCR with two 125-h.p. BTH motors and BTH equipment. About 1921 it was given the body and frame of Metropolitan Railway British Westinghouse locomotive no. 1.

The signalling system, which underwent some initial modifications to suit the requirements of the Board of Trade, was also peculiar, and the GNCR was the first tube to have automatic signals throughout on opening. There were no moving parts in the all-electric signals, which had red and green lights and were operated by d.c. track circuits. As an extra precaution, a treadle was fixed about 350/400 feet beyond each signal, to be struck by a brush on the last car of the train, thus providing a final clearance and unlocking the signal control relay in the rear. At Drayton Park, there were semaphore signals, worked by electric motors.

The end-to-end fare was 2d, with intermediate 1d stages. Through bookings with the GNR were not arranged until August 22, 1904, and through seasons were not available until January 1, 1905. The main-line company paid £1,800 a year for the use of the tube by its season-ticket holders in foggy weather.

About 15 million passengers were carried in 1905, and in each of the following two years, 16 million. Electric tramway competition caused traffic to fall sharply to 12½ million in 1908. It declined to just under 12 million in 1910, after which it increased slightly. (A prospectus issued in 1904 had estimated it as 23,250,000 in the first years.) Once Pearson's contractual dividend was finished, the ordinary shareholders received nothing. This sad story had interesting sequels in 1913 and again in 1935, as later chapters will relate.

Whilst the GNCR was struggling into stunted life, the vast injections of capital into the Yerkes companies were at last beginning to show some concrete results, and London was about to experience a crowded two-year period of urban railway electrification and tube openings.

CHAPTER 6

Consummation and Disillusion

1906–1909

EXPERIMENTS AT EALING

IN the summer of 1901, just after Yerkes had acquired control of the District, the finishing touches were given to the Ealing & South Harrow Railway, a five-mile, double-track extension of the District from Ealing (Hanger Lane Junction) to Roxeth (rechristened 'South Harrow' by the Railway). This line had been substantially completed some eighteen months earlier, but the penurious state of the parent company had caused it to be left unopened. Yerkes decided that he would use it as a test-bed for the new techniques proposed for the District and the tubes, and it was electrified on the 550-volt d.c. system with separate positive and negative conductor rails. Work on this proceeded during the latter part of 1902, and the generating plant used in the 1899–1900 Inner Circle experiment was transferred to the canal bank at Alperton to supply the power. The older section between Hanger Lane and Mill Hill Park (Acton Town) was also electrified so that convenient connections could be given with District services. Electric trains ran to Park Royal from June 23, 1903, and to South Harrow from June 28th.

The signalling and multiple-unit trains of the South Harrow line closely followed the pattern of those of the Boston Elevated Railway, opened in 1900, where automatic signalling, operated electro-pneumatically in combination with d.c. track circuits and polarized relays, had proved most successful. Experience gained on the Ealing & South Harrow was put to good use when equipment was ordered for the District electrification and the three associated tube railways.

THE CHELSEA MONSTER

The site between Chelsea Creek and Lots Road which had been selected for the Brompton & Piccadilly Circus Railway's power station was developed under the skilled guidance of Yerkes and Chapman, who designed a mighty generating station to supply the District, the three allied tubes and part of the London United Tramways system; at that time, it was the largest traction power station in Europe. The District even offered to supply the Metropolitan Railway, but the rival company chose to go forward with their own plant at Neasden.

Work on the foundations of two of the four 275-foot chimneys began in March 1902, and a few weeks later the timber bank of the Creek was replaced with a wall of Portland cement and Staffordshire blocks. To carry the main structure, 220 concrete piles were sunk 35 feet into the riverside clay. During 1903, erection of the 6,000-ton steel framework proceeded rapidly, and on a good day some 75 tons of steelwork could be assembled. When the dock walls and barge basin were finished about the end of the year, work began on the roofing of the main building, and clothing the steel frame with brick and terra-cotta. Late in 1904, the machinery was installed.

Not everyone welcomed the new arrival at Chelsea. The artist James McNeill Whistler was incensed at the intrusion, which he thought would completely ruin the bend of the river immortalized by Turner. R. D. Blumenfeld, Editor of the *Daily Express*, recorded in his diary Whistler's opinion of the parties responsible: '. . . they ought to be drawn and quartered.' Whistler had forgotten that the power station's predecessor on the site was a privately owned dust destructor.

The American artist had spent many happy years on the Chelsea riverside composing his nocturnes; for him, the evening mist transformed tall chimneys to *campanile*, but the four 275-foot smokestacks of Lots Road were too much. He died in July 1903, mercifully unaware of the full beauty of the completed structure, but may have turned in his grave eight years later when the Underground Group calendar carried a picture of the building in the blue of a summer evening which the *Railway News* thought 'suggestive of a Whistler nocturne . . .'.

Current was to be supplied to the four railways under terms agreed between the UERL and each statutory company and the proposed separate power stations at London Road (Baker Street & Waterloo), Kentish Town (Hampstead Railway) and Holloway (Great Northern & Strand Railway) were not required. The agreements were subsequently sealed and approved by the Board of Trade.

Lots Road came into partial use on February 1, 1905. At first the maximum output was 44.5 MW but soon rose to 57 MW at normal load, a figure exceeded by only two other power stations in the whole world. The main structure, with its two pairs of chimneys, conveyed a strong impression of the immense power generated within. Although the building was not without decoration, there was more than a hint of the coming functional style in railway architecture.

Inside, there were sixty-four Babcock & Wilcox water tube boilers on two floors, with space for sixteen more. Overhead elevators brought coal from the barges in the basin. From the boilers, steam passed through 14-inch steel pipes to eight sets of turbo-generators (not all erected until 1906). When all the railways were taking peak load in 1908, not more than five generators were needed simultaneously, so there was ample reserve. These generators were at the time the largest of their kind and each consisted of a 5·5 MW steam turbine by British Westinghouse and a 3-phase alternator. In use, the turbines, which were of a new type, exceeded the guaranteed maximum steam consumption per unit of electricity, and from 1908 onwards they were replaced by eight Parsons 6 MW turbines. They stood on massive concrete piers built up from the basement to the level of the engine-room floor, and operated with a tremendous din that seemed to vibrate the whole building.

Boiler water came from artesian wells, but the condensing supply came from the Thames through 66-inch pipes. The switch gear, by British Thomson-Houston, was housed on three galleries running the length of the north side of the building. All the high-voltage oil-switches were worked by 125-volt d.c. motors. Ashes were carried from beneath the furnaces to the barges in trucks hauled by a narrow-gauge battery locomotive. The coal bunkers had a capacity of 15,000 tons.

From the power house, the 11,000-volt three-phase $33\frac{1}{3}$ c/s current was passed through cables to Earl's Court, one mile to the north-west. Here it was distributed to 24 substations [1] on the allied railways, each of which was equipped with British Westinghouse rotary converters producing traction current at 550–600 volts d.c. Substation buildings also housed compressors supplying air for signal, point and lift gate operation and motor generators for the track circuiting current.

Lots Road was one of the first power stations in the world to rely exclusively on steam turbines for the generation of traction current. This was one of several bold innovations introduced by the American engineer Chapman. Prudently, he left enough space for vertical reciprocating engines if the turbines failed. Turbines had many advantages in a power station of this size. Their ability to use superheated steam and to turn a compact generator at high speed made them economical; they took up little floor space and provided a uniform driving torque, an important factor when three-phase alternators have to be operated in parallel.

As Lots Road was being completed, electrification was proceeding on the District and Metropolitan. The latter was first off the mark,

[1] These substations, which were not all in use until a few years later, were: Sudbury, Hounslow, Kew Gardens, Mill Hill Park (Acton Town), Ravenscourt Park, Earl's Court, Wimbledon Park, Putney Bridge, South Kensington, Victoria, Charing Cross (sunk below the level of the Embankment Gardens), Mansion House, Whitechapel, Campbell Road Bow, East Ham, Hyde Park Corner (below the surface), Russell Square, Holloway, Euston, Kentish Town, Belsize Park, Golders Green, Baker Street and London Road Southwark.

with electric trains from Baker Street to Uxbridge on January 1, 1905; Lots Road was completed the following month, and the District's Hounslow branch was electrically worked from June 13th. The District main line was ready on July 1st and by the end of the year all steam passenger working of District trains had ceased.

DEATH OF THE TITAN

In this 1905 summer of realization, Yerkes' health began to fail. He was confined to his house from the beginning of July but recovered sufficiently to chair the UERL board meeting on October 10th and the general meeting a fortnight later. In the following month, he went to New York for one of his periodical business trips to the USA, and upon arrival suffered a further attack of the kidney trouble which had been plaguing him. This was the last, and death came on December 29th. He was then 68. His London ventures were still incomplete; the District electric trains he had seen, but his tube railways he never knew. At their meeting on January 3, 1906, the UERL Board expressed their 'very high appreciation of the valuable services rendered by him to the Company, and also to the Public, in connection with the promotion of better facilities for traffic in London'. A well-deserved tribute, for without the dynamic Yerkes the three tubes might well have remained a roll of plans, and the beginning of unified control for London's passenger transport would have been deferred for a decade or more.

Yerkes held 32,000 £10 shares of the UERL at the time of his death, £5 paid. Calls of the remaining £5 per share were made in July 1906 and January 1907, but the company had the greatest difficulty in obtaining the £160,000. Eventually a judgment was secured against Yerkes' estate, and his Fifth Avenue mansion and art gallery were auctioned in April 1910 to meet the debt. The outstanding payment, with interest, was finally cleared in the early part of 1912.

NEW MEN

Yerkes' death came at a time when the UERL was within sight of completing its original purpose of raising capital and acting as 'contractor' for electrification of the District and construction of the three tubes. As there was little prospect that the general public would buy the ordinary securities of the subsidiary companies at anything other than give-away prices, the UERL was forced to remain in being as the holding company and policy-making body of the group. In this role, its directors needed both financial wisdom and railway-operating experience, and to this end, on January 3, 1906, they appointed Edgar Speyer as chairman and Sir George Stegmann Gibb as deputy chairman and managing director.

Speyer's appointment caused little surprise. As senior partner of Speyer Bros, he was intimately concerned in the fortunes of the UERL. His standing was such that six months later he received a baronetcy in the Birthday Honours.

But some eyebrows were raised in railway circles when Sir George Gibb relinquished the post of general manager of the North Eastern Railway, which he had held since 1891, to join the Underground. In *The Railway Magazine,* it was suggested that 'the pecuniary inducement must have been large to persuade him to vacate such a leading position in the railway world as the office of general manager of the North Eastern Railway confers, to accept the managing directorship of a few miles of electric lines'. Yet Gibb's previous experience had in many ways made him eminently suitable for the post; in 1901, he made a tour of American railways and closely studied their management methods, and a year later the NER was reorganized with Operating and Commercial Departments; he also adopted American ideas on the uses of railway productivity statistics as a tool of management; under his guidance, electric traction had been introduced on the North Tyneside lines in 1904, using multiple-unit trains taking 600 volts d.c. from a third rail. Gibb had served on the Royal Commission on London Traffic from 1903 to 1905, and was interested in London's special problems. He left the North Eastern a most excellently organized railway, and the Underground group was fortunate in securing his mature judgment and experience at this critical stage in its career.

CARS FOR THE THREE

As Gibb was completing his last few months with the North Eastern, the rolling stock for the Baker Street & Waterloo tube was steadily filling the depot at London Road. Assembled at Trafford Park, Manchester, the 108 cars travelled to Camden goods yard in London in trains of six or twelve, on their own wheels, hauled by an LNWR goods locomotive. The front car of each batch bore the legend ELECTRIC CAR TRAIN CONVEYED BY LNWR, and the trains ran about once a week (later once a fortnight) from September 1905 on Saturday nights. Every night, following the arrival of a train, one car body was lifted from its bogies to two trolleys on which it was hauled through the heart of London in the small hours by a team of fourteen horses. One contemporary account describes the equipage leaving Camden at midnight, arriving at St George's Circus at 1.35 a.m., then waiting until 2.10 a.m. for a gap in the tram service; by 5 a.m. the car was safely inside the yard, where it was lifted back on to its bogies. On one occasion a steam tractor was used instead of horses. It broke down at the depot entrance and blocked tram services for over twelve hours. Eventually the car was pulled into the depot by a winch, and the incident brought a claim from the LCC for several thousand pounds of lost tram traffic.[1]

It was a much simpler matter to deliver the 218 Piccadilly cars, as the depot at Lillie Bridge was on the main railway system. Delivery began in June 1906, and a month later the cars were arriving at the rate of fifteen a week. No exact account of the route followed has come to light, but it is likely that the cars travelled via Tilbury Docks, the Whitechapel & Bow and the District Railways.

Horse and trolley transport was also used for the Hampstead cars, which were moved from Camden LNWR and West Hampstead, Midland goods depots. Delivery began in September 1906 and the 150 cars were safely berthed in Golders Green by mid-March 1907. As with the other two lines, the assembly of the electrical equipment was done in the tube depot after arrival.

The very first test run of the three tubes took place between London Road depot and Baker Street early in November 1905; soon afterwards, test trains were operated regularly and there was a full dress rehearsal in the first week of March 1906. Trains ran to the full working timetable and the crews went through their drill as though the trains and platforms were filled with passengers; gatemen opened and closed their gates, announced the station names and passed the bell signal to guards, who estimated the correct station time before giving the starting signal to the driver. Up above, liftmen and booking clerks practised. Similar procedures were followed on the Piccadilly from the end of October 1906 and on the Hampstead from early in May 1907.

As a final step before the opening, much-needed publicity was obtained by inviting the Press to trial runs and formal meals at the company's expense. On March 7, 1906, journalists were taken by train from London Road depot to Baker Street and lunched at the Great Central Hotel. Edgar Speyer and Sir George Gibb made appropriate speeches, and the reporters went away happily clutching their 'handouts', the first of a long line of ready-wrapped and processed briefs. A special train for the Press left Hammersmith at 11 a.m. on December 12, 1906, stopped to pick up at Knightsbridge and Leicester Square, and then ran non-stop to Finsbury Park. On the way back, the newspapermen worked up an appetite for the lunch at the Criterion by inspecting the spiral conveyor at Holloway Road and the sub-station at Russell Square.

OPENING THE THREE

The first of the Yerkes tubes to open was the Baker Street & Waterloo, which began public service between Baker Street and Kennington

[1] From an account in *London Transport Magazine*, May 1956, by Mr J. P. Thomas, then Superintendent of the Line, later general manager, London Transport Railways.

Road [1] on Saturday, March 10, 1906. Intermediate stations were sited at Regent's Park, Oxford Circus, Piccadilly Circus, Trafalgar Square, Embankment and Waterloo. On August 5, the present southern terminus at Elephant & Castle was opened. The northern end was prolonged to a station called Great Central,[2] serving the Marylebone terminus of the GCR, on March 27, 1907, and extended still further west, to Edgware Road, on June 15th of that year.

Although work at some stations was not complete, the Great Northern, Piccadilly & Brompton was opened between Finsbury Park and Hammersmith on December 15, 1906, with intermediate stations at Gillespie Road, Holloway Road, Caledonian Road, York Road, King's Cross, Russell Square, Holborn, Leicester Square, Piccadilly Circus, Dover Street, Hyde Park Corner, Knightsbridge, Brompton Road, Gloucester Road and Earl's Court. The trains also called at Barons Court, a new District station between Earl's Court and Hammersmith. Further stations were brought into service as they were completed: South Kensington (January 8, 1907); Down Street (March 15, 1907); and Covent Garden (April 11, 1907). A short spur from Holborn to Strand, on the alignment of the original plan of the Great Northern & Strand Railway, was opened on November 30, 1907. This was a double-track line, at first worked by a two-car train shuttling in the eastern or southbound tunnel (in rush hours a spare train normally accommodated in the western platform at Holborn shuttled additionally in the western tunnel). From March 3, 1908, the all-day service was worked by an adapted single car which ran in the western tunnel. (After 1912 the normal service ran between the western platform at the Strand and the eastern platform at Holborn, a practice which has continued ever since. The eastern tunnel was taken out of railway use from August 16, 1917.)

The Charing Cross, Euston & Hampstead Railway began service on June 22, 1907, between Golders Green and Charing Cross (the present 'Strand'), with a branch from Camden Town to Highgate. There were stations intermediately at Hampstead, Belsize Park, Chalk Farm, Camden Town, Mornington Crescent, Euston, Euston Road, Tottenham Court Road, Oxford Street and Leicester Square. On the Highgate branch there were stations at Tufnell Park, Kentish Town and South Kentish Town. It was soon realized that the choice of station names was not entirely satisfacory and on March 9, 1908, Tottenham Court Road became Goodge Street and Oxford Street became Tottenham Court Road to match the contiguous CLR station. Euston Road became Warren Street on June 7, 1908. An interesting feature of the Hampstead was that it was the only tube railway to possess a station that

---

[1] Renamed Westminster Bridge Road, August 5, 1906, renamed Lambeth (North) April 15, 1917. Brackets dropped about 1928.

[2] Renamed Marylebone, April 15, 1917.

HOLBORN AND STRAND 1906-1907

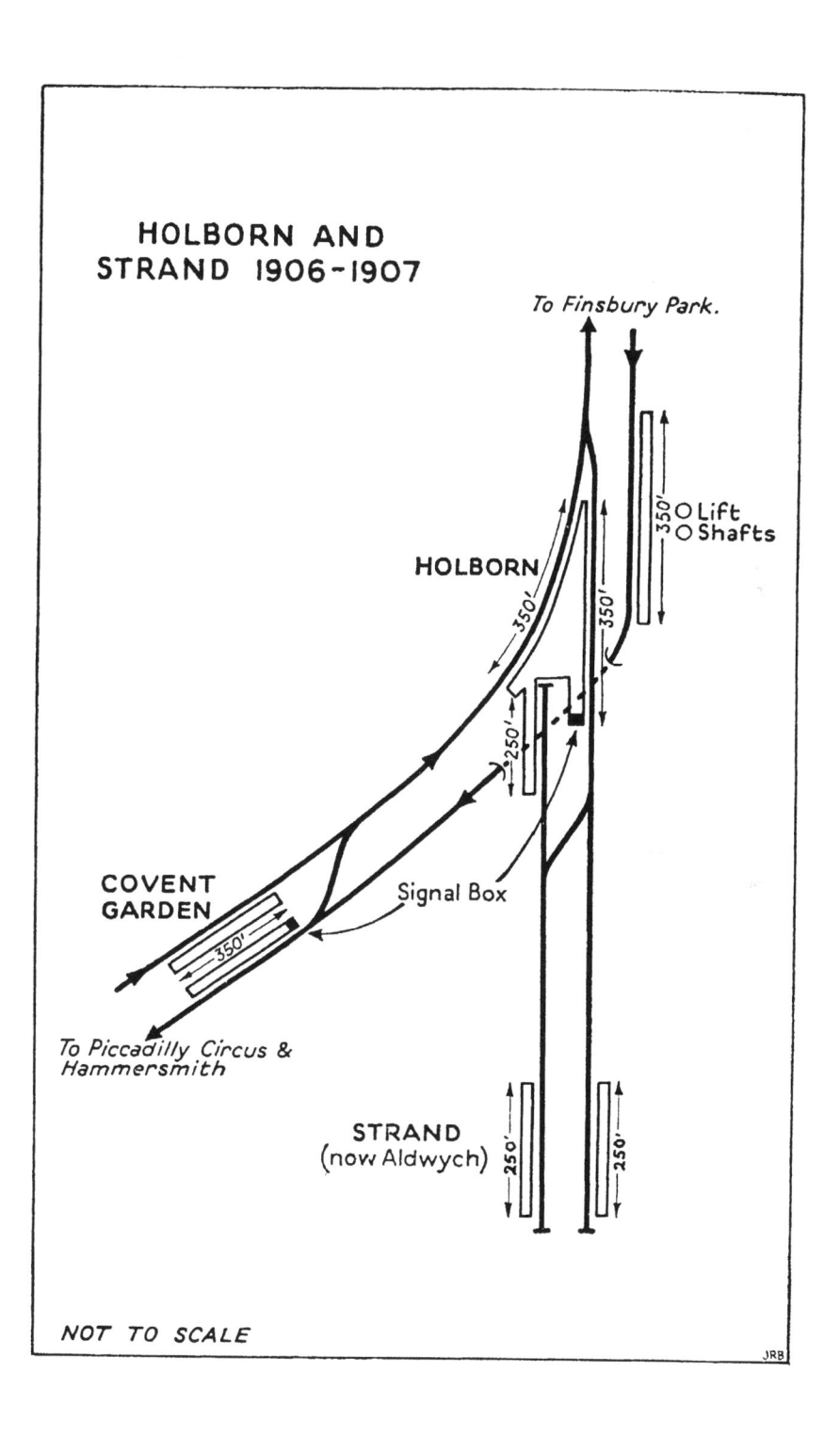

To Finsbury Park.

350' ○ Lift
     ○ Shafts

HOLBORN

350'
350'
250'

COVENT GARDEN
350'

Signal Box

To Piccadilly Circus & Hammersmith

STRAND
(now Aldwych)

250'   250'

NOT TO SCALE

JRB

was built but not opened; this was North End, between Hampstead and Golders Green, where station tunnels, platforms and stairs to the lower lift landing were finished but no vertical shafts were driven and nothing was done on the surface. A full account of this 'ghost' station will be found in Appendix 5.

The clumsy statutory titles of the three lines were, of course, too much for the man in the street. A writer in the *Evening News* (G. H. F. Nichols, or 'Quex') coined the tag *Bakerloo* for the BS&W, and this was quickly adopted by the company, who used it officially from July 1906. This move staggered the prim, anti-American editor of *The Railway Magazine*: 'for a railway itself to adopt its gutter title, is not what we expect from a railway company. English railway officers have more dignity than to act in this manner.' Mr Nokes was spared further nicknames; the other two lines simply became known as the Piccadilly Tube and the Hampstead Tube, names which were officially promulgated and stuck well enough.

Formal opening ceremonies followed traditional patterns, with some amusing variations. David Lloyd George was asked to open the Bakerloo in his capacity as President of the Board of Trade, but found excuses and the ceremony was performed by the Chairman of the LCC, Sir Edwin Cornwall, MP. Guests travelled in a special train from Trafalgar Square to Kennington Road, whence they proceeded to Baker Street and luncheon at the Great Central Hotel. Whilst they made speeches, the line was opened to the public. In his address, Sir Edgar Speyer referred to Yerkes as 'the great master mind' and hoped to have the whole system open by the following spring. Theodore Julius Hare, chairman of the BS&W, pleaded for public patronage of the railway and hoped the public would not be led away by 'that lively and young motor omnibus'. W. R. Galbraith gave details of the construction, mentioning that the tunnels driven from the top of Haymarket met those in Northumberland Avenue with a deviation of only one inch. The tunnels under Regent's Park had already existed five years without any ill effects on the trees and foliage above (this was to calm fears about Hampstead Heath).

Lloyd George came to open the Piccadilly and was given a golden controller key to start the first train from Hammersmith. Officials around him had some uncomfortable moments when it became apparent that the key would not fit, but the train was started with a normal key procured from a call examiner, whose offer to file the gold was sternly refused. The special train travelled to Finsbury Park, returning to Piccadilly Circus, where the guests alighted for lunch at the Criterion. As soon as they had cleared the station, the line was thrown open to the public, who were able to use trains already in service. At the luncheon, Lloyd George spoke of the part to be played in London transport by private enterprise and joked about the vitiated air from the tubes being blown into the streets for the benefit of the bus passengers.

Undeterred by his Hammersmith experience, the Welsh politician agreed to open the Hampstead tube. This time the gold key fitted, but the amateur driver shortly afterwards 'dropped' the dead-man's handle, stopping the train in tunnel and causing some delay. The opening was distinguished by the offer of free rides to the public, of whom about 150,000 rode on the trains between 1.15 and 8.45 p.m. Meanwhile, Lloyd George and the lucky ones were regaling themselves in the suitably decorated paint shop at Golders Green. These workaday surroundings seemed to match the sober nature of the luncheon speeches. Sir George Gibb warned that the tube railway fares must be such as to afford a proper return for the shareholders, and was followed by Sir Edgar Speyer who deplored the lack of government or municipal support for the tubes which, among other benefits, had reduced the necessity for expensive street widenings. London stood out alone among great cities in not encouraging undertakings such as this one. He strongly urged the formation of a London Traffic Board, and offered the right of purchase of his railways to the public authorities if they would supply part of the initial capital or lend their credit for raising money. Lloyd George took care not to commit himself, but offered the opinion that much might be done by municipal assistance to private enterprise in this field. He hoped that the Hampstead was not the last tube railway that London was going to see. The Underground Company seemed to be fairly sure it *was*, as far as they were concerned, as they had advertised the Hampstead as 'The Last Link' and left no doubt in anyone's mind that it represented the conclusion of their 'task'. Significantly, they made no mention at this time of the already authorized extensions to Edgware and Watford from Golders Green, although, as we shall see, a connecting bus was soon provided to Hendon.

THE YERKES TUBES DESCRIBED

The Bakerloo, Piccadilly and Hampstead tubes had a common design, and stations, rolling stock, permanent way, signalling and other equipment were uniform on all three lines. It is therefore convenient to offer a general description in which differences of detail and certain points of special interest can be mentioned.

Almost always, the route of the tube railway followed the lines of streets so as to avoid the expense of easements. This led to sharp curves, and often meant that platforms at the same station had to be sited at different levels. The severest curves were 313 feet radius on the Bakerloo, 330 feet on the Piccadilly and 462 feet on the Hampstead. Tunnel station platforms varied in depth, from 20 feet at Finsbury Park to 192 feet at Hampstead, the average being between 75 and 80 feet below the street. The point furthest from the surface was 1,900 feet north of

Hampstead station, where the tunnels were 221 feet below the Heath. Generally, the profile of each line tended to follow the surface relief. The Bakerloo gradually ascended from the Thames to Paddington, and the Hampstead climbed more steeply from the river to the Heath, rising 272 feet in its six miles. The Piccadilly tunnels were generally falling from Finsbury Park to Holborn, level to Earl's Court, then rising to the surface at West Kensington. On the Bakerloo and Hampstead, the steepest gradient was 1 in 60 and the profile of both lines made it impracticable to adopt Greathead's undulating pattern with stations on 'humps' to assist retardation and acceleration. Piccadilly stations were usually approached at 1 in 66 up, and left at 1 in 33 down.

For the tunnels, a standard internal diameter of 11 feet $8\frac{1}{4}$ inches was adopted, but this was enlarged on curves to a maximum of 12 feet 6 inches. Throughout their length, the tunnels were lit by 16-candle-power incandescent lamps sited 42 feet apart (40 feet on the Bakerloo), separately fed by a 220-volt circuit from the substations. For some months, tunnel lights were left on all the time, probably in the hope of raising the morale of the more nervous passengers. At Camden Town, on the Hampstead, was London's first running junction in tube tunnels.

The permanent way was built up in an entirely new fashion to the design of J. R. Chapman. A concrete bed 46 inches wide was first placed down the centre of the tunnel floor. Australian karri-wood sleepers, 6 feet 6 inches by 1 foot 2 inches by 5 inches, were set in sand and cement grout across the top of this bed, and the concrete then built up 3 inches between the sleepers, which were spaced 3 feet 4 inches apart (1 foot 8 inches at rail joints). The spaces under the sleeper ends and the whole road to sleeper level were filled up with washed granite cubes. The ends of the sleepers were thus virtually unsupported, and the centre portions rested on the concrete shelf. Three pieces of angle iron were nailed under each sleeper to prevent lateral movement. The ballast drained into a 3-inch agricultural pipe embedded in the concrete. Rails were bullhead, 90 lb. per yard, and usually about 45 feet long.[1] Conductor rails were rectangular in section and 85 lb. a yard. On the Bakerloo, the negative conductor was in the centre of the track, $1\frac{1}{2}$ inches above the running rails and the positive near the tunnel wall, 16 inches outside and 3 inches above the right-hand running rail. In practice it was found that misplaced collector shoes could bridge the gap between the positive conductor and the tunnel wall and so run to earth. The same conductor rail arrangement was adopted on the other two lines, but the iron tunnel segments were lined with concrete for 2 feet above ballast to reduce the risk of earthing. On the Bakerloo, polarity was reversed at points where the clearance was tight between

---

[1] The process of lowering the rails down vertical shafts and turning them into the tunnels restricted the Bakerloo rail length to an absolute maximum of 36 feet 5 inches.

the outside current rail and the tunnel wall. The conductor rails were held in vitrified earthenware insulators of special shape, which were fastened to the sleepers by malleable iron clips. The running rails rested in 30-lb. chairs and there was a ¼-inch pad of wool felt between the chair and the sleeper.

Chapman's unconventional track construction was intended to provide a certain amount of elasticity in running, economy in maintenance and reduced noise and vibration. Centre-bound track was a condition that the permanent way engineer had hitherto taken great care to avoid, and its introduction caused some flutter in the technical press, the *Railway Engineer* being particularly rude about it. The ends of the sleepers, it said, were unsupported cantilevers and would be subjected every few minutes in rapid succession to hammer blows of 3 to 4 tons; the ballast would rise, and any attempt to pack it would speedily reduce it to powder, but if the sleepers were so stiff they would not bend, 'the object of introducing all this debris into the tube becomes obscure'. The writer thought that there was great danger of sleeper movement, with consequent disaster owing to the limited clearance; what was worse, 'the precious granite cubes' would obscure sleeper movement and were 'quite useless'.

Under service conditions, none of the prophesied troubles and disasters came about and for many years the rails showed a remarkable immunity from the curse of corrugation. There *were* disadvantages and official spokesmen gradually lost enthusiasm, admitting the arrangement 'not perfect' in 1910. In a paper delivered to the Permanent Way Institution in 1935, Mr H. Windmill stated that it gave little improvement in riding comfort and many coach screws were broken. It was not perpetuated and for the new construction of the 1920s, sleepers were wholly embedded in concrete; and in the 1930s and later, concrete benches were placed below the sleeper ends, with ballast in the centre —a complete inversion of the original design.

The surface stations consisted of a brick-clothed steel framework faced on the street frontages with glazed ruby red terracotta blocks, giving an utterly drab but distinctive, cheap and hard-wearing surface. A mezzanine with large glazed arches and small circular windows under prominent hood moulds housed the lift machinery. Above this was a flat roof ready for extra stories of offices or dwellings; after over fifty years, some stations still await this development and have a stark unfinished look.

The names of the railway and station were displayed in different ways. On the original Bakerloo stations the full name of the railway was spelt out in raised gilt Roman letters above the mezzanine, and the station name, also in Roman letters, embossed in the tiles and picked out in gilt, appeared below. Piccadilly stations displayed their names in similar fashion, but the space above the mezzanine was usually blank.

On the Hampstead, there were two rows of black *sans-serif* letters each on a white strip; that above the mezzanine showed the station name, and that below, the line name. Station exteriors were illuminated by Maxim arc lamps and later, canopies of iron and glass were added to protect the entrances. On these canopies, the glass panels at right angles to the station façade usually carried the station name in white on blue and those parallel to the street bore the word UNDERGROUND.

Inside, the walls were tiled green to shoulder height, white above, and the influence of *art nouveau* was evident in the decorations and ticket windows. Sensible one-way 'passenger-flow' arrangements were adopted, and the lifts unloaded directly on to the street. All station design was by Leslie W. Green, ARIBA, who was appointed architect to the Underground company in 1903 when only twenty-nine years old. He died in 1908, just over a year after the last of his stations had opened.

For site reasons, the stations at Embankment, Waterloo, Trafalgar Square, Regent's Park, Oxford Street, Charing Cross and Finsbury Park did not have a surface building of the standard design. Embankment shared booking facilities with District's Charing Cross station; at Finsbury Park, the Piccadilly and Great Northern & City had facilities beneath the Great Northern station; Waterloo's booking hall nestled against the main-line station, and the other four had booking halls below street level. From Earl's Court to Hammersmith the Piccadilly shared surface facilities with the District (Barons Court and Hammersmith had L. W. Green façades of similar but earlier vintage). Golders Green was on an embankment and had three platforms with four faces and a two-storey, flat-roofed brick station building at road level. This bore no resemblance to the other stations and was probably without benefit of architect. The two tube stations at Finsbury Park were constructed by the GNR; both were directly below the main-line platforms and each had a booking office at the top of a short flight of stairs leading to the platforms from a system of subways which gave out to surrounding streets. Both also had a pair of hydraulically operated lifts and a spiral staircase which communicated directly with a large interchange booking hall just beneath the GNR platforms. The Piccadilly station was tiled dark green (lower half) and white (upper half), whilst the GNCR was brown and white. The Piccadilly terminus at Hammersmith was alongside the District station and had three platforms (four faces) under a low-pitched steel and glass all-over roof. Where two lines crossed, at Piccadilly Circus and Leicester Square, combined surface stations were built, with intercommunicating passages at low level. At Holborn, the eastern or southbound line of the Strand branch had its own platform, and just north of this it made a trailing junction with the eastbound main line. The western or northbound line of the branch terminated in a dead end with a 250-foot platform. All four platforms were interconnected by a somewhat complicated system of stairways and passageways. Low-level subways were provided for interchange with the Central London at Oxford Circus and Oxford Street, and with the CSLR at Euston, King's Cross and Elephant & Castle.

Minor variations in the internal layout of stations were dictated by site conditions and the varying positions and levels of the running lines. There were from two to four lifts at each station, according to the amount of traffic expected, and usually one 23-feet diameter shaft held two lifts, but there were some 30-foot shafts accommodating three. The 'double' stations at Piccadilly Circus and Leicester Square had eight and five lifts respectively; at Piccadilly there were four for each line, but at Leicester Square all five were used for either line as the lower landing was between the two platform levels. After 1907, when all the lines had opened, there were 140 lifts in service—35 on the Bakerloo, 60 on the Piccadilly and 45 on the Hampstead; all were electrically powered and were supplied by the Otis Elevator Company. Each lift car held about seventy passengers and travelled at speeds up to 200 feet a minute (generally at 100–120 feet per minute), with automatic acceleration, deceleration and positioning at landings. The lift motors were fed by an independent 575-volt circuit from the substation. Entrance gates were worked by hand, and passengers left the lift from the opposite gates, which were pneumatically worked. Lift shafts were clustered and usually descended to a landing alongside the tunnels about 15 feet above platform level. At South Kensington there was a difference of 18 feet between the levels of the east and westbound tunnels and the lifts made separate calls at each level. One of the lift shafts at Holloway Road contained an experimental 'double spiral continuous moving track' installed by the Reno Electric Stairways and Conveyors Co. Ltd. This was a moving belt of teak slats and had a speed of 100 feet a minute; although in use on the opening day of the Piccadilly tube, it was soon afterwards closed and never heard of again. At Gillespie Road, the platforms were only 30 feet below the street and no lifts were provided, and at Embankment the tube platforms were reached from the District's Charing Cross station through a long inclined passage and stairs.

Up and Down running lines were always in separate station tunnels, 21 feet 2½ inches in diameter, containing stone-edged concrete platforms 350 feet long (291 feet on the Bakerloo.) Each station tunnel had distinctive colour patterns in the tiling to assist regular passengers to pick out their destination. The ceilings were not tiled, but plastered and whitewashed and divided into sections by rings of coloured tiles reaching over to the opposite wall. At Trafalgar Square northbound platform, there was a reminder of the Whitaker Wright regime in the form of a ceiling covered with plain white tiles in contrast to the standard decorations on the other platform. Station names appeared on the platform wall tiles in brown letters about 15 inches high and *art nouveau* WAY OUT and NO EXIT signs were also incorporated in the tiling. Lighting current came from the tunnel lighting circuits and was supplemented by emergency lights fed from the local power supply. Many of the platforms also had Maxim arc lamps connected to the lift supply; these slid down round the tunnel wall for maintenance. Every station had an 18-foot shaft, containing an iron spiral staircase for emergency use, and a 4 foot 6 inch ventilation shaft. The somewhat austere platforms were furnished with handsome large-dialled electric clocks, and garden seats, most of which still survive.

For the first time on a London tube railway, adequate provision was made for ventilation. Almost all the stations had exhaust fans, extracting 18,500 cubic feet of air a minute from the tunnels and drawing it through ducts in the stair shafts to the roof, where it was ejected. Fresh air entered at street level and passed down the lift and stair shafts. At first, all the fans were run all the time, but from July 8, 1907, some were operated only for a limited period during the day. This ventilation system had two main advantages; passengers did not meet vitiated air as they entered the station from the street; and, if there were a fire, it ensured that smoke and fumes would be sucked up through the ducts and not pass up the stairs and lift shafts. *The Tramway and Railway World* remarked '. . . the tunnels and stations are as if continually swept by a breeze, and some delicate passengers may think they are getting too much of a good thing'. It was, of course, quite impossible to please everybody.

The signalling system was something quite new for Britain. Designed to handle intensive traffic with the maximum safety, it was so successful that the basic elements have survived to the present day on all London Transport lines. Its safety record over more than half a century is impressive. Automatic signals, normally in 'line clear' position, operated electro-pneumatically in combination with train stops and track circuits had been working satisfactorily for some time on the Boston Elevated Railway. The Westinghouse Brake Co. designed and manufactured similar equipment for the Ealing & South Harrow line, using the patents of H. G. Brown, Signal Engineer of the Boston Elevated.[1] The use of d.c. in the signalling system made it desirable to provide an insulated traction current return, and this is one of the reasons for the fourth rail on the London Underground (others were: if one pole earthed because of defective insulation, the service could be maintained by allowing the other pole to assume the full voltage above earth; and on the shallow lines, the system avoided the injurious effects of earth leakage currents on nearby service pipes,

---

[1] Brown came to England to advise on the E&SH installation and later became successively managing director and deputy chairman of the Westinghouse Brake and Signal Company.

telephone cables and observatories). As an additional precaution against unwanted interference from stray currents, the track circuits were protected at each end by polarized track relays so arranged that extraneous current could never cause both relays to operate in the same manner and provide a false clear. At each signal, a mechanical connection operated a train stop with a trip arm which stood up when the signal was at danger; if a train passed such a signal it would be brought to a halt as the arm would catch the brake pipe vent cock on the motor car. A 400-foot 'overlap' of the block section beyond the signal gave ample room for a 'tripped' train to stop. Signals controlling junctions, sidings and crossovers and other signals operated from signal boxes by dwarf levers were semi-automatic, returning to danger automatically after the passage of a train. All other signals were operated by the movement of trains through track circuits.

Signal-boxes on the Bakerloo were at Elephant & Castle, Kennington Road, Piccadilly Circus, Great Central and Edgware Road—the last two went out of use when the line was extended. Piccadilly boxes were at Finsbury Park, York Road, Holborn, Covent Garden, Piccadilly Circus, Hyde Park Corner and Hammersmith; in addition, a District box at West Kensington West controlled a crossover east of Barons Court and access to Lillie Bridge depot. The Hampstead had cabins at Charing Cross, Mornington Crescent, Camden Town, Highgate, Hampstead and Golders Green. Most of these boxes controlled crossovers, and in the tunnel stations they were sited at one end of a platform, the station tunnel and the running tunnel headwall forming two of the sides.

As a result of experience on the Ealing & South Harrow, the track relays were redesigned, a neater air cylinder was evolved for the signals and the train stops were given independent air cylinders. Tube tunnel signals consisted of spectacle plates moved in a vertical plane by compressed air. They were lit by oil lamps which gave much trouble as the strong draughts were apt to blow them out and they did not burn well, requiring nightly inspection to remove crust from the wicks. Oil was probably chosen for reliability in the event of electrical breakdowns, but it is interesting to note that electric signal lamps had been in use for some time on the CSLR and had been installed from the start on the Central London and Waterloo & City. Gradually the oil lamps gave way to electric bulbs, and much later the lantern plates were replaced by colour light signals. The conversion of the d.c. track circuits to a.c. was the only other alteration made to the original principles.

An interesting feature of the signalling equipment was the illuminated track diagrams in the signal boxes. These showed the track layout painted on glass, with track circuit block sections lit by strip lamps behind. When a train entered the section, the track-circuit relay opened the lamp circuit and caused that portion of the diagram to darken. The first of these diagrams was brought into use at Mill Hill Park (Acton Town) on June 11, 1905.

Whilst similar in general appearance and design, the cars of the three tubes came from three builders—American, French and Hungarian—and had detail variations.[1] The Bakerloo was stocked with 108 all-steel cars, equally divided into motor cars, control trailers and trailers, and all built by the American Car & Foundry Co. at Berwick, Pennsylvania. They were shipped 'knocked down' to Manchester, and assembled in sheds at Trafford Park. The motor cars had Sprague-Thomson-Houston multiple-unit electro-magnetic control and two BTH 240-h.p. motors fitted to 36-inch wheel, diamond-frame motor trucks, identical to those of the standard District stock. Accelerating current was maintained through both motors by 'bridge transition' from series to parallel, thus avoiding the jerk characteristic of the 'open-circuit transition' of the CLR and GNCR motor cars. There were forty-six rattan-covered seats in the motor cars, including six transverse pairs either side. At the driving end, the floor was raised over the motor bogie, and here three steps led up to four longitudinal seats, two either side. Beyond this was a steel bulkhead with a door to the control compartment and the driving cab. At the other end of the car, a door led to a platform with hinged inward-opening lattice gates either side. Access to the next car could be gained through openings in the centre of the platforms. The trailers had similar gated platforms at each end and inside, fifty-two seats (sixteen transverse in pairs in the centre of the car). They were almost 2 inches longer than the motor cars—50 feet 2 inches over buffers. Trailing wheels on all cars were of 30-inch diameter, and trailer bogies were of the girdle type with swing bolsters and the frame suspended from springs over the axle boxes. The control trailers had driving gear on the gate platform and the driver stood behind a crude weatherboard, peering through a small square window. All cars had clerestory roofs and the fireproofed mahogany-veneered interiors were rather more austere than those of the CLR cars. Livery was scarlet below waist, cream above, with BAKER STREET AND WATERLOO RAILWAY inscribed in full.

A single steel-framed trailer was built for the Piccadilly tube by the Brush Electrical Engineering Co. and delivered in September 1905. This had fifty-four seats and was similar to the Bakerloo cars in general layout, as was a second British-built trailer soon afterwards supplied by the Metropolitan Amalgamated Railway Carriage and Wagon Co. In all, the Piccadilly stock amounted to 72 motor cars, 72 control trailers and 74 trailers, the main order being shared (108 cars each) between Les Ateliers de Construction du Nord de la France (Blanc Misseron) and the Hungarian Railway Carriage and Wagon Works (Györ). Apart from the two English cars, which were scarlet and

[1] A table of dimensions is on page 139.

cream, the livery was 'engine lake', a deep crimson lake. The cars lacked the very prominent side bulge and big rivet heads of their American sisters on the Bakerloo and their control compartments were longer, reducing the motor-car seats to forty-two.

The Hampstead had 60 motor cars, 50 control trailers and 40 trailers, again of similar design, except that the motor cars, like those of the Piccadilly, had only 42 seats. American Car & Foundry built them at Manchester, using what the handouts described as 'British materials almost entirely', and the trucks, bodies and equipment were assembled at Golders Green depot. The characteristic ACF rivet heads again appeared and a new feature was a spring-loaded buffing plate to minimize bumping as the train stopped. Destination plates were provided on the car sides to indicate whether the train was for Golders Green, Hampstead or Highgate. Another innovation, later adopted on the other lines, was the use of Mr Frood's patent consolidated canvas and pitch brake block insert instead of the traditional cast iron. After experience, these inserts were replaced by blocks of woven cotton fabric laminated with jarrah wood pegs, possessing the advantage of greater efficiency and longer life (with the indirect benefits of longer rail and tyre wear and the absence of metallic brake dust). But the improved block did not work well in the open in wet weather, and a satisfactory Ferodo all-weather block was not achieved until 1931. The Hampstead cars had a lake livery, picked out with gold lining, and both Piccadilly and Hampstead cars bore only the initials of the statutory company.

Each driver had a telephone set which could be clipped to the bare wires running along the tunnel walls, allowing him to speak to the substation attendant or the nearest station. Westinghouse air and hand brakes were standard, as was the 'dead-man' current cut-out and brake application device in the controller handle.

Depot accommodation was provided at London Road, Southwark, for the Bakerloo; at Lillie Bridge, Fulham, for the Piccadilly; and at Golders Green, for the Hampstead. The Bakerloo was the only line entirely below ground, and depot access was gained by a single-track incline leading off the northbound line just north of Kennington Road station. London Road, with its cramped, expensive site, hemmed in by steep retaining walls, had fourteen storage tracks, a small three-track car shed 350 feet long, a paint shop and a substation, all fourteen feet below street level. The tunnel to the railway and a blind tunnel used as a shunting neck were at the western corner; at the other end, a ramp, used for the delivery of the cars and stores, led out into London Road near St George's Circus. Nearby was a vertical shaft containing stairs and a ventilation duct. A signal box over the tunnel entrances controlled depot movements.

On the site of the old District steam locomotive depot and repair shops, the six road car shed at Lillie Bridge was 78 feet 6 inches wide, and almost a quarter of a mile long. The cars inside were moved by current fed through flexible cables from overhead conductors. At the southern end were workshops containing two 12-ton travelling cranes. Access to this depot was awkward and vulnerable; it was obtained via sidings east of Barons Court station, the District running lines in West Kensington station, and a single line spur into the yard. The Golders Green depot, situated on the north side of the station, was by far the most spacious as the land was cheap. The whole Hampstead fleet of 150 cars could be accommodated under cover, and south of the ten-track running shed were a three-track paint shop and a two-track machine shop. Inside, the cars were moved as at Lillie Bridge. Outside, there were four open storage sidings, and the depot also included a driving school with demonstration m.u. equipment.

Train services were ambitious from the very beginning. The Bakerloo started at 5.30 a.m. with a five-minute headway for two hours, then three minutes till 11.30 p.m., concluding with a six-minute headway in the last hour. Sunday service was from 7.30 a.m. till midnight, six minutes till 11 a.m. and three minutes after that. Piccadilly trains ran between the same hours and the basic service was $4\frac{1}{2}$ minutes, every three minutes at peak periods. The Hampstead ran from 5.15 a.m. till 12.30 a.m. At peak periods, a two-minute interval was maintained between Charing Cross and Camden Town, four minutes thence to Highgate and Hampstead and twelve-minute intervals between Hampstead and rural Golders Green. In the off-peak periods, the trunk service was reduced to $2\frac{1}{2}$-minutes interval and the branches service to five minutes (Golders Green fifteen minutes). To develop the traffic to Hendon, which was about halfway along the authorized but postponed extension to Edgware, a motor-bus service was started in connection with the trains at Golders Green. The two buses and drivers were supplied by Messrs Birch Brothers, but the conductors were employees of the tube railway. This service began on July 28, 1907, with through tickets in each direction to and from the tube stations and The Bell at Hendon. The last bus from Golders Green was at 8.36 p.m. on weekdays and 9.12 on Sundays, but later weekday buses were introduced in 1908 so that the last weekday bus became 11.59 p.m. Sunday buses ran until 10.25 p.m. from September 1909. The motor buses did not prove reliable and were soon replaced by horse buses.

The Bakerloo ran six-car trains in the peak and three cars at other times, but traffic was slow in building up and most trains were soon reduced to three in the peak and two in slack hours, the slack service being reduced from three to $4\frac{1}{2}$ minutes at the same time. Six-car trains were also used on the Piccadilly at times of pressure, but the normal formation was three cars (motor-trailer-control trailer). The Hampstead peak-hour trains consisted initially of five cars, and at other times the motor-trailer-control trailer formation was in use.

Manpower was cheap, and the company was able to afford train crews so large that at quiet times in the difficult early days of the Bakerloo they must have all but exceeded the number of passengers. The guard [1] was at the front of the train between the first and second cars, and on the gangway plate between each of the other cars, there stood a gateman, able to control both gates by means of crank handles. When all the passengers had been ushered in, guard and gateman closed their lattice gates and the bulkhead doors into the cars. This done, the rear man rang a bell through to the next gateman and so down the train until the guard was reached; *his* bell rang in the driver's cab and provided the signal to start the train. This tintinnabulation down the train was a joy to the ear and a characteristic sound of tube operation in the early days. Starting whistles were tried for a few weeks on the Piccadilly at the end of 1907 but were not considered suitable. Just before the train arrived at each station, the gateman would slide open the bulkhead doors and bellow into the car a slurred distortion of the station name. Strangers would slowly realize that *Totnacorranex* meant *Tottenham Court Road next*. Letters were written to *The Times* complaining about *Ampstid* and *Igit*. A letter writer in the *Evening Standard* in October 1907 deplored the 'awful noise' made by the gate and liftmen, pointing out that remonstrances only produced rudeness and more noise. The *Railway Engineer* remarked that 'such methods are only importations from America, where rudeness and noise from railway servants are meekly tolerated, if not appreciated'. In fact, the London travelling public were undergoing the process of learning how to travel by rapid transit railways. They were not yet used to the idea that to ensure a frequent service, station time must be kept to a minimum and that, to this end, a fair amount of pushing and shoving (by themselves and by the staff) and shouting (by the staff) were all part of the game. The bulkhead doors were normally closed whilst the train was running, but instructions were issued in July 1907 that the doors at the trailing end of all cars, except the last in the train, be kept open to assist ventilation unless passengers complained of draughts. At first this applied to warm weather only but it was later extended to the whole year. In a 1923 rule book, staff were instructed to open the *front* bulkhead doors when there were standing passengers inside, and also on the open sections, to blow fresh air into the car.

The gateman's task was rendered more difficult because passengers were tolerated on the end platforms in the rush hours, although at first this practice was officially forbidden. It is not hard to imagine the 'friendly persuasion' required to close an inward opening gate on a crowded platform.

[1] Following American practice, guards were known as *conductors* for many years.

Staff discipline was strict and men were discharged for smoking on duty, running trains past stations or altercating with passengers. Untidiness in appearance would usually mean suspension from duty without pay, and the tunic uniforms had to be worn buttoned to the neck in all weathers. Trouser bottoms had to be turned down and black boots were *de rigueur* ('brown boots neither look well, nor achieve the uniformity of appearance aimed at', warned a notice of November 1907). Before booking off, the driver was required to spend an hour examining his train, noting any defects in his log; if he missed anything, he would be suspended and lose pay. Train crews carried meal baskets and ate when they could; station staff had a mess room, and booking clerks ate in their office while they worked, cooking on a gas ring. Later, systematic arrangements were made for meal breaks and these varied from twenty minutes minimum for train crews to one hour for signalmen per ten-hour day. Drivers were the élite, and in 1909 a Grade I man received 7s 1d a day compared with the Grade I Porter's 2s 10d, and the Booking Clerk's maximum of 4s 6d. Guards (Grade I) also received 4s 6d, whilst Grade I Gatemen earned 3s 8d and the best-paid signalmen (in Camden Town and all terminal boxes except Golders Green) 5s 4d. A ten-hour day ($9\frac{1}{2}$ for drivers) and six-day week was normal, and time and a quarter was paid for Sundays (9 hours), Christmas Day and Good Friday (again with the exception of drivers, who worked an $8\frac{1}{2}$-hour day plain time). Annual leave with pay was meagre, a maximum of six days for drivers but only four for porters and gatemen. In 1909 there was a standard annual uniform issue of jacket, cap and two pairs of trousers to porters, ticket collectors, liftmen, guards and gatemen (signalmen received a waistcoat instead of a jacket). Overcoats were issued every two years to liftmen, signalmen, guards and gatemen, also to ticket collectors working on stairs or in subways.

The fares on each of the three lines varied. Bakerloo passengers at first paid a flat rate of 2d (25 tickets in a book for 4s) and dropped their tickets into a box as they passed the inwards barrier, but, from July 22, 1906, 1d to 3d fares by $\frac{1}{2}$d stages were introduced. Piccadilly fares were more complicated. Averaging 0·77d a mile, they ranged from 1d to 4d according to distance, and season tickets were available. In March 1902, Sir Clifton Robinson, managing director and engineer of the LUT, had read a paper to the Royal Society of Arts in which he foresaw through fares between tramways and railways, and it was probably through his inspiration that from a very early date, bookings were available between the Piccadilly and the London United Tramways to points as far west as Uxbridge and Hampton Court. District and Piccadilly tickets were inter-available inwards from Hammersmith, and passengers could book to Piccadilly stations from some western District stations. Hampstead fares were simple enough, graduated from 1d to 3d according to distance. All three lines had the statutory workmen's single and return tickets, available on line of issue only and issued up to 7.30 a.m. on the Piccadilly and up to 7.58 a.m. on the other two lines.

AMERICAN MEN AND METHODS

It will already be apparent to the reader that many of the features of equipment and operation bore a close resemblance to American rapid transit practice, notably that of the elevated railways in Boston, New York and Chicago. As in the USA, sharp curves (entailing heavy rail wear and speed restrictions) were adopted to avoid disturbance to private property and stations were closely spaced (which kept down the average speed). The track circuits and train stops were based on the system in Boston, and the long traffic day from 5 a.m. to 1 a.m. was as near to the American 24-hour operating day as could be reached on a railway with restricted room for maintenance. The rolling stock bore many signs of transatlantic influence, with its hard wooden 'separators' in the longitudinal seats, cross-seats in the centre of the car, 'monitor' roofs, end platforms and gates, and air brakes. Crew arrangements were similar to those in America, and features of operating and ticket practice such as the acceptance of standing passengers as a normal feature in peak hours, the use of short trains in the off-peak, and the Bakerloo flat fare had all been first used in the USA. In the official notices to the staff, the frequent use of the expression *OK* struck a somewhat jarring note. Staff had to learn to describe the running lines as *northbound* and *southbound* (*eastbound* and *westbound* on the Piccadilly) instead of the traditional British *Up* and *Down*.

All these things were introduced and nurtured by the team of American experts brought to London by Yerkes, a team headed by the brilliant mechanical and electrical engineer James R. Chapman who served for some years as general manager and engineer-in-chief of the UERL. With him were Zac Ellis Knapp (an electrical engineer from North Carolina who stayed to become naturalized and a director of the UERL), F. D. Ward (the rolling-stock engineer), S. B. Fortenbaugh (an electrical engineer lent by the General Electric Co. of America), W. E. Mandelick (who became secretary of the UERL), and G. Rosenbusch, (assistant engineer, lifts and ventilation), to mention only a few.

FOREIGN EQUIPMENT

Untrammelled by any feeling of British patriotism, Yerkes exercised a carefree economic internationalism in his shopping, buying where the price was lowest. Thus all the Bakerloo cars came from America, whilst the Piccadilly stock came half from France and half from Hungary (except for the two British trailers.) Krupps of Essen supplied many wheels and axles, and much rail came from American and German firms. The lift contract with the Otis Elevator Co. aroused particularly strong feelings in Britain. Many of the contracts were with British registered firms that were merely subsidiaries of American companies.

This policy was defended by Perks at a District general meeting in August 1904. He explained how the prices quoted by English firms were originally 30–40% higher than European prices, but eventually the English prices were lowered. (The District bought 280 cars from Europe and 140 from England.) He was reported as saying that he did not know whether the shareholders wished the directors to spend £120,000 more for what was called patriotism—'at any rate they did not consider it their duty to do so.' (Cries of 'Hear, hear!')

The picture was not wholly black. Orders were given to many well-known British firms: Crittalls, Babcock & Wilcox, British Insulated and Helsby Cables, Callenders Cables, Hadfields, Dorman Long. British civil engineering contractors made the tunnels, and the station structures were largely built by British companies. In December 1903, Perks estimated than not more than 10% of the £17 million to be spent on the four railways would be used to buy foreign goods. The undertaking provided work for many pairs of British hands, even if some of them were assembling foreign material.

A BAD PRESS

With its American finance, engineers, methods and equipment, the Underground group was a natural target for the criticism of the more chauvinistic journals. The situation had grown so bad by May 1906 that Sir George Gibb wrote to the Press, defending the group's actions and complaining of 'a perpetual shower of virulent and premature criticism'.

The District was passing through a trying time, with axle failures, primitive air-operated doors scandalously tearing off ladies' skirts, and a fare increase in September 1906, which aroused widespread and disproportionate indignation.

In its early days, the Bakerloo had very light traffic and the newspapers had fun in arranging for loads to be counted and photographs taken of empty trains. On April 4, 1906, a *Daily Mail* reporter found a maximum load of 86 on the 6.17 p.m. train from Baker Street to Kennington Road.

The technical Press also joined in, although not all papers were hostile. The *Railway Magazine*, edited by G. A. Sekon (Nokes), and the *Railway Engineer*, edited by S. Richardson Blundstone, were particularly virulent critics. Blundstone made loaded comments about the level of traffic on the Bakerloo, and called it a 'beautiful failure'. Some of Sekon's comments have already been given.

VAIN HOPES

The level of traffic on the three tubes was far below that estimated in 1905 by Stephen Sellon, MICE, Consulting Engineer to the British

Electric Traction Company Ltd. For example, the estimate of the annual number of passengers compared with the actual traffic for as late as 1909, after the lines had had plenty of time to settle down, was:

	Estimated (millions)	1909 (millions)
Bakerloo	35	28·2
Piccadilly	60	37·5
Hampstead	50	29·4
Total	145	95·1

Details of the financial estimates and results are given in Appendix 4A, but it is interesting to note that the figure of 145 million passengers per year was not achieved until 1929, and then over a much greater mileage.

These profoundly disappointing results were largely due to the proliferation of road transport facilities; the supply had temporarily outrun the demand. The electric tram had attained a high degree of technical efficiency, and motor buses were steadily replacing horse-drawn, although a thoroughly reliable chassis was not yet available. The LCC was rapidly electrifying the horse tramways; fares were low, workmen's fares very low.

The motor-bus business was open to all comers. Bus fares were tied to LCC levels on parallel sections and kept down by fierce competition on others. Only the more prudent operators were able to survive. The tubes outside the Yerkes group also suffered severely from surface competition—the CSLR and GNCR from tramways running over the same route, the CLR from intensive bus services along the east–west axis.

DEEP WATER

Thus as London at long last saw the consummation of Yerkes' plans to provide it with a network of electric railways, the UERL found little demand for the only commodity it had to sell—seat-miles. Meanwhile, expenses had to be met and interest paid.

As early as December 1905, the company had been obliged to borrow £455,000 from Speyer Bros, and by March 1906 the amount owing to Speyers and their associated companies was £620,000. In the latter month, a £1 million loan from four London banks was negotiated (£250,000 each). It was to be repaid on November 2, 1906. Interest was at Bank Rate with a minimum of 4%.

The next possible source of capital was to make further calls on the £5-paid £10 UERL ordinary shares, and accordingly the company announced in July 1906 that it was making a call of £2 10s per share, payable half on August 15th and half on September 13th. An official statement explained that in the current depressed state of the world money markets, this action was preferable to selling the securities of the operating companies at very low levels. The depression was attributed to the unsettled state of Russia, losses in the San Francisco fire and a great demand for money for trade purposes. Nearer home, the work of tube construction had taken longer than anticipated, and thus more money had been needed to pay interest during construction.

A £2 10s call on 500,000 shares should have yielded £1,250,000 (less the yield on the 32,000 shares held by Yerkes' executors), but even this very large injection of capital was insufficient to meet demands for more than a few months, and in October 1906 the £1 million bank loan was extended to May 1907. One bank required repayment, and the deficiency of £250,000 was made up, half by another bank and half by Speyer Bros.

January 1907 saw the final call of £2 10s per share, payable on February 14th. With the proceeds, the £1 million loan was repaid, but a further bank loan of the same amount was raised to form a reserve fund. A second debenture was raised on the Power House, and, with tube debentures, deposited with the banks as security.

STANLEY ARRIVES

A large proportion of the ordinary shares of the UERL was still held by interests in the United States, and when the final calls were made on the shares, the American holders (in common with others) had perforce to pay up or relinquish their original investments. With so much capital at stake they viewed with dismay the decline in the fortunes of the UERL, and no doubt thought that the company needed the benefit of American drive and experience that it had lost with the death of Yerkes.

In November 1906 Gibb began to look for a general manager, and in February 1907 came the announcement that a thirty-two-year-old tramway official, A. H. Stanley, had resigned his post as general manager of the Public Service Corporation of New Jersey to take up a similar position with the UERL. Later accounts stated that a Boston firm which was deeply interested in London's underground railways (presumably the Old Colony Trust Co.) decided to send Stanley to London, as the best man to look after its interests.

Albert Henry Stanley was born in Derbyshire on August 8, 1874, but his family emigrated to the United States when he was very young, and he received most of his education in Detroit. He was fascinated by the horse trams which passed his house, and at fourteen he obtained a post as messenger with the Detroit tramways company. At the age of eighteen he was divisional superintendent, and two years later was general superintendent of the Detroit tramways, at an annual salary of $5,000 (£1,000). This was the era of tramway electrification, and Stanley learned tramway mechanical and electrical engineering by evening study and by working in the company's repair shops. Detroit was growing rapidly, and he was continually supervising the opening of new lines.

In 1903 he heard that the tramways in New Jersey were being amalgamated, and successfully applied for a post with the new company, being appointed assistant general manager of the Street Railway Department of the Public Service Corporation of New Jersey in October 1903. He became manager in February 1904, and general manager of the whole Corporation in January 1907.

From his own later accounts of the affair, it appears that he was reluctant to leave New Jersey, and agreed to come to London only on the understanding that he would be free to return to America within one year (although the relevant UERL board minute states that he was appointed for three years from April 1, 1907). There is also the (possibly apocryphal) story that shortly after his arrival he called the principal UERL officers together, told them that the company was facing bankruptcy, and asked them to hand in their resignations to take effect in six months' time. He then succeeded in borrowing £50,000 from the banks to spend on traffic promotion publicity, but, lo and behold, when he approached the newspapers, they said that all Underground developments were news which they would be pleased to print. They were as good as their word, and the £50,000 remained largely unspent. The London newspapers began to regale their readers with stories of an unceasing series of improvements and changes in their travel facilities.

FINANCIAL CLIMACTERIC

Whilst Press relations were improved, financial problems now threatened the existence of the whole undertaking.

The situation was so bad that early in 1906 Speyer had tried to obtain the help of the LCC, suggesting that they raise £5 million for the UERL on which he would pay 4%. In return, the County Council would be given a right to purchase in twenty-one or forty years. But after almost eighteen years of power, the sands were running out for the Progressive majority on the Council, and the election of March 1907 saw the Conservatives at the helm (in the guise of Municipal Reformers). Although Speyer mooted the topic again at the Hampstead opening in June, it fell on deaf ears. Such socialistic ideas from the mouth of a capitalist were obviously not to be taken seriously.

The largest and most menacing cloud in the sky was the need to repay the £7-million-worth of 1903 5% profit-sharing notes on June 1, 1908. In November 1907 the company announced that this problem had been under consideration for some time. Plans were being prepared to extend and convert the notes, but with the serious worldwide financial crisis, it was not then opportune to implement them. In the meantime, Speyer Bros would purchase the December coupons at their face value. The contemporary UERL board minutes mentioned the possibilities of 'liquidation, petitions to wind up, appointment of a receiver'.

In February 1908 it was announced that advisory committees of independent financial experts had been appointed in London and Amsterdam to consider, in conjunction with the UERL board and Speyer Bros, a scheme for dealing with the profit-sharing notes and the financial requirements of the UERL. Serious differences of opinion arose between the American and English sections of the UERL directorate about the form the scheme should take, and short-tempered cablegrams crackled to and fro beneath the Atlantic. The London board regretted that Speyer and Co. had used their influence to prevent the formation of an American advisory committee, and they declined to believe that the American noteholders would prefer liquidation to the London scheme. The New York interests (headed by James Speyer) did finally accept the London and Amsterdam scheme, but insisted that Stanley be appointed a director of the UERL; this was duly done on April 22, 1908. The scheme, published in mid-April, provided that for each £100 of profit-sharing notes, holders would be given in exchange £40 of $4\frac{1}{2}$% bonds due January 1, 1933, and £70 of 6% income bonds due January 1, 1948. The $4\frac{1}{2}$% bonds were fixed-interest securities but the 6% income bonds resembled preference shares to some extent. They carried voting rights, and dividends depended on earnings. The '6%' was the maximum dividend, and this had to be fully satisfied before a dividend could be paid on ordinary shares. Speyer Bros guaranteed to purchase sufficient $4\frac{1}{2}$% and income bonds from the UERL on each January 1st and July 1st to provide the company with enough money to pay the interest on the publicly-held $4\frac{1}{2}$% bonds, if its own earnings were insufficient. Finally, there was an issue of £1 million prior-lien 5% bonds (due January 1, 1920) to provide working capital.

Under the Joint Stock Companies Arrangement Act of 1870, the company had to go into voluntary liquidation to carry out the scheme. A successful application was made in the Chancery Division on April 15, 1908, and Sir George Gibb was appointed temporary receiver and manager. On behalf of the UERL it was stated that proceedings were likely to be taken against the company by creditors; if they were taken, the Chelsea power house and the railways would be stopped. The appointment was essential to avoid the possibility of a stoppage.

Sir Edgar Speyer presided at an extraordinary general meeting of the UERL on May 11, 1908. He was in a sombre mood: 'It is with a feeling of profound regret that I address you. It would be idle to pretend that the result of our six years' labours has not been very disappointing.' He recounted the adverse factors—unfavourable money markets (it had been planned to redeem the profit-sharing notes with the receipts from selling the operating companies' securities); delays in completion; bus and tram competition; additional expenditure imposed by the require-

"London's Latest Tube"—reproduction of an official postcard issued by the Great Northern, Piccadilly & Brompton Railway. Three-car train (control trailer in foreground) at Piccadilly Circus station at tea time.

Interior of Bakerloo car, 1907. Reproduction of an official card issued by the Company. The reverse reads "THE BAKERLOO & PICCADILLY TUBES afford the maximum of speed, safety, comfort and convenience in all weathers."

Into the tube—the Piccadilly tunnel entrance at West Kensington, between the District tracks, April 1957.—*Alan A. Jackson.*

Albert H. Stanley, Lord Ashfield (1874–1948).

Sir George Stegmann Gibb (1850–1925).

'The Underworld'—a reproduction of a painting by Walter Bayes showing people 'taking cover' from a German air attack at Elephant & Castle station in 1917. Shelterers in 1940–45 faced more crowded, but more organised conditions.
—*Imperial War Museum.*

ments of Parliament and local authorities; fares reduced below the economic level; an unwarranted burden of rates and taxes. However, mistakes and miscalculations had been made by their traffic experts— no allowance had been made for the differences between American and English conditions (and especially in relation to New York, which was constrained geographically so that passengers made longer journeys); the undertaking was an unqualified success from an engineering viewpoint and had ample capacity for the additional traffic which would develop with increased public appreciation, population and purchasing power. He foresaw lower commodity prices and cheaper money, and even mooted the possibility of one central management and pooling of all London passenger earnings.

The company had spent about £18 million, and capital issues had been: £5 million ordinary shares, £7 million profit-sharing notes, £3·7 million tube and District debentures and shares, and £·775 million of power-house debentures. There was also the £1 million loan. For the next three or four years the UERL income would be insufficient to pay the interest on the 4½% bonds. Finally, he reminded the members that Speyer Bros had preserved the company intact against the creditors by buying the December 1907 coupons, and that 'our firms' (the Speyer group) were by far the heaviest pecuniary losers.

The meeting approved the scheme, and a further meeting on May 27, 1908, approved the voluntary winding up and alteration of the articles of association. Final approval was given by a meeting of profit-sharing note holders on June 30th, and in the Chancery Division on July 16th. The voluntary liquidation was closed on July 21st.

The company's fortunes now slowly improved, and for 1908 the deficit was £12,000 against an estimate of £54,000. The first payment on the 6% income bonds was made for the half-year to June 1910 (1%), and the dividend had risen to the full 6% by December 1912. Nevertheless, the company had escaped disaster only at the price of creating a heavy burden of fixed-interest and prior-charge securities which was to preclude the payment of UERL ordinary dividends for many a year. This in turn meant that fresh capital could be raised only by debentures and similar securities, thus increasing the fixed-interest element.

FARES CO-ORDINATED

Fares had been driven down to unremunerative levels by the severe surface competition, and Sir George Gibb took the initiative in bringing the competitors together to negotiate raising the fares to more sensible scales. A conference held in mid-June 1907 between the Underground group and the Central London and Metropolitan Railways secured agreement for the fundamental east–west fares to be raised from July 1, 1907. The Central London 2d fare from Shepherd's Bush now ended at Tottenham Court Road, and that from Bank at Marble Arch. A 3d fare applied for longer journeys, but the blow was softened by the issue of books of twelve 3d tickets for 2s 9d. The Metropolitan made corresponding increases on the Hammersmith and City Line.

Following this success, Gibb invited the railway and road service operators to a further conference. On July 22, 1907, there were assembled round one table representatives of the Underground group, the Metropolitan, CLR, CSLR, GNCR, the London United Tramways, London General Omnibus Co., and the Vanguard, Tilling and Star bus undertakings. It was agreed to form a London Passenger Traffic Conference, with each company having a vote but able to dissent from Conference decisions. The objects were to eliminate unprofitable fares and overlapping routes, and to facilitate mutual consultation.

The Conference held its first meeting on July 29, 1907, augmented by representatives from the North London Railway and more bus operators. It agreed to examine the traffic returns from various routes by sections.

The fruits of these labours, and of a conference between the underground railways on December 2, 1907, were gathered on December 15, 1907, when rail and road fares were increased. Certain fares on the District, Piccadilly and Bakerloo railways were increased by $\frac{1}{2}$d or 1d. Bus fares were also adjusted (and to some extent co-ordinated with the rail fares), by the process of cutting back 1d stages, mostly on east-west routes.

This fare adjustment was the high point of private enterprise road-rail co-ordination. Subsequently, little more was heard of the full road-rail Conference, and in December 1908 the *Railway Gazette* recorded that the Conference then existed only in name. The bus companies had withdrawn on the grounds that the underground railways had introduced accelerated services and more through bookings; road-rail competition was again as fierce as ever.

The bus companies had correctly assessed the commercial importance of underground through bookings. Such bookings, which were an essential part of the campaign to persuade the public to regard the London underground railways as one system, were the result of continued co-operation by the railway members of the Conference. However, through bookings had perforce to await the abolition of flat fares, as competitive lines could scarcely be expected to give free transfers.

The Bakerloo flat fare lasted only until July 21, 1906, and the first breach in the Central London flat fare system was made on July 1, 1907, as already mentioned. Through fares were introduced as part of the 'UndergrounD' co-operation scheme, and by the end of 1908 they were available for all of the more obvious (and some less obvious) interchange journeys. The sequence of their introduction was:

Bakerloo and 50 LSWR stations 2.9.06.
Bakerloo and Metropolitan 12.11.06.
Bakerloo and District 12.11.06.
Piccadilly and CSLR by 26.7.07.
Hampstead and Bakerloo by 9.8.07.
CSLR and District by 8.07.
Hampstead and Central London 1.9.07.
Bakerloo and Central London 18.12.07.
CSLR and Metropolitan by 2.08.
CSLR and Central London via Bank by 2.08.
Bakerloo and LBSCR (stations Norwood Junction to Forest Hill inclusive, and Crystal Palace, via London Bridge and Elephant & Castle to Piccadilly Circus only) by 14.3.08, withdrawn 12.09.
CSLR and Central London via Bakerloo 24.4.08.
Piccadilly and Central London (via Holborn and British Museum or Piccadilly Circus and Oxford Circus, or Leicester Square and Tottenham Court Road) 8.6.08.
Central London and Metropolitan by 19.6.08.
Hampstead and District (walking along Villiers Street) 1.7.08.
Bakerloo and GNCR (via Elephant & Castle and Moorgate Street, or via Piccadilly Circus and Finsbury Park) by 24.7.08.
Hampstead and GNCR (via Euston and CSLR) by 24.7.08.
Piccadilly and GNCR (via Finsbury Park) by 2.10.08.
Piccadilly and Metropolitan 1.11.08 and 7.12.08.
There were Piccadilly–District, Piccadilly–Bakerloo, Piccadilly–London United Tramways, Piccadilly–Great Northern Railway, Bakerloo–CSLR, Hampstead–Piccadilly, Hampstead–LNWR, and Hampstead–CSLR bookings from the respective opening dates or soon afterwards.

PUBLICITY CO-ORDINATED

Another aspect of the movement to integrate the London underground lines was the co-ordination of publicity.

In their very early days the Yerkes group tubes pursued independent publicity campaigns. The Bakerloo had an unusual and attractive heading to its poster frames with 'Bakerloo' in chintzy capitals having the 'O's interlinked with a small 'tube' beneath them. There were some ingenious posters, including one entitled 'The Link of London Lines', showing a station tunnel encircled by a steel ring. From the ring, chain links projected radially, each with the name of a connecting railway or tramway and the interchange station. This design also appeared as a coloured postcard, as did an interior view of a Bakerloo train, graced with the presence of gentlemen in top hats, and ladies in foamy lace.

Similarly, the Piccadilly and Hampstead had their own poster frames and poster headings. The opening of the Hampstead was accompanied by a publicity barrage, including posters, handbills, illustrated handbooks, picture postcards and maps, and a folding card which, when opened, revealed a tube train emerging from a tunnel. There was also a mock newspaper called *The Mole*, with hints on through travel by underground.

Albert Stanley soon busied himself with a scheme for joint publicity for all the London local railways, to run side by side with the new through fares His fertile imagination envisaged all the railways as one system, to be described by the single word UNDERGROUND,[1] and he persuaded the general managers of the CSLR, CLR, GNCR, District and Metropolitan to agree to this in December 1907. The word was to appear in large vertical signs outside all the stations and also over the entrances. Special treatment was given to the lettering, which was white, on a light-blue ground; each letter was separated by a dash and the initial 'U' and final 'D' were made equal in size and larger than the rest. A standard map of the whole system,[2] showing the stations boldly, with a different colour for each line, was designed and used extensively to educate the public. Large illuminated versions of the map were displayed outside stations, and there was also a monster enamelled version, so durable that some examples lasted until after the Second World War. The illuminated maps were in glass-fronted boxes about 7 feet high; along the top, 18 inches deep, was a silhouette of the London skyline, below it the map, 3 feet high, and at the base, a slogan, *Underground to Anywhere—Quickest Way—Cheapest Fare* (this had been obtained by holding a competition in the *Evening News*). A folder version of the map, measuring about 13 inches by 11 inches, had a green border and was headed 'London Electric Railways'. The map was common to all issues of the folder, but on the reverse side the issuing company gave full details of its own facilities, and as much as it thought fit of other companies' services. The folder maps were placed in 'Please Take One' tins or literature boxes fixed near the station lifts and bookstalls. Arrangements were made to insert the standard map in guides and catalogues, to have it posted on hoardings, framed and hung in public buildings, hotels and restaurants, or scattered as leaflets. Over six million of these maps were sent out in 1908–1909, the scope of distribution extending to eastbound Atlantic liners.

Other publicity of the period included the issue of a game, called *How To Get There*, which consisted of a railway map, four 'cars', a packet of tickets and a *tee-to-tum*. Based on Ludo, the idea was to move the cars over the map by the extent of the score shown on the *tee-to-*

---

[1] Not to be confused with 'the Underground group,' the term used to describe collectively the UER, the Bakerloo, Piccadilly and Hampstead tubes, the MDR, the LUT, and later additions. The group were sometimes later known as 'The London Traffic Combine'.

[2] Not the first Underground map. The MDR had long sponsored 6d and 1s maps of London and suburbs, with its own and associated lines shown boldly, and the CLR had first issued a map and guide in 1906.

tum, and the tickets embodied various forfeits or benefits including non-stop runs, lost tickets, signal stops, permission to travel one, two or three stations, and so on. When a new edition was in preparation, the *Railway Gazette* facetiously suggested further forfeits such as:
Breakdown on District, proceed on foot.
Lift sticks on CLR, lose six turns.
CSLR closed for cleaning, retire from game.
Polite conductor on Hampstead Tube—miss eight moves through shock.

Once the word *Underground* was adopted, from the beginning of 1908, to describe the whole system of the co-operating companies, Stanley made great efforts to erase the word *Tube* from official usage. Until this time, the official notices and publicity of the Yerkes lines had freely used the word, and it had also been popular on the Central London, whose stations bore it on huge vertical signs by their entrances. The staff were now enjoined to use 'Railway' after the line titles, but Stanley never succeeded in expunging 'Tube' completely—it was so neat and convenient to use.

Below ground, the space taken up by the tiled station names was put to more profitable use as a foundation for posters, and the names were shown on printed slips instead. These were difficult to distinguish from the surrounding welter of commercial advertising, and the device of displaying the station name on a bar across a red circle was evolved. There have been various suggestions [1] about the origin of this, but it may have been copied from a similar device displayed on the rocker panels of 'General' buses from about 1905. The bar and circle were adopted from 1913 on the District and Underground group tubes, the red circle being solid on signs installed up to about 1916. The Metropolitan Railway preferred to be different and used a diamond instead of a circle.

Other Stanley-inspired features introduced in 1908–1909 to attract and keep traffic included posters of high quality, 'next-lift' indicators, lift operation co-ordinated with train arrivals, illuminated direction signs in subways, line diagrams inside trains (in the clerestories), an interpreter-guide at Piccadilly Circus, co-ordinated last trains, and headway clocks to ensure regular running.

SEASONS AND STRIPS

Season tickets had a chequered career at first. The Piccadilly and Hampstead had them from the start, but they were not introduced on the Bakerloo until November 1, 1906. Local seasons on these three lines were abolished from October 1, 1908, and through seasons between them from October 14th. Through seasons to other lines continued. In explanation it was stated that seasons were used by only about 3% of the passengers and that their use for lunch-time journeys made them un-

[1] See *London Transport Magazine*, February 1955.

economic. Instead of seasons and local and through return tickets, the three Yerkes tubes began to issue 'strip' tickets, in sets of six, available at any time, in either direction, between specified stations (or intermediately). A strip of six 1½d tickets cost 8d, 2d tickets 11d, 2½d–1s 2d, 3d–1s 4d, 3½d–1s 7d, 4d–1s 9d, and 5d–2s 3d. On sale from October 1, 1908, these undated tickets were completely transferable; a single ticket could be used for two children or for a dog. A month later, to complete the scheme, strips of six 1d tickets were sold, without discount. Supported by a publicity campaign, the strip-tickets were doubtless designed to reduce pressure on booking offices, but as early as February 1909 it was realized that they could be abused, and they were withdrawn in 1916. Season tickets were re-introduced on July 1, 1911, and the Central London had them from the same date.

Strips of twelve 2d tickets at 2s had been sold by the Central London from the start, later in the form of booklets. In November 1908 the price was reduced to 1s 10d, and in January 1910 booklets of fifty 3d tickets were available for half a guinea. The Hampstead also had booklets from an early date, containing twelve 3d tickets for 2s 9d—not such a good buy as the strip tickets later introduced. The GNCR abolished its monthly seasons in January 1910 and started to sell strips of six tickets at a discount.

To meet bus competition, the Central London reduced short-distance fares from the 2d flat rate to 1d from March 14, 1909. The CSLR was also forced to make reductions in the same year.

IMPROVING THE SERVICES

As more experience was gained with the new equipment on the Yerkes tubes, it was found possible to lop several minutes from the scheduled running time. For instance, on the Piccadilly, the 38 minutes allowed for the Hammersmith–Finsbury Park journey were cut to 36 on March 11, 1907, 35 on April 29th and 33 on October 14th.

For the first three years of operation, there was a good deal of trial and error in the development of a satisfactory pattern of service, particularly on the Hampstead, where the timetable was altered nearly every month. Different ways were used to strengthen the service in peak hours. On the Hampstead, shortly after it opened, each branch had a flat service all day, but the number of cars was increased in the peaks. For instance, in June 1908, the Highgate service had two- or three-car trains in the off-peak and five-car in the peak, whilst the Hampstead section had two- and four-car respectively. Each branch had a five-minute service all day, with half the Hampstead trains continuing to Golders Green. The practice of running more frequent peak trains was not adopted on the Hampstead until November 1908. On the other hand, increased peak frequencies were early features of the Bakerloo and Piccadilly timetables, and some trains were usually lengthened as well, but in October 1908 each of these lines had three-car trains all day except for six-car trains in the evening theatre rush.

With the growth of traffic, headways were shortened. The Bakerloo had reached 27 peak trains per hour and 17 p.h. normal in December 1907, and 35 peak, 20 normal in October 1908. Corresponding figures for the Piccadilly were 30 and 20 in October 1908. From February 1909 the Hampstead had 40 trains per peak hour on the Charing Cross–Camden Town section, alternate trains serving each branch.

Train lengths increased gradually. On the Bakerloo, normal hour passengers were given three cars from October 1906, and some six- and four-car trains were introduced in March 1907, although by July of that year they had been shortened to five- and three-car. The all-three-car service of 1908 was still running in October 1909 on the Bakerloo, but in that month half the Piccadilly peak trains were made up to five cars. The Hampstead was still running some three-car peak and two-car normal trains in 1909, but the minimum lengths later became three cars in the normal and four in the peak, with some five-car trains.

The provision of control trailers and the flexibility of multiple-unit control allowed the operation of trains of any length up to the limits of the platforms.

Late-night travellers were well looked after. In December 1907 the last Bakerloo train left the Elephant at 12.36 a.m. on weekdays, the last Hampstead was at 12.45 from Charing Cross, and the last Piccadilly left the termini at 12.32 westbound and 12.27 eastbound. On Sundays, last trains were about 45–55 minutes earlier. Co-ordinated last trains at 1.00 a.m. from Piccadilly Circus (Leicester Square for the Hampstead) were introduced in October 1908.

During this era the Underground group developed the practice of non-stopping certain stations into a fine art. The first timetable embodying a series of non-stop operations was introduced on the District on December 16, 1907, on an experimental basis (the District, with its many branches, gave most scope for scientific non-stopping). The success of this experiment prompted Stanley to try the same principle on the tubes.

The Piccadilly had a non-stop train from October 1907, when a theatre train left Holborn at 11.16 p.m. nightly, calling only at King's Cross and Holloway Road and arriving at Finsbury Park at 11.26, in time to connect with Great Northern suburban trains. When the Aldwych branch opened on November 30, 1907, this train was altered to start from Aldwych (then Strand) at 11.13 p.m. (later altered to 11.28 p.m.), calling at the same stations, but it lost its non-stop characteristics from October 5, 1908.

Conditions for non-stopping seemed particularly favourable on the Hampstead line, with only half-service on each branch, relatively lightly-used stations north of Euston and the long downhill runs from the north. From July 1908, the 7.59 a.m. train from Highgate ran non-stop to Camden Town in $4\frac{1}{2}$ minutes, and a second train leaving at 8.49 a.m. was added in October. A month later the trains were altered to leave Highgate at 8.2 and $8.40\frac{1}{2}$ a.m., the first missing Kentish Town and South Kentish Town, and arriving at Charing Cross at $8.16\frac{1}{2}$, the second missing Tufnell Park and South Kentish Town, arriving at 8.55. From August 1909, three southbound trains in the morning peak and four northbound in the evening ran non-stop between Golders Green and Euston, taking 17 minutes to Charing Cross instead of the normal $19\frac{1}{2}$ (19 minutes northbound instead of $21\frac{1}{2}$). This experiment was successful, and during the trial period most trains had a clear run. A new timetable with four non-stops each way daily was introduced in September 1909, and the Hendon buses ran express in 12 minutes to connect with each of these trains. The leading car carried a large 'NON STOP' sign, and the trains were waved through intermediate stations by a man with a lamp, confirming the aspect of the starting signal. With the non-stop trains, the September 1909 timetable was advertised as 'the most frequent train service in the world'—42 trains an hour in the peak period south of Camden Town.

In October 1910, the residents of Golders Green were given a 7.16 p.m. theatre train which non-stopped Belsize Park to Goodge Street inclusive, reaching Leicester Square in 14 minutes. The return working was at 11.15 p.m., and was also non-stop. By 1913 this had become two return workings from Golders Green and one from Highgate. The rapid housing development at Golders Green made it prudent to terminate almost all the trains there (instead of half at Hampstead), and this improvement was also introduced in October 1910.

From July 1912 the Hampstead service was stepped up to the record frequency of 44 trains per hour—the highest ever reached in London railway operation. In July 1913, the non-stop operation continued all day, with alternate trains stopping at alternate stations. The stations usually non-stopped were from Goodge Street to Chalk Farm, and also South Kentish Town.

The Piccadilly had different traffic characteristics, as it ran across the central area, but some of the stations situated at the half-mile-working-shaft intervals were poorly patronized, so that a system of skip-stopping would speed up the journey and cause little inconvenience. This practice was introduced on October 11, 1909, with alternate three- and five-car trains in the rush hours. The first train passed Gillespie Road and Caledonian Road and stopped at Holloway Road and York Road, whilst the following train passed Holloway Road and York Road and stopped at the other two, and so on. From October 1910 this was developed so that trains with odd running numbers passed Holloway Road, York Road, Hyde Park Corner and Gloucester Road; those with even numbers passed Gillespie Road, Caledonian Road, Down Street and Brompton Road. The overall time from Finsbury Park to Hammersmith was thus reduced to 28 minutes. By 1915, Covent Garden had been added to the first group and Russell Square to the second, and the train numbers transposed. This form of operation applied during the two business peaks and the evening theatre peak. Between the peaks, non-stopping was confined to the four 'Deadwood Gulch' stations between Finsbury Park and King's Cross.

The Central London made less spectacular but quite solid progress. Thirty trains per hour had been reached as early as 1903, and in 1905 31 trains per peak hour were proudly announced.

A half-mile extension of the Central London was opened on May 14, 1908, with a station at Wood Lane to serve the Franco-British Exhibition at the White City. A new single-line tunnel was built so that, with the existing depot-access tunnel, a loop was formed at the western extremity of the line. The new station was at ground level at the top of the loop, and had direct communication with the Exhibition; it also formed a convenient interchange point with the trams in Wood Lane, serving Harlesden. After the opening of the new loop, there were 30 trains an hour all day except in the late evening. December 1909 saw an experimental Christmas shopping service of 35 trains per hour between 10 a.m. and 4 p.m. In 1910, normal hour trains were three cars long, two such sets being coupled to form a six-car train for peak traffic.

The GNCR and CSLR conformed to the other lines by providing a service from 8 a.m. on Sundays (instead of about 11.30 a.m.) from August 1, 1909. The CSLR and CLR had no non-stop trains until they were taken over by the Underground group. In 1913, alternate peak trains on the GNCR missed Drayton Park and Essex Road, completing the journey in $9\frac{1}{2}$ minutes.

Traffic was growing steadily throughout this era, and gave rise to periodical complaints of 'overcrowding' when a surge of traffic coincided with a three-car train, and not everyone could obtain a seat. There were irate letters to *The Times* about the Bakerloo service in January and February 1909. One correspondent, travelling in the morning peak from Baker Street to Oxford Circus, counted as many as 25 straphangers! On another occasion, there were 17, and a passenger who wanted to alight at Regent's Park could not penetrate the throng and was carried on to Oxford Circus. The guard told him that he had no right to stand so far up the car if he wanted Regent's Park. The passenger, 'like one of the brave British public, said nothing but suffered in silence'.

In January 1908 staff on the Yerkes tubes were strictly enjoined: 'On no account must overcrowding of trains be allowed . . . overcrowding may constitute a serious public danger', and in March of that year there was a conference of underground railway managers at the Board of Trade to discuss overcrowding.

On the Hampstead, heavy traffic at Tottenham Court Road caused station stops to be extended to 35–40 seconds instead of the normal 20. This was countered by the installation of additional home signals, but, in general, the tubes were demonstrating the paradox that, within limits, the lighter the traffic, the more frequent can be the service.

SALLY TO THE NORTH WEST

The year 1909 saw the unhappy result of the marriage of two unhappy schemes—the proposal to extend the Bakerloo to Paddington and the ill-fated North West London Railway.

In 1900, the Bakerloo had been authorized to extend to Paddington via Harrow Road, Bishops Bridge and Bishops Bridge Road, with a very long subway below Eastbourne Terrace to the main-line station. A 43-chain deviation authorized in 1906 made the line swing round from Harrow Road to a proposed new station site on the east side of London Street, with a subway connection running west to the Great Western station. Thus the final section of line now pointed south-east. This inept routeing precluded any further extension of the Bakerloo main line in its natural direction—west or north-west—and the Underground group wisely decided to halt construction at Edgware Road for the time being.

The North West London Railway was authorized between Cricklewood and Marble Arch in 1899. Its 1903 proposal for a Victoria extension was withdrawn, but was re-submitted in 1906 after favourable mention by the Royal Commission on London Traffic.

The Bill was passed, but it was burdened with some unusually severe penalty clauses. The promoters had openly admitted that the Brush Electrical Engineering Co. (in which the British Electric Traction Co. had a substantial holding) was sponsoring, and would raise £100,000 towards the scheme if it received the contracts for the electrical equipment, lifts and rolling stock. An agreement to this effect was thereupon scheduled to the Bill. If the Brush company were given the contracts, it would be bound to pay up, but the contract prices would be fixed by arbitration. Furthermore, the railway company was forced to provide additional interchange facilities and the powers to acquire land lasted only until August 1908; the whole line had to be completed by August 1910.

The NWLR promoters found it impossible to raise the capital, and the powers were not exercised. In 1908 the British Electric Traction group devised a scheme to make a running connection with the Bakerloo, near its Edgware Road station and to abandon the authorized section from Edgware Road to Victoria. By early 1909 an agreement had been reached for the Bakerloo to work the Edgware Road–Brondesbury rump of the North West London for 25% of the gross receipts, with a minimum annual payment of £25,000.

Parliamentary approval of these proposals was sought in the North West London Bill of 1909, and the Bakerloo Bill for the same session included extensions of time for the 1906 Paddington deviation and subway. The proposed junction at Edgware Road made it unlikely that Paddington would have a full, through service, and there were angry mutterings from Great Western headquarters, with promises that both Bills would be actively opposed. The Metropolitan also objected, on the grounds of competition for the Baker Street–Kilburn traffic.

Before the Commons committee, Sir George Gibb described a proposal for a shuttle service between Paddington and Edgware Road, with the main Bakerloo service running to or from the North West London. He thought that the Great Western should contribute financially towards the Paddington extension even if they were given only a shuttle service. The Bakerloo was not likely to receive more than 14% of the GWR suburban and outer-suburban traffic—in fact, the amount of traffic that the tubes received from the main-line termini had been much exaggerated.

The Great Western witnesses estimated that the number of passengers using a Bakerloo Paddington extension would be five to six million, against Gibb's one million. The Bill was refused on May 14, 1909, on the grounds that it did not provide the Victoria–Marble Arch link recommended by the Royal Commission. The North West London scheme was now dead, and the Bakerloo Bill was withdrawn.

The two parties interested in the tube extension to Paddington played a waiting game. The Bakerloo lay quiet until the Great Western needed the extension so badly that it was prepared to pay for it; the Great Western hoped that self-interest would compel the Bakerloo to extend at no cost to themselves.

Widths and Heights of Original Yerkes Cars

	Waist width	Height	Floor height
Baker Street & Waterloo	8 ft. 11⅛ in.	9 ft. 5⅜ in.	1 ft. 10 in.
Piccadilly (French)	8 ft. 7⅛ in.	9 ft. 5 in.	2 ft. 0 in.
,, (Hungarian)	8 ft. 7 in.	9 ft. 4 in.	1 ft. 10⅞ in.
Hampstead	8 ft. 6¾ in.	9 ft. 3 13/16 in.	1 ft. 10 7/16 in.

(Source: *Electric Traction on Railways*, Dawson, 1909.)

CHAPTER 7

Integration

1910–1914

THE LONDON ELECTRIC RAILWAY

ALTHOUGH they all had the same chairman (Sir George Gibb) and were all under the financial and administrative control of the Underground Electric Railways Co. of London Ltd., the Piccadilly, Hampstead and Baker Street & Waterloo companies were still in separate corporate existence, each with its own board of directors, accounts and obligations. It was obvious that the amalgamation of the three statutory companies would lead to economies, and a Bill prepared for the 1910 session sought to transfer the Baker Street and Hampstead companies to the Piccadilly company whose capital was to be rearranged and name changed to the London Electric Railway Company. It was proposed that the LER be released from the obligation to issue single tickets for workmen but issue instead cheap return tickets up to 8 a.m. each weekday between any two stations on all three lines from August 8, 1910. This pleased the LCC so much that they offered no opposition to the Bill, and after an effortless passage, the measure received Royal Assent on July 26, 1910.

Including an additional debenture issue to cover amalgamation costs and certain new works, the total authorized capital of the new company stood at £16,800,000. The amalgamation took effect from July 1, 1910, with Lord George Hamilton as chairman and Albert Stanley as managing director.

Stanley was also managing director of the Underground company, having replaced Sir George Gibb when he accepted chairmanship of the new Road Board in May 1910.[1]

The new undertaking operated 168 motor cars and 308 trailers over 22 route-miles of tube railway, but it was by no means an interlocked and conveniently arranged system. In the absence of a central planning

[1] Sir George retired from the Road Board in 1919 and died on December 17, 1925, at the age of 75.

or co-ordinating authority, each tube line had been built in isolation, as an independent entity; physical connections at points where lines crossed were either inadequate or non-existent. After the arrival of Gibb and Stanley, the development of joint publicity and through fares had gone some way towards persuading the public to think of the underground railways as one system, but the great need was for better interchange facilities, and there now began the lengthy and expensive process of knitting the underground lines together into an integrated system, bringing benefits both to the operator and the passenger.

CHARING CROSS LOOP

One of the black spots was Charing Cross, where the Bakerloo, Hampstead and District lines all came within hailing distance of each other. The Hampstead terminated under the forecourt of the SECR Charing Cross station and passengers transferring from it to the District or Bakerloo were faced with a 200-yard walk down Villiers Street to the Embankment. Interchange between the Bakerloo and District was little better, for it involved the use of a long, dreary subway between the two rail levels. A 1910 Bill proposed the extension of the Hampstead in the form of an 838-yard single-track loop passing beneath Villiers Street and the river, returning under Northumberland Avenue and Craven Street to join the existing western Hampstead tunnel. An interchange station on the loop would afford easy communication with the other lines. These proposals met the opposition of the Commissioners of Woods and Forests and of Gordon Hotels Ltd, who feared for the safety of the Hotel Metropole, and the Bill was withdrawn.

Objections were accommodated in a second Bill for an 880-yard single-track loop on a slightly different location, and the Act was passed on June 2, 1911. John Mowlem & Co. Ltd began work in October on a tunnel from the existing terminus which passed diagonally across Villiers Street, continued beneath the Embankment Gardens from Watergate to Cleopatra's Needle, and then made a $3\frac{1}{2}$-chain curve under the river, reaching a maximum distance of 65 yards from the bank. Returning, it passed under the District for a second time, almost at right angles. At this point was sited a 350-foot platform on the outside of the loop. North of this, the running tunnel continued parallel to Villiers Street and terminated at a junction with the existing northbound line. Work was carried out from a shaft sunk below the railway bridge near the western pavement of the Embankment. The top of the 21 feet $2\frac{1}{2}$ inches diameter station tunnel was 45 feet below the street but only 10 feet below the foundations of the District station. The running tunnel, of 12 feet 6 inches and 12 feet 9 inches diameter, had a flimsy covering of only 10 feet of clay and ballast under the river bed, and compressed air working was necessary both here and in water bearing ballast beneath the District. A large circulating area was scooped out below the District platforms, and from this, two escalators in separate shafts descended to the single-tube platform 28 feet below. A similar pair of escalators connected the area with the Bakerloo platforms, 4 feet deeper. A connecting subway was made at rail level between the two tubes, and emergency stairs placed in the working shaft. The old sloping tunnel that formerly provided the only access to the Bakerloo platforms was sealed off. When the Bakerloo escalators were opened on March 2, 1914, it was claimed that the journey time from the tube to the District was reduced from 3 minutes 15 seconds to 1 minute 45 seconds. On April 6th, the new Hampstead line station and loop were opened and later in the year the surface buildings of the District station were replaced by a handsome single-storey structure in white Portland stone, designed by H. W. Ford, the company's architect.[1] With its three separate railways, each on a different level, the combined station boasted 178 trains an hour [2] and soon became the busiest station on the Underground. Some nomenclature changes made as a result of the new works caused trouble which has still not been satisfactorily resolved. The original Hampstead terminus beneath the forecourt of the main-line station was quite logically given the same name as that station, but in 1914 it became *Charing Cross (Strand)* to distinguish it from the new combined tube and District station, which was christened *Charing Cross (Embankment)*. On May 9, 1915, some restless mind renamed the two stations *Strand* and *Charing Cross* and the original Piccadilly line *Strand* became *Aldwych* from the same day; this has led to much confusion and many thousands have since had an unnecessary walk along Villiers Street as a result.

Escalators were now well established as a more convenient and expeditious means of vertical transportation. London's first railway escalators were at Earl's Court, where two machines in a single shaft connected the District and Piccadilly platforms from October 4, 1911. These were of the step-off-sideways type, without cleats, and had a vertical rise of 40 feet. Nervous passengers ignored them and, *pour encourager les autres*, the company hired a man with a wooden leg, one 'Bumper' Harris, who rode up and down all day, demonstrating how easy it all was. Large whitewashed signs on the platform surface (THIS WAY TO THE MOVING STAIRCASE) were supplemented by a porter barking in true fairground fashion: 'This way to the moving staircase —the only one in London—NOW RUNNING!' Needless to say, in the face of all this, the lifts were soon deserted by all but the most timorous. The Central London Railway used escalators in their new station at Liverpool Street and Broad Street in 1912, and the LER installed them

---

[1] John Betjeman has described it as 'the most charming of all the Edwardian and neo-Georgian Renaissance stations', but the Villiers Street side has been sadly spoilt by later additions.

[2] 2,139 a day (1,028 District, 540 Bakerloo and 571 Hampstead).

at the new Paddington station in 1914. Lifts put in at Broad Street by the CLR in 1913 were the last to be installed as part of the equipment of a new station; all deep-level stations built after 1912 had escalators only.

OXFORD CIRCUS

At the busy centre of Oxford Circus, the Bakerloo and Central London stations frowned at each other across Argyll Street. Passenger bookings in 1909 at the Bakerloo station had exceeded 3 million and the building was inadequate for a fast-growing traffic.[1] Nearly 2½ million passengers were exchanged with the Central London in that year. A 1909 Bill to enlarge the Bakerloo station was withdrawn at an early stage, but powers were obtained on July 26, 1910, which included provision for escalators (proposals for a new station for both railways on the north side of Oxford Street had been opposed by property owners).

Work began in November 1912 and a new concourse booking hall was constructed in the basement of the original Bakerloo station. From this, a 16 feet 4 inches diameter shaft, 110 feet long, was driven to a landing 54 feet below. In the shaft were placed two 90-feet-per-minute escalators. A deep-level subway had existed to connect the CLR and the Bakerloo since the opening of the latter, and the lifts of the CLR were thus still available for those who wished to avoid the 'moving staircase'. The three Bakerloo lifts were removed, the shaft floored over and part of the space used for the new concourse. When the escalators came into use on May 9, 1914, the street-level part of the station became available for lucrative letting.

TO LIVERPOOL STREET AND PADDINGTON

Two important main-line stations were still not served by tube railways, and the links, long planned, were now established.

The Central London had first obtained powers for an extension to Liverpool Street in its 1892 Act, and these were renewed in 1909 with the curious provision that no foul air was to escape from the tube into the salubrious atmosphere of Broad Street station. Permission was given by the Great Eastern for a tube station to be built under its property in return for an undertaking that there would be no further extension north or north east. John Mowlem & Co. began work in July 1910 on 833 yards of twin 12 feet 5 inch tunnel under Broad Street in extension of the 400-foot siding tunnels east of the Bank station. At Liverpool Street, the two tubes opened out into a 25-foot diameter crossover tunnel followed by two 21 feet 2½ inches station tunnels.

---

[1] Traffic on the Bakerloo line had grown remarkably after a very shaky start. In 1910 it was carrying the heaviest traffic for its length of any electric line in London. Exchange traffic at Baker Street with the Metropolitan Railway almost equalled that with the CLR at Oxford Circus.

Beyond the 400-foot platforms there was another crossover and two sidings extending to a point below the far north-east corner of the main-line station. Ninety-pound bullhead rails were laid.

The terminal platforms were 70 feet below the GER and were linked to a subsurface booking hall by two escalators with a 40-foot rise. Two more escalators, with a 41 foot 6 inch rise, led from a landing at the south end of the tube platforms to a second booking hall below the forecourt of Broad Street station, whilst two lifts in a 75-foot shaft gave direct access from the same landing to the elevated concourse of Broad Street station. Emergency staircases in vertical shafts were provided to each booking hall. The extension was ceremonially opened on July 27, 1912, and at the luncheon, Lord Claude Hamilton emphasized that the GER were making no charges for the station site as they were taking a 'broad view' that what would benefit the CLR would benefit the GER. Public service began the following day, but the Broad Street escalators and booking hall were not opened until October 10th and the lifts to Broad Street concourse, with a small booking office at the top landing, were not in service until February 23, 1913. Through tickets were then issued between the North London Railway (stations to Chalk Farm and stations to Old Ford) and the tube. A shaft was put in for a third escalator to the GER station and used for stairs until the escalator was installed on May 19, 1925.

Paddington remained a logical objective of the Bakerloo and fresh powers were obtained in 1911. The proposed 972-yard line entered Paddington station in the opposite direction to the 1906 proposal and pointed north-west, ready for further growth. In spite of all their past misgivings, the GWR undertook to subscribe £18,000 to the cost of the extension, which they would pay in eight diminishing annual instalments. Metropolitan opposition remained strong and was exacerbated by the GWR subsidy, which was considered a breach of the famous agreement of 1865. Work began four months after Royal Assent had been given on June 2, 1911.

Construction was in the hands of John Mowlem & Co. and the extension opened on December 1, 1913. From Edgware Road, the line ran under the basin of the Grand Junction Canal, then under Praed Street, crossing it twice, finally turning north on a five-chain curve to terminate under the main-line station. The 12 foot and 12 foot 6 inch tunnels were all in London clay and no difficulties were encountered. At Paddington, stairs from London Street led to a booking hall under the cab road on the Arrival Side. The subway between the Metropolitan Railway and the GWR was altered to allow access to the LER booking hall. Two escalators and a fixed stairway in a 21 foot $2\frac{1}{2}$ inch shaft led to the platforms, 40 feet below, and 60 feet beneath the street. These were in separate tunnels and connected to the lower escalator landing by cross-passages in the fashion that was to become standard.

The LER booking office was stocked with tickets to all GWR stations as far as Reading, and the GWR issued through seasons to LER stations. Connections were worked out with remarkable ingenuity: the last tube train from Paddington at night met the last Piccadilly line trains at Piccadilly Circus and the late traveller could change to reach the last Hampstead trains at Leicester Square, arriving at Golders Green in time to catch the last MET tram to North Finchley! Consideration was also shown by arranging for tickets to be collected on the trains between Edgware Road and Paddington so that passengers with close main-line connections could make a good running start on arrival.

For the growing traffic and extension, the first orders were now placed for new rolling stock to supplement the original fleet. Brush Electrical Engineering Co. built ten all-steel motor-car bodies and Gloucester Carriage & Wagon ten motor trucks. Trailing trucks were assembled by the LER and the new cars were working on the Bakerloo early in February 1914. An outstanding innovation was the introduction in this stock of centre doors in addition to the end gate platforms. Studies had been made for some years of the problem of easing passenger flow and reducing station time, and experiments with centre doors built into existing stock had begun on the Piccadilly line late in 1912. In the new cars, the centre doors in each side swung inwards and as the gateman opened his gates, bolts locking these doors were withdrawn by electro-magnets. When the last passenger had boarded, the rubber-edged doors were closed by springs in Bardsley door checks beneath the floor. The action of closing the platform gates shot the door bolts home and a green light indicated to the gateman that it was safe to give the bell signal. Space for the doors was provided by arching the car roof, and the doors were set back to give a clear height of about 6 feet through the 2 foot 9 inch wide opening. Interior improvements included armrests and moquette velvet upholstery after the CLR example. There were thirty-six seats in each car, mostly longitudinal, and 'hygienic' enamel straps for standing support. Electric tail lamps, with emergency battery supply, were fitted for the first time and each car had two GE 212 240-h.p. motors.

A complete four-car train of two motor cars (two 240-h.p. motors per car) and two trailers was also obtained in 1914. These Leeds Forge cars had the usual gate platforms and twin swing doors in the centre of each side (quickly replaced by single wooden doors 2 feet 2 inches wide). Operating procedure for the doors was similar to that in the Brush cars. Seats were upholstered in moquette and the interiors were given what was described as a 'warm' appearance by a décor of dark terracotta and cinnamon. These arch-roofed, all-steel cars weighed 27 tons 10 cwt. (trailers 17 tons 15 cwt.).

Both batches of cars were on plate-frame bogies and both had emergency accumulator lighting now fitted for the first time (older stock also began to receive this in 1914).

OUT INTO THE OPEN

The expected north-westerly extension of the Bakerloo beyond Paddington was announced late in 1911, but to find its origins it will be necessary to return for a moment to 1906. At the LNWR Half Yearly Meeting of that year, the chairman announced that an electric railway was proposed from Euston to Watford alongside the main line, to afford relief to the congested approaches to Euston and develop a London suburban traffic (hitherto much neglected owing to the inadequacy of accommodation at Euston). The new line was to begin as a single-track balloon loop in tubes about 55 feet beneath Euston station, and the two ends of the loop would join just west of the Hampstead Road. The twin 13 foot 6 inch tunnels would then run below the existing railway with stations at Chalk Farm and Loudoun Road, rising to surface at a point between Loudoun and Abbey Roads. From there, the suburban railway would follow the main line on the north side, sweeping round to avoid the sidings at Willesden Junction. Crossing the main line south of Wembley, the electric line would run on to Bushey on the south-west side of the main. At Bushey, it would swerve west to join the Rickmansworth branch, which was to be doubled through High Street station to Watford Junction. A new spur from the Rickmansworth branch would terminate at Croxley Green, and a new curve would allow through running from Bushey and Euston to Croxley Green or Rickmansworth. These proposals received Parliamentary sanction in 1907, but the unfavourable state of the money market postponed construction until the beginning of 1909, when a start was made between Willesden and Sudbury & Wembley stations. It was then announced that the tube section would be left till last and that, pending completion of the scheme, the open-air parts of the line would be worked by steam traction.

By the end of 1911, work was well advanced and on November 18th the LNWR made it known that terminal arrangements had been reconsidered. Realizing that most of their suburban passengers would be bound for places other than Euston, they now planned to direct the traffic into three streams as it approached London: after passing Queen's Park, where there would be interchange facilities, one stream would go to Broad Street via the North London Railway, another direct to Euston on the surface, and the third to the West End and Elephant & Castle via the Bakerloo tube. This last was the result of an agreement with the LER for the construction of a 2 mile 36 chain extension of the Bakerloo from its authorized terminus at Paddington to a junction with the LNWR at Queen's Park. The idea of a tube loop at Euston was discarded and an LNWR Act of 1912 sanctioned alterations in the Primrose Hill area which would bring the new suburban electric line into the existing terminus without interfering with other traffic.[1]

It was, of course, not an entirely new idea to combine tube and suburban railways—the GNCR tube had been built to main-line size for that very reason and there had been earlier proposals [2]—but it was certainly a transformation for the Bakerloo which, like all the other existing small-diameter tube lines, had originally been conceived as a self-contained 'rapid transit' line providing shuttle service between traffic centres in the urban area. The 1911 proposals for the Bakerloo and the less ambitious 1910 scheme to project CLR trains over the new GWR line from Shepherd's Bush to Ealing (to be mentioned in a moment) are therefore of considerable importance, as they mark the beginning of what was later to become a common pattern of tube development. Traffic experts lost no time in pointing out that these schemes were almost an ideal arrangement for all parties: passengers would benefit by more convenient journeys; tube railways would acquire a new traffic; main-line companies would be relieved of increasing terminal congestion and thus save the expense of reconstruction; and, finally, street traffic round the main-line termini would be relieved as the tube trains would distribute suburban traffic all over central London. There were, of course, some snags in the arrangement, such as the different platform heights of tube and main-line stations, the small seating capacity of tube stock, the impossibility of complete integration owing to the small-bore tunnels, the extra strain on station and line capacity on the inner sections of tube railways, and the inevitability of weather and other delays on the open sections affecting the tube service.

Powers for the Bakerloo extension were obtained in the LER Act of August 7, 1912, and the LNWR agreed to lend up to £1 million in perpetuity at 4% to cover the cost of construction. Walter Scott & Middleton (tube section) and Mowlems (the remainder) began work in October 1912.

In the same session, the Metropolitan and GCR obtained powers for a branch from their main line at Rickmansworth to Watford; the LNWR opposed and Stanley appeared as a witness for their side. He explained that the first idea had been to bring the Hampstead tube to the LNWR at Chalk Farm but this had proved impossible owing to the intensive service on that tube. Although much was made of the proposed electric

---

[1] Although the LNWR gave up the idea of having its own tube railway, it is interesting to note that the new scheme included twin tube tunnels of 16 feet 4 inches diameter, shield driven and iron-lined, at Kensal Green and Primrose Hill, the latter pair almost a mile long.

[2] Notably the London, Walthamstow & Epping Forest Railway, authorized in 1894, whose 16 feet diameter tube tunnels would have taken GER suburban trains behind electric locomotives to a new terminus in South Place, Finsbury, and whose own trains would have partly used GER lines.

service to Watford, Parliament evidently thought a little competition would do no harm.

In 1905, the Great Western Railway promoted a successful Bill for a railway between Ealing and Shepherd's Bush which provided for a connection with the West London Railway and a suburban line terminus opposite Shepherd's Bush Green (adjacent to the CLR station, 32 feet below). When reviewing this line in the *Railway Magazine* for July 1905, the Rev. W. J. Scott suggested that before it was finished some way would be found to connect it to the tube railway. And this came about. The CLR Act of 1911 authorized a half-mile connection between the tube terminus at Wood Lane and the Ealing & Shepherd's Bush line, and also gave the CLR running powers over the E&SB to Ealing. Construction of the E&SB began in 1912 (the suburban terminus in Uxbridge Road was dropped), but tube trains did not use it until 1920 and we shall deal with it in a later chapter.

THE LOTS ROAD DEAL

During the four years following the formation of the LER, the Underground company took steps to consolidate and strengthen its position. The first move was to sell Lots Road power station to itself—at any rate it seemed rather like that.

A Lots Road Power House Joint Committee was established by the Metropolitan District Railway Act of 1911 and consisted of representatives of the District and LER. This Committee was authorized by the Act to purchase the power station from the Underground company and then lease it to the two railway companies. The explanation given for this odd transaction was that it was necessary to protect the position of the statutory railway companies as (in theory) the limited company owning the power house *could* part with its interests in the two railway companies and sell the power house to another company, leaving the railway companies in the cold. Powers were also sought in the Bill to permit the Committee to supply current for non-transport purposes but this was firmly refused in the face of powerful opposition from local authorities and electricity supply companies.

THE GENERAL JOINS THE FAMILY

The motor bus was now becoming firmly established in London, and before the first decade of the century had passed, it was apparent that this new arrival was to be a serious competitor for urban and suburban transport. Astute minds were at work within the Underground company, and about the beginning of 1910 negotiations had been started with the London General Omnibus Co. Ltd, the capital's main bus operator. By the end of 1911, the air was thick with rumours of an imminent amalgamation between the two companies. Meanwhile, in its Act of 1911, the LER had obtained powers to acquire, provide and work buses, and the Underground company was busy altering its articles of association to allow it to carry on business as a bus and cab proprietor and manufacturer. As a sign of the way the wind was blowing, it may be mentioned that from September 1, 1911, the subsidized buses which had been maintaining the Golders Green–Hendon connection were replaced by LGOC vehicles with the same through booking facilities. The rumours of the pending merger caused a sensational rise in LGOC Ordinary Stock, and the price touched 240 before the excitement was ended on January 19, 1912, by the publication of an official statement of the proposed terms. For each £100 of LGOC Ordinary Stock, the holder would receive £105 of 6% UERL Cumulative Income Debenture Stock, plus £105 of UERL 6% Income Bonds, plus 100 'A' shares of the shilling denomination. The *Railway Gazette* of February 2, 1912, stated that opinion was divided on the financial merits of the offer, but the General could only expect dividends of 12% and over if there were little or no competition. With several new bus companies being floated and the UERL threatening to run its own buses, it considered the General shareholders would be well advised to accept.

And this they did. At the beginning of February, the LGOC announced that the majority of their shareholders had accepted the offer.

By this shrewd and far-sighted financial manoeuvre, the Underground company had neatly removed a most dangerous source of competition and brought into its family a lusty profit earner which could support the poorer relations[1]; it had avoided the otherwise inevitable establishment of its own bus fleet and could now contemplate the prospect of a vast integrated transport system based on through bookings and connected services.[2] Sir Edgar Speyer had cause to be well pleased with his achievement and with his clever young lieutenant, Albert Stanley, from whom he had undoubtedly received much help and stimulus.

The financial benefits of the merger were soon apparent, and for the second half of 1912 the Underground company was able to pay the full interest on the 6% Income Bonds for the first time. The public benefited by the introduction of new bus services from Underground stations and road-rail through bookings, though it was many years before the latter took full shape.

Fears of competition from the reliable 'B' type buses (introduced in October 1910), particularly on new routes from Underground railheads, prompted the British Electric Traction-controlled Metropolitan

[1] The LGOC dividend for 1913 was 18% and the *Railway Gazette* rightly described it as 'the most valuable jewel in the regalia of the Underground Electric Railways Company of London'.

[2] One of the first examples of fully integrated working was the bus station opened in April 1914 at Hammersmith for LGOC service 33, with its booking offices for the Underground railways and a covered way to the railway platforms.

Electric Tramways to float a bus subsidiary and order 350 Daimlers, but an amicable agreement between the Underground company and the BET was reached before the end of 1912, as we shall see in a moment.

There were also rumours of working arrangements between the Underground company and the Metropolitan Railway and between the Underground and the LCC, whose tramways were a serious rival to many Underground services. Nothing more was heard of the talks with the LCC, but at the Half Yearly Meeting of the Metropolitan on January 29, 1913, Lord Aberconway, the chairman, said that although amalgamation had been ruled out, there was an understanding with the Underground company towards the elimination of unnecessary competition and working together in mutual interest.

THE METROPOLITAN BUYS A LOSER

The Metropolitan Railway was at this time about a little empire building of its own. The luckless Great Northern & City, with its full-size tubes, had been an utter failure; the expected through running with the GNR had never come about and the electrification of the LCC tramways in its area had had a damaging effect on local traffic. In March 1912, at the Half Yearly meeting, Lord Lauderdale, the chairman, wistfully remarked that his company were watching the UERL-LGOC amalgamation with 'sympathetic interest', but later in the year it became apparent that sympathy and interest obtained in another quarter, for the Metropolitan Railway Bill of 1913 proposed to acquire the GNCR, extend it to a junction with the Waterloo & City, and make a junction with the Metropolitan at Liverpool Street. It was suggested that a short line like the GNCR was very vulnerable to surface competition and linking with other lines was essential if it were to become self-supporting. As arrangements had been made regarding a GNR-Metropolitan partnership in the GNCR, the GNR offered no opposition, but there was a considerable furore from City property owners, excited by the fact that both the proposed link lines would have to come near the surface, one to join the shallow Metropolitan line and the other to pass over the top of the CLR. The Metropolitan emerged with powers to take over the GNCR as a going concern from June 30, 1913,[1] and build a 500-yard extension to Lothbury only.

In the following year, a joint GNR and Metropolitan Bill sought to transfer the GNCR from the Met. to a Joint Committee of the two railways and again to link the GNCR with the Met. near Aldgate, also to have an exchange station, but no junction, with the Waterloo & City. The GNR proposed to build connections between their up and down slow lines at Finsbury Park and the GNCR at Drayton Park. Now the winds of opposition blew from the North London Railway, which saw itself being squeezed out of the GNR suburban zone. Certain conditions

[1] The GNCR was worked by the Metropolitan from September 1, 1913.

inserted by Parliament to protect the rights of the NLR were not acceptable to the promoters and the Bill was withdrawn. The Met. was left holding a GNCR unadorned and had to make the best of a bad job.

Its efforts were not entirely unfruitful. By the end of 1913, the overall journey time had been reduced to $10\frac{1}{2}$ minutes ($9\frac{1}{2}$ for semi-fast trains) and the rush hour service was increased to two minutes in the mornings and $2\frac{1}{2}$ minutes in the evenings (slack hour interval was four minutes). In the middle of 1914, it was announced that passenger loadings had increased and a small profit had been made. From September 24th that year a useful economy was made by closing Poole Street power station and supplying the line from Neasden through a new substation at Drayton Park. Asserting its main-line complex, the Met. introduced First Class accommodation on February 15, 1915, mainly for the benefit of GNR season-ticket holders. The GNCR thus became the first and only tube railway to offer more than one class of accommodation, and the facility lasted until March 31, 1934. Some rebuilding of cars for both First and Third Class was begun in 1915.

MORE ADDITIONS TO THE FAMILY

Whilst the Metropolitan was embracing the GNCR, Speyer and Stanley were also active. On November 19, 1912, London learnt that the Underground company proposed to acquire both the Central London and City & South London railways and that a new company, called the London & Suburban Traction Co., in which the Underground company was to have a very substantial interest, would take over control of the important tramway systems of the London United and Metropolitan Electric companies, including the bus fleet of the latter, which was to pass to the LGOC.[1] Sir Edgar Speyer pointed out that the object of these arrangements was to improve London traffic facilities and 'develop traffic to work best for shareholders'. Consolidation and co-ordination of traffic facilities, he blandly observed, were just what the 1905 Royal Commission on London Traffic had thought desirable. The *Tramway and Railway World* had already remarked [2] that it seemed to be Sir Edgar's ambition, in default of the appointment of a London Traffic Board by the Government, 'to constitute himself and his directors as a *de facto* Traffic Board'. Behind Sir Edgar, always ready with advice and ideas, and with acute perception of the whole London

---

[1] The only remaining privately owned tramway system in London, that of the South Metropolitan Electric Tramways and Lighting Co. Ltd., fell into the hands of the London & Suburban Traction Co. in May 1913 and two years later the Underground Co. put aside all pretence and took over administrative control of L&ST and its subsidiaries. Even so, the British Electric Traction Co. retained a large holding in L&ST until November 1928 when it sold its shares to the Underground Co. who were naturally anxious to "prevent them falling into unfriendly hands" (Lord Ashfield in 1929).

[2] September 21, 1912.

transport scene, stood Albert Stanley, whose 'ability, energy and tact in completing difficult and complicated negotiations' were commended by Speyer at the Underground company's Annual Meeting in February 1913.[1] Stanley's ability and contribution to the national weal were recognized in the Birthday Honours of June 22, 1914, when at the age of 39, he became Sir Albert H. Stanley.

The offers to the Central London and City & South London shareholders were made late in 1912. For each £100 of Ordinary Stock, the CSLR shareholders would receive £40 of LER Preference Stock and £25 of LER Ordinary Stock. This was reasonable, as the pioneer company's financial position had been shaky for some years owing to severe tramway competition. It was impossible to improve the service and attract more traffic without drastic modernization which would require capital expenditure far beyond the company's reach. The only way out was consolidation with the larger concern and the shareholders accepted this. When announcing the offer on November 19, 1912, Speyer revealed that powers would be sought to enlarge the tunnels so that LER trains could run over the CSLR via a new junction near Euston.

Although its traffic had slumped badly in the face of motor-bus competition,[2] the Central London was in somewhat better shape than the CSLR. Holders of CLR Ordinary Stock were offered equal amounts of new 4% Guaranteed Stock with a contingent right to 40% of higher earnings if earned for three successive years. The *Railway Times* thought that as the CLR had only paid 3% for some time, this would raise the dividend by 1%, but it was doubtful whether the directors had secured adequate terms. Despite this opinion, the shareholders accepted.

Both railways were taken over from January 1, 1913, and whilst the companies retained their separate statutory identities, the lines were gradually shaped into integral parts of the Underground system. A leaflet issued by the Underground in January 1914 smugly reported of the CLR: '. . . the advertisements at the stations have been straightened up to give a brighter and tidier appearance and care has been taken to indicate thoroughly the names of the stations by the adoption of the bull's eye design of nameplate repeated at frequent intervals. . . .' It added that the CLR lift gates were being converted to compressed-air operation, that the line had been resignalled with the latest automatic equipment and that the fare from one end to the other had been reduced from 3½d to 3d.

[1] At this Meeting (February 1913), it was also announced that most of the company's securities and shares were now in British hands and the two remaining American directors had therefore resigned. Speyer took the opportunity to recall that the great projects of tube construction and the District electrification were only made possible by the availability of American and other foreign capital.

[2] From 45 million passengers in 1902 and 1903 to 38 million in 1911 and just under 36 million in 1912.

In accordance with Sir Edgar's promise, powers were duly obtained in 1913 for the opening out of the CSLR running tunnels to a maximum of 13-feet diameter (station tunnels to a maximum of 30 feet). An LER Act of August 15, 1913 empowered that company to run over and work the CSLR and construct new lines to join the two railways at Euston.

The CLR also had a Bill in the 1913 session embodying proposals that had been worked out before the take-over. Much to the chagrin of Thames Valley local authorities, who had been given strong hints of extensions towards Twickenham, Sunbury and Chertsey, the original plans had been whittled down by the Underground Co. during the take-over negotiations. The Act of August 15, 1913, sanctioned a 2½ mile extension from Shepherd's Bush to Gunnersbury station, LSWR, via Goldhawk Road, Stamford Brook Road, Bath Road, Chiswick Common and Chiswick High Road, reaching the surface just before Gunnersbury station, where a junction would be made with the LSWR. This opened up the possibility of through tube running to Richmond and beyond.[1]

Another CLR Act in the following year authorized escalators at Shepherd's Bush, Bank and Oxford Circus, and a new interchange station at Holborn Kingsway to replace British Museum. None of the works authorized for the CSLR or CLR in 1913 and 1914 had been started when war broke out, nor was the Ealing line by any means ready for CLR trains, although the twenty-four motor cars to work over it were ordered from Brush in July 1912—more will be heard of these cars in the next chapter.

The Underground empire building stimulated a new flush of through bookings. A start was made on February 1, 1914, when 'T-O-T' (Train-Omnibus-Tram)[2] tickets were introduced. These first issues were available via Paddington (Bakerloo) on LGOC routes 7 and 32, via South Kensington (Piccadilly) on LGOC route 49, via Wood Lane (CLR) on LGOC route 66 and via Shepherd's Bush (CLR) on LUT tramcars. In the following month, these tickets could be purchased for through journeys via Stockwell (CSLR) and LGOC routes 20 and 20S, and soon afterwards there were TOT bookings via Highgate, via Hammersmith, via Golders Green, via Finsbury Park and via Clapham Common. Then on July 28, 1914, there came the first through bookings between the

[1] The LER also had its eyes on the LSWR outlet, and in its Act of August 15, 1913, obtained powers for a connection between the Piccadilly line at Hammersmith and the LSWR Kensington–Richmond branch, with the intention of running Piccadilly trains to Richmond.

[2] 'T-O-T' was used as the title of a publicity leaflet issued free to the public of which there were 33 issues between May 17, 1913 and the final one of August 15, 1914. The staff magazine published from October 1922 until 1933 was called 'T-O-T Staff Magazine' and there was also a T-O-T Benevolent Fund.

LER and the LCC tramways, via Highgate, where an additional entrance in Highgate Hill had been added in 1912.

In April 1914, the LER booking offices at Elephant & Castle, King's Cross, Oxford Circus and Tottenham Court Road began to issue CLR or CSLR tickets as well as their own, and the other lines' booking offices at these points were stocked with LER tickets. The CLR and CSLR booking offices at Bank were similarly rationalized. The two surface station buildings of the CSLR and the LER at Euston, in Eversholt Street and Drummond Street, were closed on September 30, 1914, and all traffic diverted through the booking hall and lifts under the main-line station.

London was to wait another twenty years for a unified public transport system, but at last a beginning had been made.

CHAPTER 8

The Tubes in Wartime
1914–1918

WHEN war came, London looked beneath her streets, like a nervous person looking beneath the bed. In the first week of August 1914, police were sent to search the disused CSLR tunnels between King William Street and the Borough. Not finding any arms, spies or gunpowder, they departed, leaving the Underground staff to seal off the deserted tubes.[1]

As mobilization proceeded, the tubes became thronged with soldiers crossing from one railway terminus to another. To ease this movement, the company allowed all men in uniform to travel free until October 1st.

At first there was some decline in evening and pleasure travel, but by the end of 1915 it was clear that the war was bringing the tubes a great boom in traffic. The increased activity within and across London was caused by constant troop movements, leave travel, reductions in road services owing to staff shortage and withdrawal of vehicles for war purposes, and more travelling by well-paid workers enjoying the high level of employment. Another contributory factor was the 'dim-out' enforced after dark as a precaution against air attack—people naturally preferred travel in the well-lit tube cars to slow bus and tram journeys through darkened streets. In the long winters of 1915–1916 and 1916–1917 many thousands of new passengers were gained.

TAKING COVER

Others made their first acquaintance with the tubes in a different fashion. After a Zeppelin raid on Antwerp late in August 1914, some precautions were taken in London and by the end of the year the

[1] This little piece of comic opera followed a suggestion by G. A. Sekon (Nokes), editor of the *Railway and Travel Monthly*, who also proposed searches of the Pneumatic Despatch Co. tube from Euston to the General Post Office and the disused Metropolitan Railway tunnel beneath St Pancras station.

Underground company had made preparations to give shelter in its tube stations to people caught in the streets at the start of an air bombardment. The first of the twenty-five Zeppelin and aeroplane raids on the London County area came on the night of May 31, 1915, after some curtain-raisers in the Thames estuary. In September 1917, there were six air raids in the space of a month, a concentration then unprecedented. The attacks were of such a persistent nature that they led to the first mass use of tube stations for sheltering. The sequence began on the night of September 4th and after this there was a break until the 24th, when it was estimated that some 100,000 people were sheltering in tube stations. The crush was very near danger point at some places and a particularly alarming feature was the complete blockage of many emergency staircases. On the night of the 25th, there was another raid and some 120,000 went down below. The original idea of using the stations to shelter those caught in the streets was swept aside—the frightened crowds arrived before any warning was sounded. They came ready to spend the whole night on the platforms and stairs, bringing their bedding, their babies, their dogs, cats and birds, their life savings. They came in their thousands on the next two nights, though no bombs were dropped. The safe working of the railways was jeopardized and passengers were inconvenienced. Shelterers even invaded trains and sat in them as long as they ran, finding this less monotonous than sitting about in the stations.

The outcome of this was that from September 28, 1917, access to stations was forbidden to all but genuine passengers unless an air-raid warning had sounded. After the warning, shelterers were allowed in only if unencumbered with animals and baby carts. On the evening of the very day this notice was issued, there was a fierce rush at Liverpool Street after a warning had been given of impending attack. The police were unable to stem the torrent of people [1] and a Russian Jewess was killed in the crush. There was no raid that night, but further 'moonlight' raids occurred on the nights of September 29th and 30th and October 1st. On the night of February 17, 1918, the total sheltering reached 300,000, the highest figure of the war. But London had only two more raids to come and the last was on the night of May 19, 1918.

SPEYER AND STANLEY ATTACKED

The war forced Sir Edgar Speyer from the London scene. His foreign-sounding name attracted attention, and in the hothouse atmosphere of wartime prejudice and distortion whipped up by the popular Press, accusations of disloyalty and suggestions of treachery were levelled at the unfortunate financier. In May 1915 he decided to resign his chair-

---

[1] '. . . people of the poorer classes, mostly aliens, women and children' according to the *Railway Gazette* of October 5, 1917.

manship of the Underground company in favour of Lord George Hamilton and on May 17th he wrote to the Prime Minister asking permission to resign his Privy Councillorship and requesting the revocation of his baronetcy. Mr Asquith replied that the King was unwilling to agree, and added: 'I have known you long and well enough to estimate at their true value these baseless and malignant imputations upon your loyalty to the British Crown.' But after further attacks in November, he joined his brother in New York. He died in Berlin in 1932.

Sir Albert Stanley offered his services to the Government at an early stage and in 1916 he was appointed Director-General of Mechanical Transport at the Ministry of Munitions. When Lloyd George formed a new Government in December 1916, Stanley was chosen as President of the Board of Trade and his posts in the Underground company were taken over by C. W. Burton. Strange as it may seem, Stanley also had to face the strident accusations of jingoist prejudice. A question was asked about his real name in the House of Commons on June 26, 1918, and the reply revealed that the name 'Stanley' had been assumed by his father, whose real name was Knattriess, when he emigrated to the USA. It was stated that there had been no foreign blood in the family for many years. Sir Albert served as a Minister until May 1919, when he resigned in favour of Sir Auckland Geddes. During his period of office, he had the unpleasant task of imposing a Coal Restriction Order which caused a substantial reduction in the Underground rail services.[1]

THE COMMON FUND

Before he took up his post in the Government, Stanley produced a 'war baby' with the assistance of the Board of Trade. This was the Common Fund, authorized by the London Electric Railway Companies' Facilities Act of July 29, 1915. Most railway companies had been under Government control since August 5, 1914, but the District was the only one of the Underground lines to come within the scope of the Government Order. There was therefore some embarrassment when railwaymen began to receive war bonuses, for although the tubes were not getting the Government subsidy, the staff naturally expected the bonus. Early in 1915, the Underground company approached the Board of Trade and other Government Departments, but it was soon established that the Government had no desire to take over the tubes. An ingenious proposal then made by Stanley gained the sympathy of the Board of Trade, who promised to give the necessary facilities for legislation. On the strength of this promise, the war bonus was paid to

---

[1] The Coal Restriction Order (366 of 1918) required a reduction in consumption of at least 15%, and from May 1, 1918, fifteen tube stations were closed after 11.30 p.m. From May 5, 1918, four stations were closed all day on Sundays and another two closed half the day on Sundays.

the Underground railway staff. Stanley's plan was to establish a common fund of the District, the LER, the CLR, the CSLR and the LGOC, each company to pay its surplus into this fund every half-year after meeting certain agreed standing charges. The fund would then be distributed in agreed proportions among the five companies. If any company's revenue was not sufficient to meet the standing charges, the fund would make up the deficiency, and the share-out would not be affected. The strong, in other words, were to assist the weak.

As part of the arrangement, it was decided that each company would retain the whole of its traffic receipts for credit to its account, including through fares to other Common Fund companies' destinations. This paved the way to a simpler and more effective system of through tickets.

As always, the LCC was suspicious of these financial manoeuvres within the Underground group, and suggested that any agreements between the companies should be submitted to the Board of Trade. It also wanted a time limit of three years, and thought that there should be a special form of accounts for the LGOC and its contributions to the Fund. As so often before, Parliament had little sympathy with the LCC, and the only provision inserted in the Bill in deference to the Council was that copies of all agreements should be deposited at the Board of Trade, which could publish them at its own discretion. The two-page Act, which contained the seeds of a unified London transport system, was followed by an Agreement establishing the Fund retrospectively from January 1, 1915. For 1915, the agreed distribution was: CSLR 2%, CLR 20%, LER 26%, District 12%, LGOC 40%. From January 1, 1916 onwards, it was: CSLR 6%, CLR 20%, LER 30%, District 12%, LGOC 32%. An immediate result was a slight increase in the small dividends paid on the ordinary stock of the tube companies and a decrease in the dividends paid on the bus company's ordinary stock.

EXTENSION AND IMPROVEMENT

When the war began, the Underground company was involved in several capital works and was making detailed preparations for others, including the important task of modernizing the CSLR. The war slowed down this activity, but a surprising amount was completed during these troubled years.

By far the most important was the connection between the Bakerloo and the LNWR suburban lines at Queen's Park, culminating in the projection of tube trains right out to Watford, almost 21 miles from the Bakerloo's southern terminus at Elephant & Castle. It was 2·09 miles from Paddington to Queen's Park, and trains ran to Kilburn Park (1·59 miles) from January 31, 1915 and into the still unfinished Queen's Park joint station from February 11, 1915. Work was behind schedule owing to labour shortage and wet weather.

On a general gradient from south to north of 1 in 200, the new line passed through London clay in twin tunnels of 11 feet 8¼ inches internal diameter (12 feet on curves, 21 feet 2½ inches at stations) and rose to the surface a quarter of a mile east of Queen's Park station at 1 in 48. There were intermediate stations at Warwick Avenue (escalator rise 35 feet), Maida Vale (38 feet), and Kilburn Park (34 feet), all with two 90-feet-per-minute escalators either side of a fixed staircase. Warwick Avenue and Maida Vale both had booking offices below street level, but the latter station also had a one-storey surface building faced in red terracotta. At Warwick Avenue, the two stairways from the street had their upper ends protected by glass screens. The commodious Kilburn Park booking hall was at ground level and had elevations similar to Maida Vale.

Maida Vale, not opened until June 6, 1915, had an all-female staff, the first station in London to be so blessed. It was true that there was a male stationmaster, but he was shared with three other stations. The women consisted of two ticket collectors, two porters, two booking clerks and two relief ticket collector-booking clerks. These ladies wore broad-brimmed blue felt hats and blue tunics with silver braiding and white metal buttons. Later they were given long overcoats, which concealed most of their black-hosed legs. The *Railway Gazette* expressed approval, and stated that it was 'preferable to employing hobbledehoys'.[1]

At Queen's Park, a new station had to be built on the north side of the LNWR platforms, the tube lines coming between what were later to be the up and down LNWR electric lines. There were two island platforms with the tube lines at their inner sides, the outer sides being reserved for the North Western electrics. Beyond the station, there was a 355-foot car shed with two centre pit roads and two outer roads, the latter being the running lines, designed to connect later with the LNWR suburban lines. By passing the running lines through the car shed, extra space was made available for car storage. East of the station, alongside the ascending ramp on the north side, there was another car shed, with two roads, 758 feet long, and space for thirty cars. Both sheds were not completed until late in 1915.

Lots Road supplied power through three 1·2–MW rotary converters in a substation at Kilburn Park. The line opened with a 2½-minute rush-hour service (3½ minutes in slack hours) and the 6·8-mile run from Queen's Park to Elephant & Castle occupied twenty-five minutes. Through fares and season tickets were available to and from LNWR suburban stations. At Queen's Park, McKenzie, Holland and Westinghouse installed upper quadrant semaphore signals with operating

---

[1] As the war progressed, employment of women became commonplace. In 1917, women replaced gatemen on trains, men remaining as front and rear guards.

motors directly applied to the arm—they were not replaced by colour lights until some forty years later.

Owing to wartime difficulties, the rolling stock provided for the extension was a mixed bag consisting of:

One complete four-car train (two motor cars, two trailers) built in 1914 by the Leeds Forge Co. and described in Chapter 7.

Ten new motor cars by Brush, built 1913–1914 and also described in Chapter 7.

Ten trailers of 1907 gate stock from the Hampstead line.

Three 1906 motor cars from the Piccadilly line.

Seven motor cars converted from 1906 Hungarian trailers by the LER at Golders Green. These had single 2 foot 2 inch centre doors on each side in addition to end gate platforms, and retained their clerestory roofs.

Three months later, the LNWR had completed its suburban lines from Queen's Park to Willesden Junction (New Station) through new tunnels at Kensal Green. Tube trains were projected over them every fifteen minutes from May 10, 1915, and terminated in the centre roads of the new suburban station at Willesden Junction. Until the LNWR power station at Stonebridge Park was opened in February 1916, current was supplied from Lots Road. The intermediate station at Kensal Green was not opened by the LNWR until October 1, 1916.

By 1916, the LNWR had electrified its suburban lines between Willesden and Watford (they were already in use by steam trains [1] which had connected with the tube service at Willesden). Although the LER-LNWR Joint Stock for the through service to Watford had been ordered in 1914, there was now no hope of delivery as the manufacturers were fully engaged on war work. Late in 1916, the two companies agreed that the Watford service be operated provisionally with Underground tube stock, and trial runs began early in November 1916 when speeds of up to 40 m.p.h. were reached. On April 16, 1917, a weekdays-only tube service was begun over the 20·8 miles between Elephant & Castle and Watford Junction, the journey taking 61 minutes. At the same time, LNWR electric trains [2] started a rush-hour service between Watford Junction and Broad Street via Hampstead Heath. The LNWR new suburban lines continued to be used by Euston–Watford steam trains at weekday peak hours and on Sundays. Additional steam trains ran on weekdays between Willesden (New) and Euston, connecting with Baker-

[1] The LNWR lines had been opened for steam working as follows: north of Kensal Green tunnels—Harrow, and Croxley Green branch June 15, 1912; Harrow–Watford High Street and Junction February 10, 1913.

[2] The full LNWR electric service on the new lines did not start until July 10, 1922, when electric trains began to run from Euston and Broad Street to Watford via Queens Park. The Watford High Street–Croxley Green branch was electrified on October 30, 1922, and the Rickmansworth branch on September 26, 1927.

The Bakerloo reaches the open air—a modern photograph showing a 1938 stock train climbing the 1915 ramp at Queen's Park. On the left is one of the two running sheds built here in 1915 to serve the extension; on the right, the down LNWR suburban line from Euston and further right, the LNWR main lines.—*Alan A. Jackson.*

Big Ben and Little Len—a contrast between the tube and main line loading gauges at Harrow & Wealdstone station on the LNWR suburban lines, first used by Bakerloo trains in 1917. This photograph, taken in 1954, shows a 1938 stock tube train and 1927 LMSR multiple-unit compartment stock.—*Alan A. Jackson.*

The Ealing & Shepherd's Bush Railway—A Central London Rly. train from Ealing approaching East Acton soon after the opening of the line in 1920. The train is composed of 1903 motorcars and 1900 trailers.—*London Transport.*

Early days at Edgware—S. A. Heaps' pleasant buildings for the terminus of the Edgware extension, soon after the opening in 1924. The feeder bus on the right is on route 140, Pinner Green via Wealdstone. There is a notable absence of passengers—much of the immediate surroundings is still pastureland.—*London Transport.*

loo trains. The basic tube service to Watford was 15 minutes, widened to 30 minutes from June 1, 1918. All LNWR suburban stations except Carpenders Park (closed December 31, 1916, not reopened until May 5, 1919) and Stonebridge Park (destroyed by fire January 9, 1917, reopened August 1, 1917) were served by tube trains, and these two stations were also included when they reopened. Watford trains were relieved of some local traffic by passing Maida Vale and Regent's Park (and additional stations in the rush hours). Traction current at 630 volts d.c. came from the LNWR power station at Stonebridge Park.

A notable aspect of the new service was the rolling stock. The Bakerloo trains were made up with two motor cars of new design and three Piccadilly line trailers in the formation M-T-T-T-M. The twenty-two new motor cars were those built by Brush in 1914–1915 for the Central London Railway extension to Ealing and kept in store pending completion of that line. They were the first all-enclosed tube cars and had two GE 212 240-h.p. motors giving a free-running speed of 40 m.p.h. BTH automatic electro-magnetic contactor control equipment was another new feature. Two additional collector shoes were fitted to the motor bogies, one on each side, to make the cars suitable for four-rail operation. As there was insufficient clearance to fit the positive shoes in the trailer truck of the motor car, the shoes were attached to the truck of the adjacent LER trailer and connected to the motor car by jumpers. At the trailing end of the motor car there was an all-enclosed vestibule fitted with single swing doors each side, under the supervision of a gateman who controlled the single central doors on each side of the car remotely, as on the 1914 stock. The open end platforms of the LER trailers were fitted with a ramp which led from the interior of the car to a superimposed raised step necessary to help passengers reach the platforms of the LNWR stations. This refinement was not practicable on the motor cars, with their closed-in vestibules and low roofs, and passengers travelling in these cars had to step up and down the considerable difference between the platform heights of the two railways—it was just as well that hobble skirts and long, tight corsets had passed out of fashion. The thirty-two seats of these clerestory-roofed motor cars were upholstered in red 'Pegamoid' and were nearly all transverse. Car ceilings were white and the walls light brown; exterior livery was lake, with white above waist and gold lettering. The fifty-two rattan seats of the trailers were re-upholstered in moquette, but some cars went into service before this was done. Trailers were also white above waist, making the Watford trains easily distinguished at stations.

Towards the end of 1917, some rush-hour trains to Watford were made up to six cars, the rear car being reserved for certain stations only. A Sunday service was inaugurated on July 6, 1919, and in that year some trains of LER gate stock were to be seen on the Watford run, as the CLR cars were withdrawn in batches to be fitted with plate-frame motor bogies by Cammell Laird (the original equalizer-bar motor bogies had proved unsuitable for sustained high-speed running).

During the war the Bakerloo also benefited by the provision of improved interchange facilities at Baker Street. Two 90-foot-per-minute escalators with a 31 foot 3 inch rise were installed in a 16 foot 4 inch shaft between the tube platforms and a new circulating area below the through Metropolitan lines, which were reached by stairs. These escalators were opened on October 15, 1914, but the original lifts maintained service to and from street level.

Construction of the Post Office Tube Railway, authorized in 1913, went on in spite of the war. This line had been recommended by a Departmental Committee appointed by the Postmaster-General in 1909 to consider the question of carrying mails across London. It had been authorized by the Post Office (London) Railway Act of 1913, which proposed a tube railway from Paddington station to the Eastern District Post Office in Whitechapel Road via the Western District Parcels Office in Barrett Street, the Western District Office in Wimpole Street, the West Central District Office in Museum Street, Mount Pleasant Post Office, the General Post Office and Liverpool Street station. Driverless trains, controlled from the stations, were to operate on a one-minute headway over 2-foot gauge double tracks in 9-feet diameter tunnels. Mowlems had begun the tunnelling in October 1914 and, in spite of labour shortages and questions in Parliament, continued until 1917 when the 6½ miles of tube were virtually completed. During the war, the empty tunnels were used to store art treasures from London museums. The equipment of the line was not taken in hand until post-war prices had fallen, and it was not opened until December 1927.

MATTERS OF POWER

The varied and interesting generating machinery of the CSLR, with its three- and five-wire d.c. transmission, fell into disuse when arrangements were made to feed that railway from Lots Road from June 13, 1915. Three new substations, converting 11,000 volts a.c. to 600 volts d.c., were erected at Stockwell, Elephant & Castle, and Old Street. The Hampstead line substation at Euston was also enlarged to deal with the CSLR, and the original Bakerloo substation at London Road was replaced by the new equipment at the Elephant. After this change, the CSLR had central positive conductor rail with negative return through the running rails.

A further economy was to be made by supplying the CLR from Lots Road, but this move was delayed owing to wartime difficulties in obtaining extra generating plant for the main power house. One 15-mW set was installed at Lots Road in 1915, but a second set could not be had until after the war.

OVERCROWDING

The increased traffic brought by war conditions in London was carried with growing difficulty, and by the end of 1917 the situation had become so bad that there was much criticism in the Press and in Parliament. Factors limiting tube capacity at this time were several: the slow loading and unloading of the end-platform gate stock, slow attendant-worked lifts and restricted train lengths owing to shortage of rolling stock. As has been seen, new types of rolling stock were emerging, but nothing more could be done during the war. Some improvements had been made to lifts. The first landing-operated lifts were installed at Piccadilly Circus on January 8, 1914 (two lifts), and were quickly followed by two more at the same station and three each at Euston and Great Central, the latter with the now familiar dome-covered controls. This system provided a more convenient control of passenger flow and, as it was developed, a substantial economy in staff. During the war, many of the old CSLR and CLR lifts were improved and standardized with LER lift practice. Some of the CSLR hydraulic lifts were replaced by electric lifts from LER stations where escalators had been installed. The restriction of train lengths could not be relieved until extra stock could be built. In December 1917, the position was grim:

	Trains per hour	
	Peak hours	Slack hours
Bakerloo:		
Watford	4 six-car trains	4 five-car trains
Queens Park	24 five-car trains	16 three-car trains
Piccadilly:		
Main line	1 six-car train and 23 five-car trains	20 three-car trains
Aldwych	15 one-car trains	15 one-car trains
Hampstead:		
Hampstead and Golders Green	19 five-car trains	12 three-car trains
Highgate	19 five-car trains	12 three-car trains
Central	11 six-car trains and 13 five-car trains	20 three-car trains

To cope with the crowding, various measures were devised. In May 1917, barriers were erected 3 feet from the edge of the northbound Bakerloo platform at Oxford Circus opposite the points where the second and third cars of a train were to be stopped. After the alighting passengers were clear, a sliding bar was moved to allow the waiting hordes behind the barriers to assault the train. This barrier system was later installed at Charing Cross (Bakerloo) (November 1917) and Piccadilly Circus (Bakerloo) (Spring 1918). Two CLR stations, Oxford Circus and Tottenham Court Road, were also equipped in March and April 1918. A similar system, involving platform queues opposite each pair of train gates, was tried at Knightsbridge early in 1918 and was introduced at Holborn and Dover Street in June of that year. The queue system proved better than the barrier plan and was started at seven CSLR stations (Bank, Elephant & Castle, Borough, London Bridge, Moorgate, Kennington, Oval) and three Bakerloo stations (Piccadilly Circus, Oxford Circus, Charing Cross) between September 1918 and March 1919. For both systems, over fifty extra staff were needed to control and marshal the crowds, and the overall saving in station time was small as drivers lost time in trying to stop their trains in the precise positions required. These barrier and queue systems lasted at least until late in 1919, but were at best a palliative, introducing a measure of order into near-chaos.

As the war drew to its end the overcrowding became even worse, and more will be heard of it in the next chapter.

CHAPTER 9

Growing and Grafting

1919–1926

A FRESH START

THERE were four major authorized works outstanding when the war ended: the extension from Golders Green to Edgware; the extension of the CLR from Shepherd's Bush to Gunnersbury, LSWR; the junction between the Piccadilly line and the LSWR at Hammersmith; and, most important of all, the modernization of the CSLR and its connection with the Hampstead line at Euston.

Before 1914, a fully equipped tube railway could be provided at about £600,000 a mile, but rising costs had brought this to almost £1 million by 1919. A private enterprise could not now obtain capital at less than $5\frac{1}{2}\%$; tube railways were returning a modest 2%. In 1919, Parliament authorized the LER and CSLR to borrow a further £2½ million, mainly for the CSLR modernization and the Euston junction, but the two Acts were passed in an atmosphere of uncertainty. Rumours of transport nationalization were current and the difficulties of the London transport situation were very much a topic of the day.

Public discontent with overcrowding had been growing for some time and an increase of fares on April 6, 1919, brought matters to a head. A committee of London MPs met in April to consider the transport situation and a month later heard evidence from Frank Pick, Commercial Manager of the Underground company. Pick explained that the serious congestion on the tubes and District Railway arose from a shortage of buses and a shortage of rolling stock. New cars could not be had during the war and maintenance had become so difficult that forty had been withdrawn from service. The Group's railway traffic had increased by 69% over 1914 so that, despite increased costs, fares had been kept low, averaging only one-third more than before the war, which compared with the main lines' increase of 50%.

As an outcome of the MPs' discussions, a Select Committee on Transport (Metropolitan Area) was appointed on May 29, 1919, under the chairmanship of W. Kennedy Jones, MP for Hornsey. Pick appeared before this committee and said that the Underground's policy was to 'seek large numbers of passengers at low rates of fares'. Working expenses were 47% of receipts in 1913 but had risen to 59% by 1918, necessitating the recent fare increase. The company would have no objection to the establishment of a London Traffic Board, and indeed had tried itself to form a pool with the LCC Tramways, but the negotiations (in 1915–1916) had broken down. The committee gleaned from other Underground witnesses that Sir Edgar Speyer was still a large shareholder,[1] and that the UERL, in its role as 'contractor' for the construction of the three tubes, had spent roundly £11 million in cash but received £15,220,000 in stock. The Select Committee reported in July, proposing a Greater London Traffic Authority covering the Metropolitan Police District.

In September 1919 a Ministry of Transport was formed, with Sir Eric Geddes, deputy general manager of the North Eastern Railway, as Minister. Very quickly, the Ministry set up an Advisory Committee on London Traffic, consisting of representatives of the Ministry, the London area county councils and the City of London. Like its predecessor, the committee recommended the formation of a Greater London Traffic Authority. This should be a permanent body under the Ministry of Transport with permanent members enjoying salaries higher than those of comparable civil servants. The Report (published in March 1920) noted that tube passengers had increased from 69 million in 1905 to 147 million in 1914 and 266 million in 1919; over the same period rides per head per year in London on local railways other than main-line companies' services had increased from 157 to 350. It was thought that some tube railway projects were required in the public interest, but Government or municipal assistance might be needed as new lines would not provide an adequate return on capital invested.

Sir Eric Geddes announced in November 1920 that a London Traffic Authority Bill was being drafted, but four months later he told the Commons that the Bill had been dropped, explaining rather weakly that the subject was highly contentious and there was need for economy in public funds. With some candour he added: 'I deplore my inability to deal with it... I cannot see how to do it.' Once again the unification of London's transport had been put off. At least the period of uncertainty had ended—the Underground men knew that the initiative was with them again.

During 1921, the number of unemployed rose to almost two millions and Lloyd George's Coalition Government produced a Trade Facilities Act intended to encourage works that would relieve unemployment

---

[1] Speyer Bros' shares (probably about £1,000,000) were purchased by S. B. Joel on behalf of Messrs Barnato Brothers at the end of 1919 (*Tramway & Railway World*, January 10, 1920).

by offering a Treasury Guarantee of capital and interest on the money raised. Stanley, now Baron Ashfield of Southwell,[1] saw his opportunity, and on October 26, 1921, presented to London MPs a £6 million development scheme to qualify for the Guarantee. His proposals were:

Modernization of the CSLR and its connection with the Hampstead line at Euston.

Extension of the Hampstead line to Edgware.

A junction between the CLR and the LSWR at Shepherd's Bush. (This was authorized in the Central London & Metropolitan District Companies' (Works) Act, 1920, and replaced the earlier and more expensive Shepherd's Bush–Gunnersbury scheme. It was intended to operate CLR trains to Richmond via Studland Road Junction and Turnham Green.)

250 new tube cars.

A third 15-MW generator for Lots Road.

He also suggested that a further £3 million could be spent in constructing the authorized District extension from Wimbledon to Sutton and in providing a new bus fleet. If the Government guaranteed interest on the £6 million, he hoped to raise the capital and provide two years' employment for 20,000 men. Ashfield included a proviso that his buses be secured against competition for ten years and the Government turned the scheme down as it was thought unlikely that Parliament would condone statutory protection against competition.

A fresh scheme was then submitted, omitting the bus-protection condition and the CLR-LSWR connection. This found favour and in March 1922 the Treasury agreed to guarantee the principal and interest on a loan of £5 million for fifty years. The capital was raised by a public issue of $4\frac{1}{2}$% debentures, which was fully subscribed, and a symbolic beginning of the new programme was made at Highfield Avenue, Golders Green, on June 12th, when the MP for Hendon turned the first sod of the Edgware extension.

Lord Ashfield could now see the way ahead, but he was seriously concerned that the competition by other bus operators might eat away LGOC profits and so endanger the health of the Common Fund.

CENTRAL TO EALING

Before describing the new works of the 1922 Programme, we must return to 1920 and record the extension of the CLR to Ealing Broadway. Henry Lovatt & Co. (Wolverhampton) had completed the Ealing & Shepherd's Bush Railway for the GWR and it was opened for freight traffic (via Kensington, Addison Road) on April 16, 1917. Stations and electrification were added in 1919–1920 and a public service of tube

[1] New Year Honours, 1920. On resigning the Presidency of the Board of Trade in 1919, Stanley had returned to the Underground group as chairman and managing director.

trains at a basic headway of ten minutes ran over the 4¼ miles between Wood Lane and Ealing from August 3, 1920, following a Directors' Inspection and formal luncheon on July 28th. East Acton, the only intermediate station, was of the wooden shack type; built to serve the LCC's Old Oak housing estate, it is in fact in the Metropolitan Borough of Hammersmith. Two other unpretentious stations, North Acton and West Acton,[1] were added on November 5, 1923. Earthworks included a 40-foot cutting at Victoria Road, Acton (a widening of the existing GWR cutting), and a 20-foot-high embankment near East Acton station. On the section between Wood Lane and North Acton, provision was made for two additional tracks.

There were two unusual features. The original CLR line from Shepherd's Bush station to Wood Lane, for depot access only, ran west from the station then sharply north in a single tunnel, and when the passenger service was extended to Wood Lane in 1908 it had been necessary for the new up line to be taken across the original tunnel in order to obtain an easier curve and form a loop with a station in Wood Lane. This meant that the tube lines north of Wood Lane had to be crossed back again to permit normal left-hand running with the GWR freight trains on the Ealing & Shepherd's Bush. A flyover was therefore built west of the Wood Lane road bridge at the point where the freight lines came in from the West London Railway. The second item of interest was the signalling, an early British installation of three-position upper-quadrant semaphores controlled by a.c. track circuits. Indications were: danger—horizontal (red light); caution—45° (yellow light) and all clear—vertical (green light). Caution was displayed after the train had cleared one section, and the vertical position was assumed when two sections ahead were cleared. The equipment, by the McKenzie, Holland & Westinghouse Power Signal Co., included train stops and fog repeaters. Automatically operated semaphore arms had pointed ends in accordance with American practice.

The extension was worked with the 1914 motor cars already described, plus two more, taken out of store in 1920, and standard CLR trailers. Six-car trains ran at rush hours, but west of Wood Lane three cars sufficed in the off-peak. Seven-car peak-hour trains were introduced in 1922 over the whole of the CLR, but those with 1903 motor cars proved to be under-powered and the six-car formation was resumed until 1926 when certain cars were re-motored.

Ealing now enjoyed a City route that was speedier than either the District or the GWR, and some relief was afforded to the former. As the Underground was at first obliged to levy full GWR fares on the extension, traffic was disappointing, but after a few years, housing estates and factories appeared and results were more encouraging. Power came

[1] Handsomely rebuilt in 1940 with a lofty glass-sided booking hall on the road overbridge.

from the GWR plant at Park Royal, which had been opened in 1906 to supply the Hammersmith & City Railway, and the substation was at Old Oak Common. In 1936, Park Royal closed and current was then taken from the Metropolitan Electric Supply Co.

EDGWARE'S TUBE

As the prewar powers had been kept alive, work on the Edgware extension was started as soon as the Treasury Guarantee was announced. The scheme was already twenty years old. As described in Chapter 5, the Edgware & Hampstead Railway Act had been passed in 1902. A year later, the Watford & Edgware Railway Act authorized a 6¼-mile extension of the E&H from Edgware to Watford High Street, just west of the Colne, following earlier schemes for steam lines between the two towns (1864 and 1897).

In 1907 and 1909 the Watford & Edgware Company came to Parliament seeking extensions of time, light railway powers and authority to amalgamate with the Edgware & Hampstead Company. The 1907 Bill was rejected, as Parliament saw no prospect of capital being raised, and the 1909 Bill was probably withdrawn because the LNWR route to Watford had become a practical possibility for a tube extension. The W&E powers finally expired in 1911, but the company was nominally alive for many years after that and some land on the route was retained. Eventually the London Passenger Transport Board registered the W&E as a limited company in 1934 (hitherto it had been a statutory company only) solely for the purpose of winding it up, and that was done early in 1935. Part of the W&E route was to have been used by the abortive Bushey Heath extension from Edgware as we shall see in later chapters.

As authorized by its 1902 and 1905 Acts, the Edgware & Hampstead line kept to the north-east side of Golders Green Road (then Hendon Road) until a point immediately south of Brent Bridge, where it crossed over that road and proceeded in a direct line to the Midland Railway at Colindeep Lane, passing north of Hendon Park. Beyond here, the route to Edgware was as later built. Deviation powers were obtained in 1909 and 1912 bringing the route south of Golders Green Road about ¼ mile west of Golders Green station. On July 1, 1912, under the LER Act of that year, the E&H was formally taken over by the LER and at last a serious beginning was made. The ten-year delay had nearly doubled the cost of the proposed line, as the Hampstead Railway had caused intensive development at Golders Green, and the land required in that area, now commanding high prices, was criss-crossed by new and projected roads. By 1913 most of the route to Edgware had been bought and fencing had begun when the outbreak of war once again deferred construction.

The contracts placed in 1922 were awarded to Charles Brand & Son

(Golders Green–Hendon) and the Foundation Company (Hendon–Edgware). By the summer of 1923, good progress had been made with the first section and the 1·67 miles to Hendon were opened on November 19th, after an official opening by Sir Philip Lloyd-Greame, President of the Board of Trade and MP for Hendon. The stations at Brent and Hendon Central were a pleasant attempt by S. A. Heaps to evolve a new suburban style. Green's ruby-tiled elevations were obviously unsuited to a semi-rural landscape and it was decided to provide something better than the shack stations until then usual for incipient suburbs. The booking halls were in sober Georgian style, afforded some dignity by Portland stone porticos of coupled Doric columns. Over all was a pyramidical roof of Italian tiles with wide eaves over the red and purple sand-faced brick walls. The spacious halls (Brent was 2,500 square feet, Hendon, 3,250) had oak passimeter ticket booths standing on a floor of large black and white non-slip tiles. Passimeters had first been installed at Kilburn Park, on the Bakerloo, on December 16, 1921, and after their adoption on the Underground were taken up by other railways, notably the LNER. Basically, they were a free-standing ticket-issuing booth equipped with automatic devices and other arrangements to facilitate ticket issue and eliminate the need for separate ticket collectors when traffic was light. An American invention, their name derived from the devices which counted or metered the passengers as they passed either side through flipper arm turnstiles situated beneath the ticket windows. At this time ticket examination on entry to platforms was generally punctilious and as the passimeter clerk 'cancelled' the ticket on issue, he replaced the inwards ticket inspector. As time went by, the word 'passimeter' came to be applied to any free-standing booking booth of the type described; of recent years, the metering devices have been dropped and at busy stations the number of booths has been reduced.

Both Brent and Hendon Central had 350-foot island platforms between the northbound and southbound lines, partially sheltered by wood and glass awnings supported by central pylons of lattice steel, in the manner of District stations built just before 1914. At Golders Green, the two-road station of 1907 [1] was enlarged by the addition of another road and by increasing the number of 370-foot platform faces from four to five. This arrangement permitted terminal working from either direction without interfering with the through service. To compensate for the loss of siding space caused by the reconstruction, a new siding was built in 1923 near the tunnel mouths, one end of it terminating in a blind tunnel alongside the running tunnels.

Leaving Golders Green, the new line crossed Finchley Road on an

---

[1] An additional entrance in Finchley Road for exchange traffic with the Metropolitan Electric Tramways (with a generous canopy extending over the whole width of the pavement) had been opened on December 18, 1911.

88-foot girder bridge, passing on to a brick viaduct through a district which had been almost covered by houses as early as 1912. After the viaduct, the line passed over Golders Green Road on another girder bridge, into cutting, and on to a viaduct into Brent station; in this section several new houses had to be demolished. The valley of the Brent was crossed on a brick viaduct 300 feet long and 24 feet high, followed by a climb of 1 in 50 to Hendon Central through a cutting planted with ornamental shrubs.

All-electric track-circuited signalling with train stops and two-indication colour light signals (the first of their kind on open sections of the Underground) was supplied by the Westinghouse Brake & Saxby Signal Co. Signal cabins were provided at the north ends of both Hendon and Brent stations, the latter to control passing loops to be built later.

A train service of about 3½-minute headway was operated from the start to encourage traffic development. Both gate and standard stock [1] ran at first, but gate stock was soon withdrawn north of Golders Green as wet weather was unkind to its non-metallic brake blocks. The 7·7 miles from Hendon to Charing Cross were covered in 25½ minutes. The fare was 6d. Hendon Central station was situated south of the old village and at first stood in lonely glory amid fields.

Miners working towards Hendon from Colin Deep dug twin tunnels at 60 feet a day beneath the Burroughs, using rotary excavators. The section under the Midland main line was tackled with pick and shovel, and the 11 foot 8¼ inch tubes were embedded in concrete at that point. About ¾ mile north-west of the tunnels, at Colindale, the line passed into a steep-sided cutting retained in concrete, and then continued without major earthworks to Edgware through prettily timbered countryside. Just before the terminus, the embankment of the GNR Edgware branch was pierced. Two girder road overbridges, at Colindale and Burnt Oak, contained provision for a third track, and land for this was taken throughout. At Edgware, the island platform and two tracks were completely roofed over and used for car storage at night. Alongside the station, a car shed 402 feet long and 69 feet wide contained four inspection pits and had accommodation for twenty-eight cars. East of this shed, four open sidings brought the total available storage space to seventy-six cars. The intermediate stations at Burnt Oak and Colindale had island platforms; Colindale resembled Brent and Hendon, but Burnt Oak, surrounded by fields, and connected to the main Edgware Road by a new 320-yard approach drive, had only a small wooden booking hut. Little or no traffic was handled at this point until the LCC developed its Watling Estate around the railway between 1926 and 1931, and a permanent station building in Heaps Georgian style was completed in August 1928. At street level Edgware had a graceful court-

[1] See pp. 187–9.

yard, 100 feet by 69 feet, flanked on three sides by the single-storey station building with its Doric columns and Italian-tiled roof. The west wing served as a shelter for bus passengers and a cycle store, the east contained shops and the main portion the entrance to the booking hall, flanked by small pavilions. The courtyard was to be a station for the feeder bus services, but immediate provision of these was hampered by the poor state of the surrounding country roads.

Signal cabins were sited at Colindale, for a reversing siding, and at Edgware, for a scissors crossover outside the station and the points into the car shed. Telephones were fitted to home signals and the usual tunnel telephone wires installed under the Burroughs. Signalling was identical with that of the first section.

A unique feature was the Burnt Oak substation, the largest automatic substation then existing in Britain and the very first for traction purposes. It contained two 1·2-MW rotary converters and space for a third, and was controlled from Golders Green. In December 1930, an additional substation was opened at Hendon, and this had the first steel tank rectifier installed for railway use in Britain. At first, the power came from Lots Road, but after the end of 1933, both substations were supplied from Neasden.

A building strike delayed the completion of the stations and they were not finished by the opening day, August 18, 1924. Burnt Oak could not be opened until October 27th.

The village of Edgware, comprising some thousand or so inhabitants, was now placed within thirty-five minutes of Charing Cross, with a train every ten minutes. From Colindale, there was a five-minute service to Charing Cross. During peak hours all trains ran to Moorgate, via the new Euston connection, every eight minutes, and a four-minute Charing Cross service was available by changing at Golders Green.

A NEW LIFE FOR THE VETERAN

The Edgware extension was not a mere projection of the old Charing Cross, Euston & Hampstead Railway, but part of a new system of tube railways formed by joining the Hampstead line at Camden Town to a modernized CSLR at Euston. Like the extension, the Camden Town Junctions and the modernization of the CSLR were financed by Government Guaranteed Stock and had long since been authorized by Parliament. Work therefore began in the summer of 1922, as soon as the Treasury Guarantee had been confirmed.

Burrowed by Mowlems' Welsh miners, the new junctions between Camden Town and Euston, CSLR, had no less than seventeen connections between new and existing tunnels, and 1,666 yards of new tunnel were made. The new line, in twin tubes of 11 feet 8¼ inches diameter (12 feet 6 inches on curves and 25 feet at junctions), began 81 yards west of the CSLR station tunnel at Euston, underneath the Hampstead

To Edgware.
CAMDEN TOWN
To Highgate
Signal Box

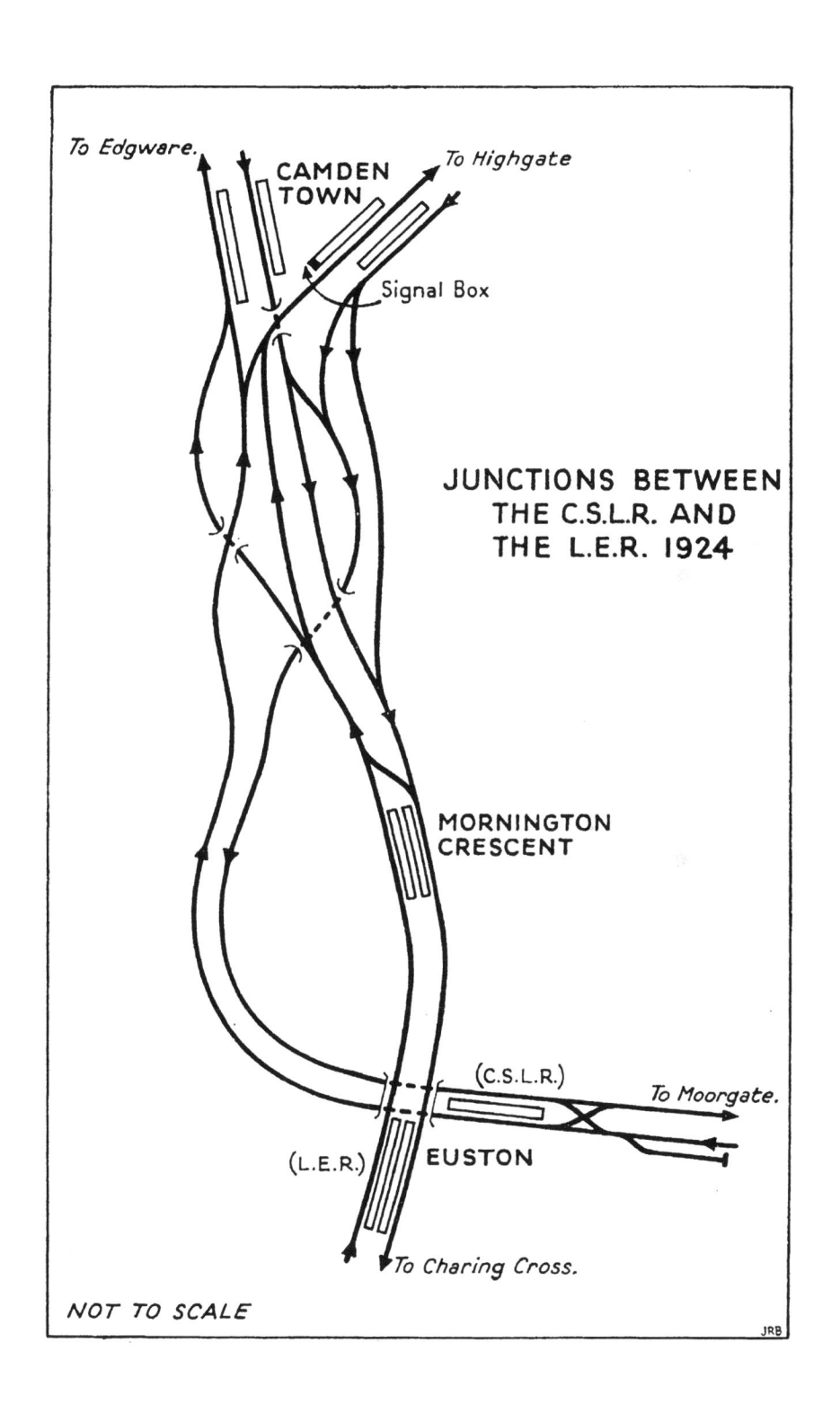

JUNCTIONS BETWEEN
THE C.S.L.R. AND
THE L.E.R. 1924

MORNINGTON CRESCENT

(C.S.L.R.)
To Moorgate.
(L.E.R.) EUSTON
To Charing Cross.

NOT TO SCALE

JRB

line, and ran to a point 1,330 yards north, where six tunnels now twisted and clustered like amorous worms, making junctions in such a way that trains could proceed to or from either the Edgware or the Highgate branch on to or off the CSLR without following any conflicting paths. The gradient became as steep as 1 in 40 uphill northbound and 1 in 30 downhill southbound. Six Greathead shields removed about 80,000 tons of spoil and the work was reached by three shafts, two at Mornington Crescent and one at Ampthill Square, where a pocket of waterlogged grey sand required compressed-air working for about 100 yards. Hampstead line traffic continued uninterrupted throughout the job, but the actual junctions between old and new tunnels were made in the night hours. A 48-lever frame (38 levers operative) was installed in the existing signal box at the south end of the Highgate platform at Camden Town with an illuminated diagram and train describers to enable the signalmen to set up the correct path for an approaching train. A power frame and diagram were also placed in the old CSLR box at Euston. The new connection was statutorily part of the LER, although for practical purposes it formed an extension of the CSLR to Camden Town.

Internal diameters of the old CSLR tunnels varied from 10 feet 2 inches (Borough junction–Elephant & Castle) to 10 feet 6 inches Euston–Moorgate and Elephant–Clapham Common) and 11 feet 6 inches (Moorgate–Borough junction). A successful experimental enlargement to standard dimensions (11 feet $8\frac{1}{4}$ inches on the straight) was made in 1914, and in 1919–1920 a short length of tunnel near Stockwell was enlarged to work out procedure with a shield. Most of the old segments in each ring were removed, the earth behind scooped out and the segments then replaced, with new segment pieces, the result being an irregularly shaped tube of approximately standard diameter. On the southern section the work was done with annular shields through which the tiny locomotive-hauled trains could pass, and some use was made of shields north of Moorgate. The 10 foot 2 inch section was rebuilt with entirely new segments throughout, and the 11 foot 6 inch stretch was enlarged by hand.

To allow the work to proceed at a good pace, the line was closed between Euston and Moorgate after traffic on August 8, 1922, and it was hoped that with twenty-four-hour occupation, completion of the enlargement and the Camden Town junctions would coincide. Another factor that influenced the decision to close this section was the need for compressed-air working near the Fleet River between Euston and King's Cross. Blue and cream LGOC buses ran every five minutes (three minutes in rush hours) over the closed route and offered through booking with Hampstead line stations north of Euston and all CSLR stations south of Moorgate. With the exceptions of Borough (closed July 16, 1922) and Kennington (closed May 31, 1923), the Moorgate–

Clapham Common section remained in use between 6.30 a.m. (8 a.m. Sundays) and 7.40 p.m. (first and last trains from either terminus).

The buses were withdrawn on Sundays from January 1, 1923, but extended to Clapham Common on weekdays at the same frequencies, operating as an ordinary bus route instead of calling only at CSLR stations. The latter development much peeved the LCC, who strengthened their tramway service along the Clapham Road to do battle with the intruder.

For about a fortnight in May 1923 the line south of Oval was closed, and when it was reopened on May 15th, weekday trains ran only between 7 a.m. and 7.40 p.m. Compressed-air working was necessary in water-bearing strata between Oval and Stockwell where the line passed below the Effra Brook, and during the summer and autumn of 1923 temporary crossover tunnels were constructed at South Island Place and Portland Place. Using these, single-line working started from October 22nd on one or more of the short sections thus created, whilst the engineers used compressed air in the parallel tunnel. Sunday trains were entirely withdrawn after October 7th and replaced by buses.

The brave attempt to run a nearly normal train service on the southern section whilst work proceeded came to grief on November 27, 1923. On that evening, a northbound train struck some poling boards projecting from the tunnel roof just north of the Elephant, beneath Newington Causeway. The driver stopped his train as 2 cwt. of sand and ballast cascaded on to the last two cars, pouring through a gap where two segments had been removed. With commendable presence of mind, he removed the loose pieces of timber from the track, pulled forward beyond the gap to make a quick assessment of the damage, and then resumed his journey, coasting into Borough station after current had been cut off by a short-circuit. Meanwhile a further 500 cubic yards of ballast tumbled into the tunnel, filling it completely for a length of 43 feet. Above, there was a large subsidence in the roadway which had exploded a gas main as it had formed. Shuttle services were worked on either side of the blocked section for the rest of the evening and on the following day, but the whole line was completely closed after traffic on November 28, 1923, until the modernization was completed. To meet the extra load, the bus service was increased to a one-minute headway. The Ministry of Transport's Inspecting Officer of Railways conducted an accident Inquiry and found that greater care could have been exercised in view of the known nature of the subsoil and the fact that trains were running. He recommended that in future, tube tunnelling in waterlogged strata should not be disturbed by the running of trains.

The northern section was reopened to public traffic on April 20, 1924 (Easter Sunday), and through trains ran from Moorgate to Hendon, using the new Camden Town junctions. City Road station, which had closed with the other northern section stations after traffic on August 8, 1922, was not reopened, as traffic was insufficient to justify the expense of reconditioning the station and extending the platforms.

The southern part of the line came back to life on the morning of December 1, 1924, when the Hon. Marian Stanley, Lord Ashfield's daughter, drove a special train before public service began in the afternoon. Trains of the new standard stock provided a peak-hour service between Clapham Common and Edgware (52 minutes, fare 10d) and a basic service between Clapham Common and Camden Town, whence half of the twenty-six trains an hour ran to Highgate and half to Golders Green. To mark the occasion, the CSLR was officially renamed 'The City Railway'. Overall speed was $3\frac{1}{2}$ m.p.h. faster than the old CSLR, with a saving of $6\frac{1}{2}$ minutes in running time between Clapham Common and Euston.

The 44 CSLR locomotives and 161 trailers were withdrawn, many of the cars finding a new lease of life as contractor's huts, poultry houses or garden sheds, at least two surviving (at Watlington and Hampton Court) until the present day.[1]

Apart from the enlargement of the running tunnels and the provision of new rolling stock, a number of stations were modernized. Not all this work had been completed by December 1, 1924, but the general plan was to equip all but seven of the stations with escalators.[2] All platforms throughout the line were lengthened to 350 or 360 feet, and station tunnels were re-tiled. Many detail improvements were made, and the opportunity was taken to carry out some much needed realignments, notably at Kennington and Elephant & Castle.

At Clapham Common, an entirely new station was built above rail level ready for the reopening. It had a passimeter booking hall, 90 feet by 40 feet, excavated below the Clock Tower with a cream terra cotta rotunda above the stairwell as the only surface structure. Two 83-foot-

---

[1] Two locomotives and one trailer were kept for museum purposes. One locomotive, numbered 1, but not the original no. 1, went to the Science Museum, South Kensington, on April 25, 1923, and is still exhibited there. Locomotive no. 36 (Crompton, 1900) was reconditioned in 1938 after a long sojourn at Lillie Bridge, and was then placed on exhibition at Moorgate (Metropolitan) station. It was damaged by a bomb in 1940 and has since been dismantled, certain parts being retained by Messrs Crompton Parkinson Ltd. A 'Padded cell' trailer, no. 10, gathered dust at Lillie Bridge sheds for many years, but was saved from destruction in 1937 by the efforts of Mr V. Boyd Carpenter. After reconditioning, it was sent to York Railway Museum in January 1938. The CSLR steam locomotive, an 0-4-2ST built in 1922, survived as a works engine and became London Transport L.34.

[2] Mostly for site reasons, lifts were retained at Euston, King's Cross, Angel, London Bridge, Borough, Elephant & Castle, and Kennington. Bank retained most of its lifts, but three escalators at the connected CLR station were opened on May 7, 1924.

long Waygood-Otis escalators, either side of fixed stairs, had a rise of 41 feet. The ascending escalator was the first installation in Britain of the new cleat-comb, straight forward step-on and step-off type. Clapham Road (which became Clapham North on September 13, 1926) was also given a passimeter booking hall and two escalators, but was not finished until May 29, 1926. Stockwell depot became a working site and the hydraulic lift was used for removing spoil. The old station building was completely demolished and the new structure was ready for the reopening. New platform tunnels of standard LER type replaced the old single tunnel, with its island platform. Two reversible Waygood-Otis escalators, flanking fixed stairs, led up to a 3,000 square feet passimeter-equipped booking hall, whose exterior was clothed in cream terracotta with black dressings. Oval had a new booking hall, as at Stockwell, and two escalators (in service May 29, 1926). Kennington was a working site and operations continued from there for some months after the reopening whilst the junctions with the new Charing Cross link were completed. Eventually reopened on July 6, 1925, this station retained its original lift shafts, booking hall and dome, but the usual passimeter was installed and the walls were re-tiled. At Elephant & Castle, a new façade in the 'Stockwell' style replaced the old. Borough was another working site and was not reopened until February 23, 1925; it had a new booking hall, resembling the others. At London Bridge, both sets of lifts were modernized, but the original buildings remained unaltered apart from the installation of the passimeters. Work here was not completed until October 1927. Alterations at Bank were few, but Moorgate was given two new escalators (one reversible) of 53-foot rise, and a new 3,575 square feet booking hall below street level; the latter was opened on July 2, 1924, the escalators on the following afternoon. Old Street was provided with a brace of escalators and a new surface booking hall, all opened on August 19, 1925. Much work was done at the Angel, where a fine new passimeter booking hall was completed early in 1924. High-speed lifts were installed, reducing the street to platform time to ninety seconds. These lifts were controlled at slack periods by the booking clerk, using buttons in the passimeter booth. The original island platform was lengthened to 350 feet and retained, as were those at Euston, Clapham Road and Clapham Common. Very little change was made at either Euston or King's Cross.

Standard LER track and signalling were fitted. The CSLR had been resignalled with a.c. track circuit automatic signals (without train stops) in 1919–1921 in preparation; new power frames of LER type were now fitted at Euston, Angel, Moorgate, London Bridge, Stockwell and Clapham Common (London Bridge and Stockwell were normally switched to automatic), and the remainder of the sixteen CSLR signal boxes were closed.

In its first year, ending December 1925, the rebuilt CSLR began to earn its working expenses and charges in respect of old capital, even making some contribution to the interest on the new capital. Traffic increased by 32% between the first and last months of the year.

A SURREY STRUGGLE

As so much capital had been sunk into the CSLR reconstruction and the new junctions, it would have been unrealistic to imagine that the southern terminus would remain at a point so near to Central London. The planners of Electric Railway House saw no reason why the fresh green meadows of Surrey should not be made to yield a traffic harvest to rival that they hoped to gather from those of Middlesex.

In the autumn of 1922, a CSLR Bill was deposited, for an extension to Morden, where the tube railway would have a junction with the authorized, but still unbuilt, Wimbledon & Sutton Railway. In the same session, an LER Bill provided for an extension of the Hampstead from Charing Cross to meet the CSLR at Kennington. South of Waterloo, this new link was to have a series of Camden Town-like junctions with the Bakerloo. Heady with dreams of tube trains from Sutton to Watford and from Sutton to Edgware or Highgate via either Charing Cross or Bank, Lord Ashfield despatched this little bundle to Parliament with hardly a glance over his shoulder at the shadow of Waterloo. In 1923, the Underground issued a map showing the new lines to Sutton, through an area which the *Railway Magazine* of February 1923 described as a 'district somewhat inadequately served by railways'.

The Wimbledon & Sutton project was within the Underground family. Originally authorized to private promoters in 1910, its powers were obtained by the District (which had been back stage) in 1913. Land acquisition, alterations at Wimbledon station and fencing were all begun before the war; nothing more had yet been done, though the powers had been kept alive.

As soon as the 1922 proposals were published, a great furore arose at Waterloo, and early in 1923 Sir Hugh Drummond, the LSWR chairman, told shareholders that it was 'a very serious matter'—the new lines would form a wedge into their hallowed territory. Sir Herbert Walker went to Parliament to defend the rights of the new Southern Railway, and said they would raise no objection if the tube extension were terminated at Tooting, but anything further would be regarded as an invasion of their territory as defined in the Grouping Agreement. He thought the threatened competition all the more regrettable in the light of the previous co-operation between the LSWR and the Underground group. If the SR were allowed to build the Wimbledon & Sutton, they would grant the tube running powers into Wimbledon and Sutton via the Tooting & Wimbledon line. Sir Herbert thought that had the Wimbledon & Sutton been built before 1914, it would have gone to the Southern at the Grouping.

Early in July 1923, the Lords Committee rejected the whole CSLR extension, but by the time the Committee met again at the end of the month, the rivals had achieved a compromise. The SR had agreed to tube extension to Morden, within an easy stone's-throw of the Wimbledon & Sutton alignment, but with no junction, and in return, the W&S powers were to pass to the Southern, who would in addition to building this line, also reopen and electrify their Streatham, Tooting & Wimbledon section, closed to passengers since December 31, 1916.

An SR Bill for the W&S was now prepared, including powers for connection with the Southern lines at either end. The District was to have running powers, and the SR was not to oppose any future connections between the W&S and the District at Wimbledon, or the provision of suitable terminal accommodation for the District at Sutton should these facilities ever be required. With the 1924 Act, the W&S passed to the Southern, who constructed it as a third-rail electric line and opened it from Wimbledon to South Merton on July 7, 1929, and throughout on January 5, 1930.

When announcing the compromise, Counsel for the SR stated that it had been made at considerable sacrifice to their own interests; this was proved after the tube extension was opened, when the SR discovered that in 1927 the new line had deprived them of some four million passengers. Considering the issue in isolation, one wonders what prompted the Southern to accede to such a brutal intrusion into their suburban area after they had succeeded in having the Bill rejected. Possibly they obtained a secret assurance from Ashfield—'this is the last territorial claim I have to make'.

The LER Bill of 1923 was also opposed by Waterloo. Of the four proposed link lines between the Bakerloo and the Hampstead, the SR objected to one which passed beneath their terminus (the connection from the Bakerloo towards Kennington) as it would interfere with their own proposal for a deep-level terminal loop for Western Section suburban trains. The LER withdrew the offending line, thus considerably restricting the scope of the junctions (there would have been little point in sending Hampstead trains to the terminus at Elephant & Castle without a corresponding diversion of Bakerloo trains to Kennington and beyond). So it was that South London's 'Camden Town' never came about. The authorized works (Charing Cross to Kennington and Clapham Common to Morden) received Treasury blessing under the Trade Facilities Acts.

CHARING CROSS TO MORDEN

Work on the new tunnels between Charing Cross and Kennington started on April 22, 1924—the first tube railway construction in central

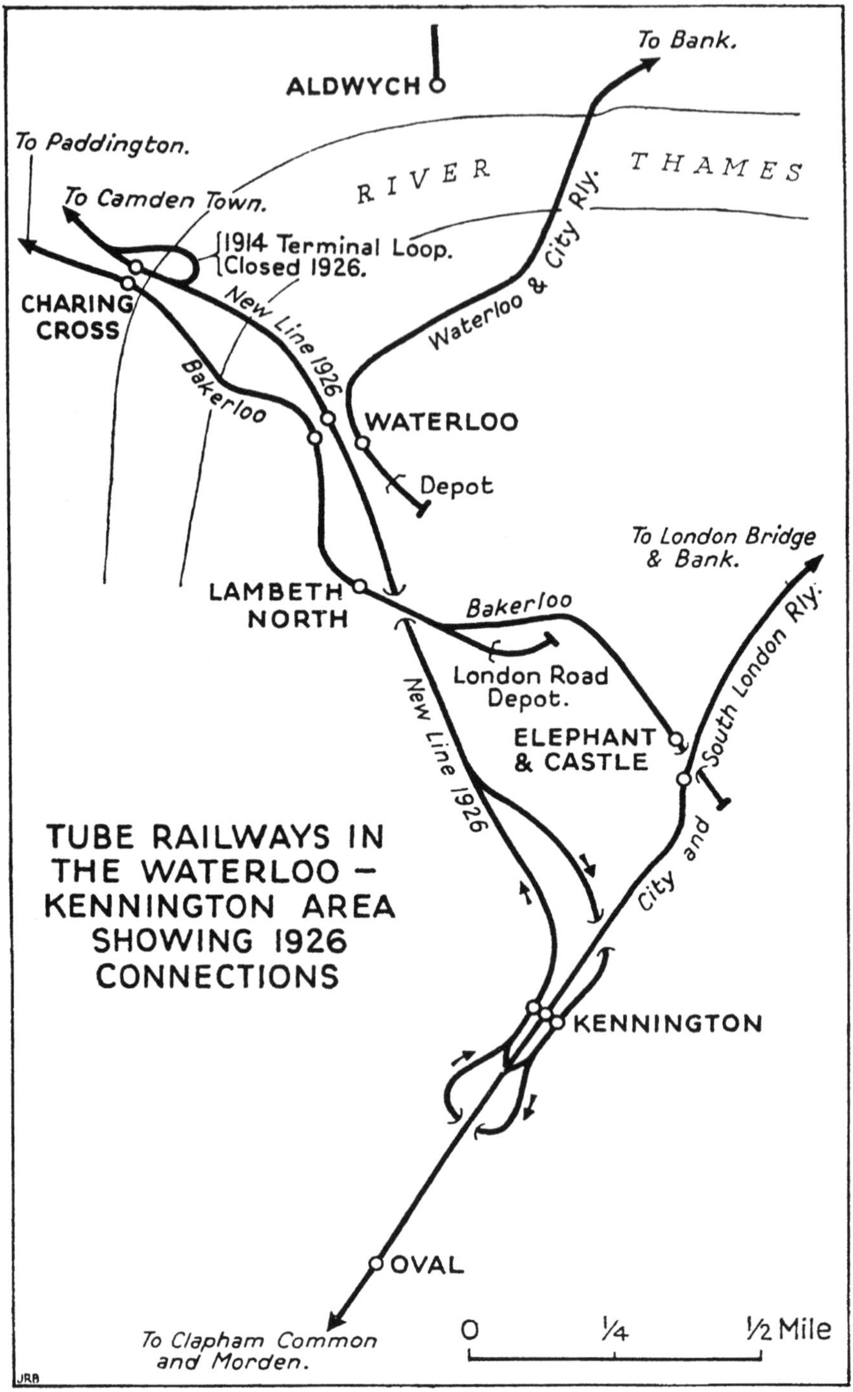

ALDWYCH
To Bank.
To Paddington.
RIVER THAMES
To Camden Town.
1914 Terminal Loop. Closed 1926.
CHARING CROSS
New Line 1926
Waterloo & City Rly.
Bakerloo
WATERLOO
Depot
LAMBETH NORTH
To London Bridge & Bank.
Bakerloo
London Road Depot.
New Line 1926
ELEPHANT & CASTLE
City and South London Rly.
TUBE RAILWAYS IN THE WATERLOO – KENNINGTON AREA SHOWING 1926 CONNECTIONS
KENNINGTON
OVAL
To Clapham Common and Morden.
0
¼
½ Mile
JRB

London for ten years. The eastern (or southbound) tunnel began under Villiers Street, leaving the terminal loop at its neck, and the western tunnel broke off from the south end of the single-platform Charing Cross station at the point where the terminal loop and the station tunnel met. The southbound tube almost at once opened out to 21 feet 2½ inches for the new southbound platform; it then proceeded under the river bed, piercing the loop at an oblique angle. At the points of junction and crossing where old and new tubes came together, the old tubes were plugged with concrete and work went on under compressed air. Whilst this was going forward, Highgate trains were terminated at Strand and only the Golders Green and Edgware trains used the single platform at Charing Cross, running through the northbound tunnel in both directions. This diversion began on January 25, 1926. In the same month, Mowlems started to scoop out a new circulating area for the tube station immediately below the District lines at Charing Cross.

Under the river, the twin tubes ran parallel to Hungerford Bridge, and then turned south into a new station at Waterloo adjacent to the Bakerloo platforms. A shunting neck was proposed between the new running tunnels at Waterloo but was never built. Beyond, the new line took a direct route towards Kennington, passing below the Bethlem Hospital (now the War Museum) and the Lambeth Workhouse (now Lambeth Hospital). Just before reaching the CSLR, the tunnels diverged, the southbound one passing under the City line, curving south, and then rising parallel to it into Kennington station, where cross-platform interchange was available with the southbound trains from Bank. The northbound tunnel also curved south until parallel with the northbound City line, and in Kennington station, the parallel northbound platforms afforded convenient interchange. Kennington station now had four platforms more or less at the same level. South of the station, the new tunnels diverged (after throwing off connections to each of the City tunnels) and descending, curved to meet each other beneath the City line—a large terminal loop stretching as far south as the edge of Kennington Park. A new shunting neck was built between the City line tunnels south of Kennington.

The four Bakerloo lifts at Waterloo were supplemented by three new escalators in a shaft that had to be built in water-bearing strata under compressed-air, with the additional complication of elaborate underpinning of the main-line station. The first of the escalators was not ready until July 29, 1927, the other two on October 12th. At the top landing, beneath the main-line concourse, a new passimeter booking hall was constructed. From this, passengers reached the Southern Railway by using escalators installed by the LSWR in 1919 to give improved access to the W&C. The Kennington loop also required compressed-air working.

CONNECTIONS AT KENNINGTON 1926

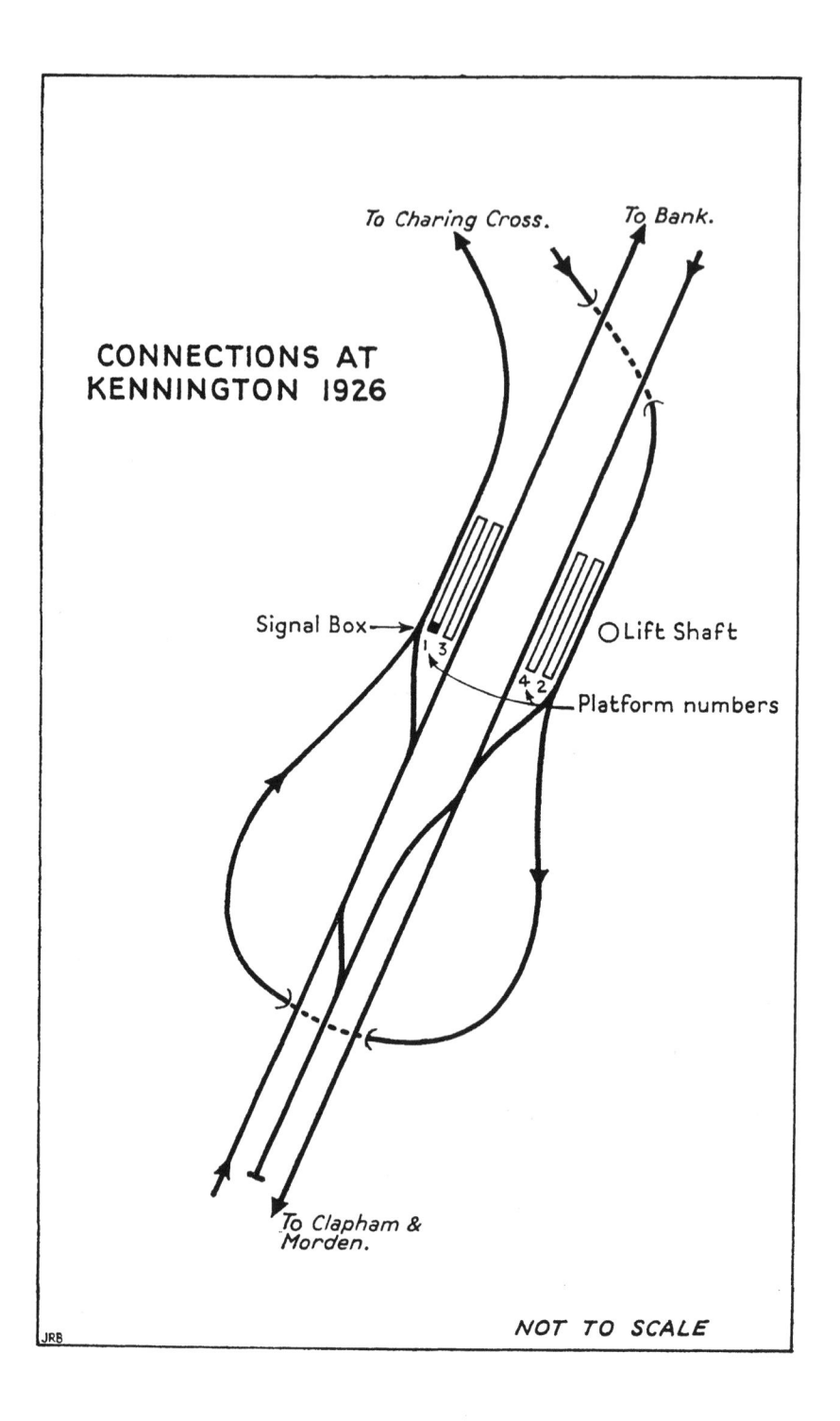

To Charing Cross.
To Bank.
Signal Box
Lift Shaft
Platform numbers
To Clapham & Morden.

NOT TO SCALE

At Kennington, a new 31-lever power locking frame was fitted in a cabin at the south end of the northbound Charing Cross line platform; this controlled the crossover roads between the Charing Cross and City lines and the crossovers for the shunting neck south of the station. All platforms on the new section were 350 foot. The distance between Waterloo and Kennington, 1·19 miles, was the longest between two tunnel stations at this time. Power came from Lots Road through a new substation at Lambeth North, which was equipped with two 1,500-kW rotary converters and remotely controlled from Charing Cross. Trains began to run between Charing Cross and Kennington on September 13, 1926, the same day as the Morden extension opening.

The first sod of the Surrey extension was cut at Nightingale Lane, Clapham Common, on the last day of 1923, and the first shafts were sunk at that point and at Trinty Road, Upper Tooting, during the following spring. The tunnelling contracts for Clapham to Tooting Broadway were awarded to Charles Brand & Son, and those for the remaining section to the Foundation Co. Ltd. By the end of the summer, work was proceeding in earnest, ten rotary excavators pushing forward at the rate of 1,400 feet a week. Bad weather in the winter of 1924–1925 delayed the construction of the car sheds at Morden, and work on the station buildings was hindered by labour troubles. Much difficulty arose between Tooting Broadway and Colliers Wood, and at Tooting Broadway, Balham and Merton owing to the presence of waterlogged strata. Tooting Broadway station was built in Woolwich Beds and pockets of water-saturated sand up to 20 feet deep—the site, which could not be changed, as it was an important traffic centre, was virtually an underground pond.

There were seven stations, about $\frac{3}{4}$ mile apart; Clapham South, Balham, Trinity Road (Tooting Bec),[1] Tooting Broadway, Colliers Wood, South Wimbledon (Merton) and Morden. The closeness of the stations reduced the speed of operation and increased running costs without adding much to revenue (the streets above were well served by buses and trams). The now familiar confusion of Underground geography was again apparent; 'South Wimbledon' was not in Wimbledon, but in Merton High Street.

Influenced by contemporary European styles, the architecture, by Adams, Holden & Pearson was angular, uncompromising and *modernistic*. The aim was to make the stations stand out from other buildings in the thickly populated areas through which the line passed on its way to the fields at Morden. To this end, the stark elevations of reinforced concrete and Portland stone were arranged to obtain the maximum benefit from floodlighting. Over the entrances, there were very large windows incorporating the bar and circle symbol, flanked by flagstaffs bearing models of the symbol; the canopies had fascias of blue enamel

[1] Renamed Tooting Bec, October 1, 1950.

and bronze. At night, floodlights concealed in the canopy bathed the frontage with a white glare. From December 1926, searchlights were installed on the station roofs to shine upwards at 45° and draw attention to the new line.

Morden's platforms were in the open air, just below street level, but the intermediate stations each had two escalators with fixed stairs between, leading straight down from passimeter booking halls, all of which were at street level except Balham and Trinity Road where the halls were under the road. A third escalator was provided at Tooting Broadway in October 1928. Tunnel stations had 350-foot platforms approached by decelerating inclines of 1 in 60 for 600 feet with accelerating dips of 1 in 30 for 300 feet at the other end. Running and station tunnels were of the now standard dimensions of 11 feet $8\frac{1}{4}$ inches and 21 feet $2\frac{1}{2}$ inches, and the average depth below street was 40 feet. At Morden, the line was in a concrete-walled cutting containing three 400-foot platform roads (five faces) roofed by a 100-foot arched span. Stairs led up to a large octagonal, top-lighted booking hall through which the passenger passed out to a 1,600-square yard forecourt bus station. Below this forecourt and the London–Epsom Road, the railway continued for $\frac{1}{4}$ mile to reach the car sheds. These had five 426-foot bays and were 300 feet wide. The twenty roads accommodated 140 cars, all over inspection pits. There was open siding space for another 41 cars on the east of the sheds and for 69 more on the west. At the rear, there were overhaul shops, stores and offices.

The tubes ended a short distance north of Morden station, but before reaching the open air, the trains passed through reinforced-concrete cut-and-cover tunnels. This last section, south of Dorset Road, was to have been open cutting, but tunnel construction was finally chosen owing to the presence of water near the surface.

Power at 11 kV, 50 c/s, 3 phase a.c., was taken from the County of London Electricity Supply Company through a switch house at Colliers Wood into substations at South Wimbledon (three 1,200-kW rotaries) and Balham (automatic; three 1,500-kW rotaries), which supplied the conductor rails at 600 volts d.c.

Track was the standard 95-lb. bullhead rail, with fibre pads inserted between the chairs and the Jarrah wood sleepers. Clapham Common signal box was converted to automatic working and a 15-lever power frame was installed at Tooting Broadway for working the 425-foot shunting neck south of that station. A 31-lever power frame was fitted in a cabin at the north end of Morden station to operate all points and signals in the area except the scissors crossover and points leading to the shunting necks near the depot, which were controlled from a mechanical frame nearby. Other signalling was of the standard automatic pattern with train stops.

Normal train service on the extension was $4\frac{1}{2}$ minutes Morden–

Golders Green via Bank, but in rush hours there were additional trains to Highgate via Charing Cross. In the evenings, a six-minute service ran to Euston via Bank, half of these trains continuing to Golders Green; this became a four-minute service at theatre time. North of Tooting there was a joint 2½-minute service to Euston and a 9½-minute service to Highgate, all via Bank, plus some Highgate trains via Charing Cross. Further Charing Cross trains were available by changing at Kennington.

Some 15,000 free tickets were issued to local residents for use after 3 p.m. on the opening day, to or from Bank or Leicester Square. Lieut-Colonel J. T. Moore-Brabazon, MC, MP (now Lord Brabazon of Tara), formally opened the extension on the morning of September 13, 1926, in his capacity as Parliamentary Secretary to the Minister of Transport. He drove the train from Clapham South to Morden, and joined the other guests at a lunch in the car sheds, taken at tables decorated with red and white carnations. In his speech, he described the extension as a 'Test Tube'—if it were well supported and earned the dividends required, it would be possible to build further extensions; there was no question of a Government subsidy for the tubes—they could not see their way to use money raised by Imperial taxation for the benefit of one city, and transport facilities must pay for themselves. Lord Ashfield, wistfully glancing through the doors at the open meadows, said that the extension needed 14 million passengers a year if it were to pay; the postwar programme founded on Government guaranteed debentures was at an end [1] and no more money was available for new construction.

The labour troubles of 1926 had delayed the opening well beyond the original target of July, and Balham station was not ready for the opening day—it came into use on December 6, 1926.

A poster advertized the new line as 'London's traffic puzzle—another piece put right', and showed a jigsaw puzzle map of London with the south-west sector completed (the south-east part was disarranged, and the map wisely showed no further north than Gillespie Road on the Piccadilly line). Naming the line caused some difficulty, and after correspondents to *The Times* had suggested such monstrous tags as *Edgmor Line*, *Medgeway*, *Mordenware Line* and the *Edgmorden Line*, the two sections were kept separate and officially known as the *Hampstead and Highgate Line* and the *City Railway*. Some years later, the clumsy title, *Edgware, Highgate and Morden Line* was used.

A new enterprise, Morden Station Garage Ltd, a subsidiary company of the Underground Group, opened a 7,600-square foot garage opposite the station on January 31, 1927. This building, in Underground car shed style, was designed to house motor cars and cycles for those

[1] One major work did in fact remain unfinished at this time. This was the vast new station at Piccadilly Circus, to be described in the next chapter.

who wished to drive in to the station from the surrounding area. Single-deck buses ran from the station forecourt to gather traffic from Worcester Park, Cheam, Burgh Heath, Wallington, Sutton, Mitcham and Banstead. Existing services were already linking Morden with Epsom and Leatherhead. Clustered around the new terminus were three streets of cottages, three or four large villas and the Crown Inn. All around, as far as the eye could see, were fields and parklands.

NEW ROLLING STOCK

The first postwar cars to be delivered to the tube railways were those designed for the Watford–Bakerloo service. This LER-LNWR Joint Stock had been ordered from the Metropolitan Carriage, Wagon & Finance Co. in 1914, but that firm had grimmer things to make in the four years that followed. Delivery began in March 1920, and the cars replaced the borrowed CLR stock on the Elephant & Castle–Watford trains. Specially designed for the long run out to Watford, the new cars were formed into 12 six-car trains in the sequence M-CT; M-T-T-M, a motor car being allowed in the middle of a small loading gauge tube train for the first time. Four-car trains ran outside peak hours.

The Watford Joint stock was considerably heavier than the older stock, as more substantial steel plate had been used in an effort to reduce vibration at speed. A six-car train weighed 161 tons 2 cwt. empty. The motor cars had thirty-six seats and the trailers forty-eight, all upholstered in red and black rep, and mostly transverse. There were single doors at the ends and centres of each car, and these swung inwards under the eye of the gateman. When the last passengers had entered and the centre doors had closed themselves for the last time, the gatemen locked them remotely from the end platforms and unlocked them at the next station as they opened their end doors. A red light indicated to the gateman that the centre doors were open and this changed to green as soon as they had swung closed. This door system did not work very happily and some unpleasant accidents brought it to the notice of the Ministry of Transport. The floor of the cars was $4\frac{1}{2}$ inches higher than that of the original tube stock and its level was a compromise between the heights of the tube and LNWR suburban platforms. The extra height allowed wheels of 42- and 32-inch diameter instead of the previous 36 and 30.

Each motor car had two BTH GE 212 240-h.p. motors, and the new trains could reach 45 m.p.h. without much difficulty. The automatic relay multiple-unit control of BTH design was similiar to that fitted to the Ealing cars.

Inside, the new stock was light and pleasing, with no clutter of straps (the standing passengers had to support themselves as best they could on small handles at the corners of the seats, or against the stanchions of the draught screens). The colour scheme was dark cream and white, and there was ample light from opal ceiling bowls and globe side lights between the large windows. Electric heaters were fitted. A final touch was provided by the neat hat and parcel racks over the windows, which added a subtle 'long distance' flavour. External livery was standard LNWR; chocolate below waist, white above, with white arch roofs.

Like most cars built as a compromise for both urban and suburban traffic, the Joint stock was not wholly successful. In 1931, when the cars had almost reached the end of their effective working life, a *Railway Gazette* reader timed their evening peak stops on the congested section between Trafalgar Square and Baker Street. At Oxford Circus the longest stop of an air-door train was 41 seconds, but one Joint stock train lingered as long as 87 seconds (a reasonable maximum station time for the tube railways is 30 seconds). The lavish use of staff along the train was a further disadvantage.

In 1919, twenty trailers and twenty control trailers were ordered for the Piccadilly Line. Built by Cammell Laird & Co., these cars were the first tube stock to have air doors throughout. The forty-four longitudinal seats in each car were covered with 'Pegamoid' leatherette. Bodies were all-steel arch roofed, with centre double sliding doors separated by a pillar, and single door at the ends. A feature of the stock, which bore a remarkable internal similarity to the contemporary District 'F' stock, was a complete absence of straps—standing passengers were given vertical and horizontal stanchions. The air doors had sensitive rubber edges, and when depressed by contact, a clutch operated, reversing the action of the door. So, if your friend was dallying, or you were just feeling good-natured towards a pretty stranger, you banged on the rubber edge and the doors obligingly swung open again. Not surprisingly, this system was soon abandoned. Air doors reduced the train staff to three—driver and front and rear guards: the latter had to watch for little yellow semaphore arms which stuck out by the centre doors of each car, for when all these were out of sight, the doors were all closed and the starting bell could be rung. Short-sightedly designed for tunnel operation only, with simple trucks and bodies inadequately weatherproofed, these cars had a brief working life and all were out of passenger service by 1939.

The new trailers enabled the running of six-car trains on the Piccadilly to relieve peak-hour overcrowding; this was done by converting twenty French gate stock motor cars to air-door operation so that they could haul the new trailers. All ten six-car trains were in service by the middle of 1923, increasing the line's seating capacity by 1,760. Displaced gate stock was drafted to the Hampstead and Bakerloo lines to make up further six-car trains.

A large new fleet was required to replace the locomotive-hauled trains of the CSLR and to work the new extensions to Edgware and Morden. This was to become the 'standard' tube stock for many years, and cars were ordered in batches between 1923 and 1934 with no substantial variations in design.

Before the new fleet was ordered, five railway-car builders were invited to design specimen 48-seat trailers on a basic 'shell' of standard dimensions, with two sets of double 4 foot 6 inch air doors either side. In addition, a 44-seat control trailer was built by Gloucester to the design of Underground staff. Attention was given to noise reduction, and the scientist A. M. Low produced some curious cowlings for the wheels and also suggested smaller windows and sound insulation in the bodywork. The car builders' design staff showed plenty of imagination in arranging the interiors of the five cars, and very lavish upholstery and fittings were provided on a scale not seen before or since on tube cars. The Leeds Forge, Cammell Laird and Birmingham cars were especially attractive; the MC&W car had a fluted rubber floor and the Cammell Laird was distinguished by diffused trough lighting requiring twice as many bulbs as normal. The Underground's own car was more austere and, not surprisingly, was the design chosen for the whole fleet. The six cars were ready at the beginning of 1923 and were shown to the Press on the Piccadilly line on February 3rd. In August they were moved by road to Golders Green and put to work on the Hamstead line.

The first orders for the standard cars were placed in March 1923—191 cars (81 motor cars, 35 control trailers and 75 trailers) from the Birmingham, MC&W and Cammell Laird companies at a total cost of £710,000. Whilst not exactly silent, the new cars were a great improvement on the gate stock, and the *Tramway and Railway World* graciously noted that 'conversation in normal tones is now possible'. On the trailers there were two sets of double air doors on each side, each 4 feet 6 inches wide. They lacked the sensitive edge of Cammell Laird stock, but the closing of the last 6 inches was at a pressure low enough to cause no injury and allow the doors to be forced apart to remove clothing or limbs. The motor cars had a set of double doors, with a centre pillar leaving an effective opening of 5 feet 7 inches, supplemented by single swing doors at the far end for the guard, whose compartment was separated from the main saloon by an arched bulkhead. The sprung seats (30 in the motors, 48 in the trailers and 44 in the control trailers) were covered in grey and dark brown moquette of a lozenge-pattern, a style also used extensively in the Underground group's road vehicles for many years. Eight pairs of transverse seats were fitted in each trailer, and the motor cars had four pairs in addition to the usual longitudinal seats. Lighting was of a much higher standard than hitherto and shades induced a restful atmosphere. The steel sides were lined with asbestos facings to reduce tunnel noise. Roofs had a clerestory interrupted by arched sections over the doors and punctuated by air scoops either side. Livery was a smart Post Office red below a black waist band and a yellowy cream above it; the roof gutter had a black edge and the edges of the window posts were also black; doors and cabs were red all over. The interior panels were cream above the waist and dark green below, whilst the woodwork was a dark mahogany shade.

Front bogies of the motor cars carried two nose-suspended 240-h.p. motors driving the 3 foot 4 inch wheels through reduction gears. All bogies were of the plate-frame type, and car frames were raised over the motor trucks which had a wheel base of 6 feet 11 inches. Trailer bogies had 2 foot 8 inch wheels on a 5 foot 7 inch base. Immediately behind the cab and over the motor bogie was a 10 foot 8 inch long control compartment with the equipment either side of a central gangway. The all-electric equipment included automatic acceleration. Brakes were Westinghouse pneumatic (later converted to electro-pneumatic). As was by now usual, automatic electro-magnetic contractor control was fitted.

The Hampstead line received the first deliveries late in 1923, and in May 1924 a further order for 52 motor cars, 50 trailers and 25 control trailers was placed with the MC&W, Birmingham, and Cammell Laird companies. At the end of 1925, 48 motor cars were ordered from Cammell Laird and 5 trailers and 67 control trailers from MC&W. All cars ordered between 1923 and 1925 had recessed windows and thick waist bands, forming a type group with only insignificant variations.

ACTON WORKS

After the war, with large arrears of maintenance and prospects of a bigger car fleet, it was decided to establish a central repair depot to undertake heavy overhauls and replace the local facilities of each line at Golders Green, Lillie Bridge, Ealing Common, Shepherd's Bush, Stockwell and London Road. To this end, some market gardens just south of Acton Town station were purchased in 1920 and a Works on this site was opened in December 1922. The new shops took District, Watford Joint, Bakerloo [1] and Piccadilly cars for periodic overhauls, and in 1924 500 men were handling sixteen cars a week. The layout was arranged so that cars passed progressively through the works from stripping to final repainting in the most efficient and economical fashion. Between 1925 and 1927, extensions were made, capacity increased to forty cars a week, and the progressive handling scheme was speeded up by the use of moving belts in some shops. A car could then be overhauled and repainted in seven days, and each car entered works after 60,000 miles (75,000 for trailers).

On March 27, 1927, a single-line tunnel with a gradient of 1 in 50 was opened at King's Cross to connect the northbound City line and

[1] Bakerloo and Watford Joint cars reached Acton via Willesden High Level, Kensington (Addison Road) and Earl's Court until an alternative wholly-owned route was available in 1939 via Finchley Road and Rayners Lane.

eastbound Piccadilly line tunnels and enable cars from the Hampstead and City lines to reach Acton Works. This 1,000-foot tube was constructed by Walter Scott and Middleton, and was authorized by a Ministry of Transport Order, the only tube line so sanctioned. By October 1927, Acton was overhauling the stock of all lines including Central London cars, which were routed via Ealing Broadway and Ealing Common. Golders Green continued to deal with light repairs. Improvements carried out there in 1925 had increased the repair roads from two to five, a 15-ton overhead crane and traverser were installed and a new four-road car shed to accommodate four eight-car trains was added at the same time.

BOOSTING THE POWER

By 1919 the load on Lots Road power house had increased above that of 1914 by nearly 18%, and a further 15-MW turbo-generator was installed in 1921 with four more boilers, bringing the boiler total to 68 and the maximum capacity to 78 MW. During the 1921 coal strike, sixteen of the boilers were converted to oil burning and the oil brought up from Thames Haven by barge, but this equipment was removed after the strike. Another 15-MW generator was added in 1925 and four more boilers were put in between 1922 and 1925, some with emergency oil-burning equipment. Apart from the new Hampstead and City line extensions, extra load was introduced on March 18, 1928, when the CLR began to be fed from Lots Road. At this time the CLR rotary converters were adapted for the change from 25 to $33\frac{1}{3}$ c/s.

The 1926 General Strike drew attention to the vulnerability of the Underground's centralized power supply with its non-standard frequency. The Air Raid Precautions Sub Committee of the Committee of Imperial Defence recommended a long-term conversion to 50 c/s so that power could be taken from other sources in an emergency, but although short sections of the Underground had been tied to other sources from time to time, a sense of urgency was lacking until 1938, when steps were taken to convert the tramway power station at Greenwich.

RESHAPING THE HEART VALVES

The increased traffic brought by the new extensions showed up the inadequacy of many of the original central area stations.

A particularly busy point was Tottenham Court Road, where the Central London and Hampstead lines exchanged traffic. It had received early attention from the signal engineers in an attempt to speed the flow of trains on the Hampstead side, and from 1920 onwards, 'Hustlers' had been used there to the same end. Hustler procedure was first introduced at Victoria, District Railway, in December 1919 when an employee stood on the platform in a prominent position, armed with stopwatch and siren. Thirty seconds after the arrival of each train he would sound his siren as a signal for the platform staff to make a determined attempt to get the train away. About 1923, a mechanical Hustler was devised and connected to the signalling circuits at certain stations, and the modern equivalent, the 'Audible Interval Indicator' has been in use at Oxford Circus, Bakerloo, since September 1935. A 1914 Act had authorized escalator subways leading to the Hampstead booking hall at Tottenham Court Road and this hall was to be enlarged to serve both lines. In the summer of 1923, John Mowlem & Sons received the contract and a 5,000-square foot booking hall with two passimeters was constructed beneath St Giles Circus. The old CLR ticket hall was closed and the station building reduced to an entrance. From the new hall, three escalators in a 26-foot shaft led to a circulating area whence stairs served the CLR platforms, and two more escalators gave access to the Hampstead line. The upper escalators came into use on September 29, 1925, and the lower flight on February 1, 1926. Eight lifts were closed (the intermediate landing used part of the space formerly occupied by the CLR lifts).

New subways were built in 1920 at Charing Cross to improve access to the tube platforms from the street and the District. A new booking office was opened on December 6, 1920, and new exits and a bridge over the District were added in 1921. This station was now the busiest on the Underground (nearly 33 million passengers were handled in 1925) and its complicated maze of passages and stairs was made clear to lost and confused customers by Traffic Guides, first introduced in 1925. At rush hours, to help speed the escalator movement, a device called the *Stentorphone* was installed at the foot of the Hampstead line escalators in 1921. This gave forth in thin and scratchy tones: PLEASE KEEP MOVING: IF YOU MUST STAND, STAND ON THE RIGHT: SOME ARE IN A HURRY, DON'T IMPEDE THEM! The passengers' replies are not recorded. This little toy was also installed at Oxford Circus and it consumed one record a week.

Oxford Circus was the second busiest station at this time and a reconstruction designed to provide room for 50 million passengers a year was begun in October 1923. The Bakerloo booking hall was extended beneath Argyll Street to form a combined hall for both railways, measuring 130 feet by 60 feet. The old CLR station was converted to an entrance. A new escalator shaft 110 feet long and 17 feet in diameter, with two comb-type escalators, was built to connect the new booking hall with a landing just above the CLR platforms. These escalators were ready on July 5, 1925, and the new booking hall was completed on August 16th. In 1929-1930 the Bakerloo escalators were replaced by modern machines of the comb type.

At Trafalgar Square, two comb-type escalators replaced the 20-year-old lifts from April 13, 1926—the space occupied by the lifts was used to enlarge the booking hall. A new subway to Cockspur Street was constructed by the Westminster City Council and opened on September 28, 1928.

Three escalators in a 27-foot shaft were built at Bank CLR station and replaced the five lifts from May 7, 1924. These were the first triple flight of escalators. One of the old lift shafts was converted to a staircase to improve low level interchange between the CLR and the City line. Another CLR station to receive attention was Shepherd's Bush, where lifts were replaced by two escalators on November 5, 1924. Bond Street's lifts gave way to two escalators on June 8, 1926, and the booking hall and street-level frontage were reconstructed in 'Morden' style.

Many minor improvements were made in the first half of the 'twenties, particularly to lift equipment and signs. Illuminated train destination indicators were installed in July 1926 on the platforms at Leicester Square, Tottenham Court Road, Charing Cross, Kennington, Euston, Bank and Moorgate, and also at Waterloo and Morden when those stations were opened. These helped Morden–Edgware [1] line passengers to unravel the several routes and destinations now available.

FARES AND FINANCES

Small fare revisions had been made on September 1, 1917, November 1, 1918, and February 1, 1919, and, as mentioned earlier, to help pay for increased running costs, another increase was made on April 6, 1919. Further rises came on September 1, 1919, and in 1920 statutory authority was obtained to charge $1\frac{1}{2}$d a mile (2d up to one mile) for a limited period. The Bill encountered strenuous opposition from local authorities and an attempt to abolish workmen's fares (described by Lord Ashfield as 'an old-fashioned dole to working men') was unsuccessful, although the latest issuing time for these fares was cut back to 7.30 a.m. on the LER and CSLR. The minimum fare now became $1\frac{1}{2}$d in most cases, though some of the old penny fares survived.

In the 'twenties the LCC pursued an aggressive policy of cheap-fare facilities on its tramcars, and when the 2d midday tram fares were introduced on April 19th and May 3, 1920, the Underground was forced to follow suit, and issued cheap midday tickets on the CSLR from the summer of 1921. On December 1, 1921, the LCC re-introduced penny minimum fares and the General bus fares had to come down too, much against the grain. Tube fares were not reduced, the official reason being that it was policy to force short-distance traffic on to surface routes. Even so, fares above $1\frac{1}{2}$d were reduced to approximately a penny a mile up to 4d, and fares above 4d were given increased distance value, from January 1, 1923.

Through road-rail ('T-O-T') season tickets, monthly and quarterly,

[1] Although not officially introduced until 1934, this phrase is adopted henceforth as a useful short description of the Hampstead/CSLR amalgamation.

Edgware, rail level, 1925—the two-platform terminus, with its overall roof, is on the left, the four-road car shed and storage sidings in the centre and on the right. Both original gate stock and standard stock are in evidence.—*London Transport*.

Hendon Central, 1924—intermediate station design on the Edgware extension. Standard stock (motor car 517 at front) on a Charing Cross train. The signal box has since been taken out of use and partly demolished.—*London Transport*.

Enterprise at Morden—Dr Charles Holden's Portland stone modernity contrasts with the ungainly buses on feeder service. In the foreground, the Station Garage (June 1927).
—*London Transport.*

were introduced in 1923. The first ones were valid from May 1st via Golders Green and from June 1st such tickets were available via Hammersmith. From July 1st they were in use via Boston Manor, and via Ealing Broadway, from August 1st via Shepherd's Bush, from October 1st via Turnham Green and via Walham Green, from November 19th via Putney Bridge and via Northfields, from December 1st via Finsbury Park and via Highgate, and from December 1, 1924, via Clapham Common. When Morden station opened on September 13, 1926, T-O-T tickets were sold on a number of routes through that point.

In 1925, the special powers obtained in the 1920 Act expired and the old penny minimum fare came back to all tube railways from January 1st, some penny fares covering as many as three stations. As a further gesture, and to encourage long-distance riding, fares between 7d and 1s 4d were reduced by 1d or 2d from April 1, 1925.

Dividends were still pathetically small, although the 1918 return was inflated by the wartime traffic. In that year the LER paid 2% (its highest to date) and the CLR went back to 4% for the first time since 1906. Even the proletarian CSLR managed 2%. In contrast, the Associated Equipment Company, an Underground subsidiary established in 1912 to supply buses to the LGOC, paid 44% on its £1 million capital, an even better earner than the General, thanks to lucrative Army lorry contracts. When things settled down again after the war, it was the General which supported the Common Fund and paid the fat dividends, and this in spite of increasingly serious competition from 'pirate' and other bus operators.

Lord Ashfield constantly bemoaned this competition at the postwar Annual General Meetings. The number of 'pirate' buses, many of them competing on General routes, rose to nearly 500 in 1924, a year in which there were as many as 105 one-man bus firms working in London. In February 1926, when the 1924 London Traffic Act and LGOC absorptions began to take effect, 662 'pirate' buses were on the road. Ashfield maintained that the Underground was forced to drop all ideas of co-ordination within its organization (i.e. restricting its buses on routes parallel to its railways and tramways) and had to run the buses to the economic limit to counter the competition. As we shall explain in the next chapter, some relief was given by the Traffic Act, and towards the end of the 'twenties, complete co-ordination of London's passenger transport gradually became a hard possibility.

By 1925, the great postwar flush of capital expenditure on tube railways was almost over. Thirteen million pounds, mostly raised on Treasury Guarantee, had been spent, and thousands of unemployed given useful work. At the Annual Meeting of 1925, Ashfield remarked, '... if we are to strive, it is in the hope of reward, and if we are to be denied that reward ... the time has surely come when we must stay our hand ... until this expenditure becomes self-supporting ... we feel we cannot possibly venture upon any fresh schemes of expenditure. The forward policy . . . must become a conservative policy.' There were pressing demands for new tube railways (notably to Ilford and the new LCC out-county estate at Becontree; north of Finsbury Park; and in south and south-east London), but 'to all these projects we must turn a deaf ear'. He explained that new lines could be built only if competition were 'brought within reasonable and fitting limits'.

In 1925 and 1926 the tubes lost much traffic to road services and were as usual carried by the General, despite the new extensions and reconstructions. There was a very urgent need for a line north of Finsbury Park, but no hope of raising the £4 million or so needed for this unless more Government assistance were made available.

CHAPTER 10
Piccadilly Progress
1919–1933

BLOWING OUT THE CORK

SO far, the routes chosen for underground extensions had been mainly influenced by considerations of traffic prospects and railway politics. Public pressure for new facilities had played a negligible part.

A profoundly different situation obtained north of Finsbury Park. As related in Chapter 5, the Great Northern Railway had in 1902 secured restrictive covenants which effectively bottled up the two tube railways terminating there. Most of the passengers arriving by tube wished to travel to points further north or north-east. It was possible to continue northwards by the GNR suburban services to Hatfield or Hertford North, but the stations were mostly sited too far west of the Green Lanes axis to be convenient. Many travellers to Harringay or Wood Green, and all those to Tottenham and Edmonton, had to change at Finsbury Park and complete their journeys by tram or bus.

Agitation for a tube north of Finsbury Park had begun as early as 1919, when the local Advisory Committee of the Labour Exchanges suggested a tube extension to Hertford. The proposal was frustrated by the Great Northern veto and Underground indifference, but local Members of Parliament helped to keep the subject alive by questions to the Minister of Transport, citing the congestion at Finsbury Park and the need to relieve unemployment in their constituencies. They received the usual stone-wall answers that passed the buck to the railway companies, but in November 1923 came the more encouraging answer that the companies were jointly studying the improvement of railway communication in this area, and that electrification was contemplated at an early date.

After an unsuccessful approach to the Ministers of Labour and Transport in December 1922, the Middlesex Federation of Ratepayers' Associations assembled a 30,000-signature petition and presented it to the Parliamentary Secretary to the Minister of Transport (Colonel

Ashley) on June 14, 1923. (It had been organized by the general secretary, J. W. Pardoe, who laboured indefatigably for an extension from 1920 onwards, and unearthed the veto in the GNR 1902 Act.) The petition referred to the overcrowding at Finsbury Park as 'a grave menace. Here converge two tube services and numerous bus and tram services, from the City and West End, and the available outlets are utterly inadequate'. The leader of the deputation told Colonel Ashley that they proposed to urge local MPs to vote for rejection of the LNER's private Bill, which was down for a second reading later that day, unless the railway company gave a satisfactory undertaking about the removal of restrictions on a tube extension. The debate lasted nearly three hours, and although the Bill received its second reading, the LNER was left in no doubt about the Commons' displeasure with the veto clause.

A further deputation attended the LNER on February 15, 1924, and was told that electrification tenders had been invited and that powers were being sought for GNCR-GNR connections at Finsbury Park. (These powers were obtained in the LNER 1924 Act and were subsequently revived in 1936, and in other Acts up to the BTC Act of 1956.)

Meanwhile the LNER had found some difficulty in piloting its 1924 Bill through Parliament. On March 5th, at the second reading stage, Mr Robert C. (later Lord) Morrison, Co-operative member for Tottenham North, moved the rejection, asserting that the 'LNER had treated representations for removal of the embargo with contempt'. It was 'a bargain made twenty-seven years ago between two railway companies neither of which now exists, to be used to deprive half a million people of reasonable travelling facilities for all eternity'. He was supported by the Conservative member for Hornsey, who asserted that deplorable and scandalous travelling conditions existed 'because the GNR had wangled into the 1902 Bill the obnoxious clause which reserved for them the whole of the area as a sort of Tom Tiddler's ground'.

Forced into an indefensible position, the LNER spokesman in the House (Colonel Wilson) promised that if the Bill was passed (which it was) the LNER would either announce its electrification scheme or waive the veto before the following October, although it would retain its rights to oppose an extension Bill in Parliament. In fact, examination of electrification proposals dragged on until November 1925, when the LNER regretfully concluded that it could not proceed. Robert Morrison returned to the attack by moving the rejection of the LNER's 1925 General Purposes Bill, saying that this was the only way to teach them a sharp lesson.

The London Traffic Act of 1924 set up the London and Home Counties Traffic Advisory Committee, empowered to appoint certain of its members to hold public inquiries. At the request of the Minister of Transport it delegated five of them (including Mr John Cliff) to inquire into the alleged inadequacy of travelling facilities to, from, and within certain areas in the north and north-east of London. The chairman was Sir Henry Maybury (chairman of the Advisory Committee) and hearings began on October 19, 1925.

The main cause of complaint was the confusion in the roadway at Finsbury Park during the evening peak when homegoing passengers coming from the Piccadilly and GNCR lines had to board a northbound bus or tram, the latter involving crossing the northbound stream of road traffic along Seven Sisters Road. Counsel for the local authorities spoke of 'pandemonium . . . something very like a free fight', and witnesses told of 'shocking congestion, many accidents, lynching' and being knocked down. Extracts from the Press were read, to the effect that 'men and women fight like rugby footballers for means to reach their homes', 'accidents occur daily', 'clothes are torn, and fainting girls and women are so common as to pass almost without comment'.

Mr Dryland, County Engineer and Surveyor of Middlesex, confirmed the danger to prospective tramway passengers from passing traffic, and the pushing and fighting to board trams.

The tramway situation was bedevilled by the multiplicity of authorities involved (for instance, the LCC Tramways were responsible for this section of road, but much of the service was provided by MET cars). The LCC policy was that passengers should make the whole journey from central London by tram, so that short-workings were not viewed with enthusiasm, and the police were happier without them. J. K. Bruce, general manager of the LCC Tramways, stated that a tram layby in the park had been discussed but was dropped because of police objections. There was in existence a third track near Stroud Green Road especially for short-working cars, but the practice of having regular short-workings had been abandoned because of allegations that this increased the congestion. Tram services were too frequent and too many to have separate queues in the roadway.

Mr Bruce produced a table of tram loadings showing that between 5.30 and 7.00 p.m., 125 trams left Finsbury Park for Manor House, with 8,308 passengers, of whom 54% had been on the cars on arrival. Corresponding figures for the buses were 172, 8,313 and 64%.

There was no unanimity amongst the 'complainant' witnesses about the alignment of a tube extension from Finsbury Park. Each naturally wanted it to serve his own area, but the most popular suggestion was: Manor House–Harringay–Wood Green–Palmers Green–Enfield, with Southgate as a runner-up.

Frank Pick, assistant general manager of the Underground group's railways and the LGOC (and assistant managing director of the Underground Co.), had more definite ideas about the course of the extension. He revealed that at Manor House, bus and tram traffic to and from Tottenham, etc., was 15,000 eastbound and 14,000 westbound, but to and from Harringay and Wood Green in Green Lanes it was

42,000 southbound and 43,000 northbound (presumably he included traffic to and from Moorgate, etc., via Mildmay Park). He concluded that the total of 85,000 passengers per day showed that there was a case for a tube railway to Wood Green. It would have to continue some distance further in order to rise to the surface to reach a depot. Such an extension had not been surveyed in detail, but the LER engineers had sampled the subsoil and made a rough estimate of the cost.

Pick was careful to point out that the Underground group did not have the resources to build the extension—he was merely advising the tribunal 'as to which tube we think they must encourage the building of by someone...'. To be able to raise fresh capital it was essential to pay reasonable returns on existing capital. Such returns were not at present being earned, and could not be earned until 'a measure of protection is introduced into London'. Traffic handled at Finsbury Park (LER) had dropped from $19\frac{1}{2}$ million passengers per annum in 1919 and $20\frac{3}{4}$ in 1920 to $14\frac{1}{2}$ in 1923, 1924 and 1925 (estimated). On bus route 29, competition from independent operators had forced the LGOC to put on extra buses, so that the LGOC carried 8 million passengers on this route in 1918 but 23 million in 1925 (with about a one-minute joint service); much of the increase had been extracted from the trams and the underground. The existing Piccadilly line service from Finsbury Park (24 six-car trains per hour) had ample spare capacity.

Hearings ended on November 26, 1925. The Report concurred with the view that the traffic demand for services to and from Finsbury Park came from *between* the various LNER lines, and sagely noted that in the districts covered by the inquiry there was considerable room for housing development. As far as the Finsbury Park problem was concerned, it recommended:

(i) That the LER be invited to give serious consideration to extending the Piccadilly line to Manor House in the immediate future, and that it should examine the proposal for a tube/tram/bus interchange at this point.
(ii) That the LER be invited to explore the possibility of extending the Piccadilly line to Wood Green or Southgate and to advise the Minister of the result.

In subsequent discussions with the Ministry of Transport the Underground group adhered to the view that an extension was impossible until its finances were on a sounder basis, but the LER quietly went ahead with the task of surveying the route in detail and acquiring the land. The site of what is now Turnpike Lane station was up for auction on May 25, 1927, and it is likely that this opportunity was taken. In September 1928, Southgate UDC gave conditional permission for a trial bore at Chase Side. By February 1929, Lord Ashfield was saying, 'it certainly cannot wait long', and when yet another deputation waited on the Minister of Transport in December 1929, the solution to the financial problem had been found, and the Minister was able to tell them that the LER was seeking parliamentary powers for a Cockfosters extension.

Two months after the election of a second Labour Government in May 1929, the Development (Loan Guarantees and Grants) Act was passed in an attempt to stem the rising tide of unemployment. Under this Act, the Treasury could guarantee interest for up to three years or *grant* interest for up to fifteen years on capital raised by public utility undertakings that exercised statutory powers for work that would help to relieve unemployment.

This enabled the Underground group to re-examine the financial prospects of a Finsbury Park extension, and the Government pressed them to make a quick decision. An application for assistance was submitted on October 29th. The scheme, which would cost roundly £13 million, included:

(i) A $7\frac{1}{3}$-mile extension of the Piccadilly line from Finsbury Park to the Middlesex hamlet of Cockfosters, via Wood Green, the Arnos Grove estate, and Southgate (surfacing south of Arnos Grove).
(ii) The widening of the District Railway to four tracks from Hammersmith to Northfields ($4\frac{1}{2}$ miles) to allow the projection of Piccadilly line trains to the latter point.
(iii) The reconstruction of many central area stations.

This was followed by an application for parliamentary powers in the London Electric, Metropolitan District, Central London and City & South London Railway Companies' Bill of 1930. As the Bill would contribute substantially to the relief of unemployment, it was 'certified' for quick passage.

Meanwhile there were some rumblings of discontent from local residents about the proposed route, which passed through undeveloped territory north of Southgate, and left built-up areas such as Enfield unserved. There was also some unsuccessful pressure for a station at Harringay, between the proposed Manor House and Turnpike Lane stations.

LER *versus* LNER

The hearing before the Commons select committee began on March 26th, and Frank Pick gave evidence for the LER. He had obtained up-to-date figures showing that in 1929 roundly 27 million passengers were handled at the two Finsbury Park tube stations, of whom 78% travelled by bus and tram to or from Manor House. Of this Manor House traffic, 75% travelled in the Harringay direction.

The northern part of the extension had been planned to run as nearly as possible halfway between the Great Northern main line and the Hertford loop line. The district was fairly well developed as far as Wood Green, but hardly at all from Southgate onwards. The Cockfosters projection was needed to reach the depot—inadequate depot accommodation had always been a bugbear on the Piccadilly—and it was essential for the cars to come to the surface for ventilation and to be cleaned. Furthermore, it was only fair that the LER should be allowed to continue for some distance on the surface, at one-fifth the cost of tunnel construction. There should not be great competition with the LNER as that railway served the City and the LER served the West End. The tube extension would in fact create development which would bring fresh traffic to the LNER(!).

The LNER were in strong opposition, contending that an extension was needed to Manor House only. Their Counsel spoke of the new railway 'coming into our territory and taking our traffic', which provoked Pick to reply, 'I don't know what reason you have for calling this "your territory". The fact is we are going into this region where the North Eastern have done nothing, and are doing nothing, to meet the traffic needs of the people.'

Sir Ralph Wedgwood (chief general manager of the LNER), well-versed in the commercial tactics of the Underground group, correctly prophesied that what are now known as Southgate, Oakwood and Cockfosters stations would be developed as railheads to bring in passengers from a wide area.

On the section of the LNER that would be affected, the number of passengers had declined by 21% since 1922. Receipts were £275,000 per annum, of which £110,000 would be lost if the tube were built. There was *no* bottleneck at Finsbury Park—road congestion there could be remedied by a Manor House extension.

The LNER's 1924 electrification scheme was then disinterred and dusted. At their meeting on March 21st the LNER board of directors had decided to electrify their suburban lines to Edgware, High Barnet, Alexandra Palace, Welwyn Garden City and Hertford North, provided that the LNER received an adequate Government grant and satisfactory terms could be arranged for City terminal accommodation. Would it not be better to have 60 miles electrified for £4,350,000 than a Cockfosters extension for £5,655,000?

Pick was recalled on April 1st, and was now able to give details of the agreement with the Treasury. For fifteen years they would pay 3% each year on the amount so far expended on works and equipment. Of the total cost of roundly £13 million for the whole scheme, the 3% grant would apply to £11,255,000 (these figures were later reduced, see below) or £4,275,000 for the Cockfosters scheme.

In his closing speech, Counsel for the LNER averred that approval of the Bill would result in the directors of the LNER meeting in a state of gloom. This dire prospect did not deter the committee from finding the preamble proved, but they added a rider that was well suited to April 1st—that the LNER should not be discouraged from going forward with their electrification scheme. No doubt the LNER found this most gratifying.

The battle continued before the Lords committee, where the hearing began on May 21st. Leaving nothing to chance, Pick had arranged for photographs to be taken of the turmoil at Finsbury Park from the roof of a shop, and for his observers to count the LNER traffic at Hornsey, Wood Green, Palmers Green, Bowes Park and New Southgate. The latter produced an estimate of 12 million passengers a year—less than that handled by Golders Green alone. To a pertinent question about the population that would be served at Cockfosters, Pick replied that there would be 25,000 people there in a short time.

In a manoeuvre reminiscent of 1924, the LNER asked that the tube powers should be withheld until December 31, 1930, and should not be exercised if the LNER had commenced electrification by then. The committee were not impressed, and took only three minutes to find the preamble proved.

Royal Assent was given on June 4, 1930, and the Act included:

New Railways
Finsbury Park–Cockfosters (7 miles 5 furlongs 4·2 chains).
Siding between Hyde Park Corner and Down Street (1 furlong 9·85 chains).
Widenings of MDR
Acton Town–Boston Manor.
At Hammersmith.
Station Reconstruction (of various magnitudes)
At Knightsbridge, Hyde Park Corner, Dover Street, Holborn, Russell Square, King's Cross, Warren Street, Kentish Town, Edgware Road, Marble Arch, Chancery Lane.
Subway Connection
Between Bank and Monument stations.

Capital for the new works was raised by the issue of £8·45 million of LER 5% Debenture Stock and £850,000 of CLR. £5 million was issued in July 1930 (at 97·5) and the remainder in October 1930 (at 99·5).

ON TO COCKFOSTERS

As soon as the Royal Assent was received, 'notices to treat' were served on the owners of properties to be demolished, and by early September 1930 the shops on the site of Wood Green station were already down to first-floor level. Soon afterwards, hoardings were erected round the working sites and the contractors began to sink the 12-foot shafts. The first shaft was at Turnpike Lane, which was the construction headquarters for the whole tunnel section, with a large dump of cast-iron segments. There were nine working sites—at Finsbury Park tennis courts, Manor House, Woodberry Grove, Colina Road, Turnpike Lane, Wood Green (Lordship Lane and Pellatt Grove), Nightingale Road and Bounds Green. Altogether 22 shields were used, working in 32 sections (which involved 19 end-on junctions).

The tunnelling contractors, Charles Brand & Son, subcontracted the removal of the clay spoil to the St Mary's Wharf Cartage Co., Paddington, who acquired a fleet of thirteen 'Sentinel' D.G.6 type steam wagons with three-way hydraulic tipping gear. These six-wheeled, solid-tyred monsters were painted navy blue, with 'Excavators' written very large in yellow along their sides, and the authors remember them with nostalgic affection, chuffing solidly to and fro, between the sites at Manor House, Turnpike Lane and Wood Green and the Lea Valley dumps. Some of the wagons brought back sand and gravel for making grout and concrete. Tunnelling went on for twenty-four hours a day except for a break from 2.30 p.m. on Saturdays to 10.00 p.m. on Sundays. Average progress was 51 rings (85 feet) a week, or about one mile of single tunnel per month. Modern aids included pneumatic picks, and 'swinging-arm' segment erectors for the station tunnels, with a telescopic arm operated by compressed air. On March 19, 1931, the new tunnels met the old near Finsbury Park, and by October 1931 the tunnelling was complete, except for the Wood Green and Bounds Green station tunnels, which were driven by the end of the year.

At Bounds Green Road, the New River passed below the roadway in a brick tunnel, and the railway tunnel was to be 25 feet below the river bed. Iron segments were used to reinforce the river tunnel, and the flow was meanwhile diverted through a steel tube 90 feet long which had been erected at Park Avenue, Wood Green, and floated along the river towards its position.

Work was also going ahead on the open section north of the tunnel mouth at Tewkesbury Terrace, for which Sir Robert McAlpine & Sons Ltd were the main contractors. This involved some substantial engineering work, including a 175-foot girder bridge over the North Circular Road, a 95-foot 6-inch girder bridge to carry Bowes Road over the railway, a quarter-mile viaduct across Arnos Park (which took three million bricks), and further girder bridges at Waterfall and Valley Roads.

There was to be an isolated half-mile section of tunnel at Southgate, and work began at Chase Side on June 22, 1931. North of Southgate was another brick viaduct, whilst between Enfield West and Cockfosters, bulldozers and 'steam navvies' were busy clearing the site for the running depot.

By January 1932 the main constructional work to Arnos Grove had been completed, and the laying of track, installation of signals and construction of station buildings began.

The Underground group announced that the 4·47-mile Finsbury Park to Arnos Grove section (with intermediate stations at Manor House, Turnpike Lane, Wood Green and Bounds Green) would be opened on Monday, September 19, 1932. Thirty thousand free return tickets to Piccadilly Circus were issued to local residents for use after 10.00 a.m. on the opening day.

There was no official opening ceremony, but on Friday, September 16th, the Underground entertained representatives from Southgate, Wood Green, Tottenham and Stoke Newington councils, MPs, County Councillor Pardoe and local pressmen. They inspected the new railway with J. P. Thomas, operating manager, in attendance to explain the technicalities.

Stations on the extension opened from 11.00 a.m. to 5.00 p.m. on Sunday, September 18th, to sell season tickets. The next day, the first train left Arnos Grove at 5.23 a.m. Trains were well loaded throughout the day, and in the evening there was 'standing room only' on many journeys. Stations were beflagged and there was a display of flowers at Arnos Grove. At night, the beams of publicity searchlights at Bounds Green and Wood Green met in the sky.

The 2·38-mile section from Arnos Grove to Enfield West, with an intermediate station at Southgate, was opened on Monday, March 13, 1933, again without official ceremony. Local residents were given free tickets for a return journey from Enfield West or Southgate to Piccadilly Circus after 10.00 a.m. on the opening day. The final ·81 of a mile section to Cockfosters opened on Monday, July 31, 1933. For both these extensions, the practice was again adopted of opening the new stations on the previous Sunday to issue season tickets and give information.

TRACKS AND TUNNELS

Of the 7¾-mile extension, about 4½ miles were in tubes. Running tunnels were of standard 11 feet 8¼ inches internal diameter except on curves of 20 chains or less, where a diameter of 12 feet applied. At Tewkesbury Terrace, the entrance to the south-bound tube was a bell-mouth to alleviate the sudden increase in air pressure as trains entered at speed.

The line fell from Finsbury Park to Turnpike Lane, and then climbed in a series of 1 in 60 gradients to Cockfosters. At Tewkesbury Terrace the track was 48 feet higher than Finsbury Park; between Southgate and Enfield West the line climbed 65 feet to reach a height of 275 feet above sea level. A notice in the Enfield West booking hall claimed: 'This station is at the highest point in Europe in a direct line west of the Ural Mountains of Russia.'

Manor House, Turnpike Lane and Bounds Green were sited on humps, but the severity of the prevailing gradients did not allow this elsewhere.

The running rails were of 95-lb. bull-head type, in 90- or 60-foot lengths (the previous standard lengths were 42 feet in tube and 45 feet in the open); the conductor rails were 130 lb. per yard, rectangular section underground, and 150 lb. flat-bottomed in the open (instead of 85 and 100 lb. respectively). In the tunnels the conductor rails were welded into 600-foot lengths. The Jarrah-wood sleepers were supported at each end on concrete benching, and concreted in. Shingle was used to fill the centre drainage channel and was also spread in a 2-inch layer over the benching and sleeper-ends. Creosoted fir sleepers were used on the open sections.

At Manor House, Turnpike Lane and Wood Green, the station tunnels were of 23 feet 2½ inches internal diameter to allow wide platforms for the expected heavy traffic, but at Bounds Green and Southgate they were of the standard size of 21 feet 2½ inches. All tube platforms were 385 feet long to cope comfortably with a seven-car train. Starting signal repeaters, headway and ordinary clocks, telephones and signal equipment were neatly housed in the tunnel headwalls. A 2-foot anti-suicide pit [1] accessible from arches beneath the platforms was provided between the rails at each deep-level station.

Station tunnels and circulating areas were faced with biscuit tiles up to ceiling level, relieved by coloured tiles round the borders of advertisement panels and passageways. The colours adopted were: Manor House—blue; Turnpike Lane—orange; Wood Green—green and biscuit; Bounds Green—red; Southgate—orange.

Bulbs in small rectangular glass fittings, at 7-foot intervals, were placed on the undersides of soffits where the ceiling joined the tunnel walls.

A NEW ARCHITECTURE

Charles Holden was generally responsible for the design of the station buildings [2] and the style was completely different from his Morden line stations.

[1] Tube stations were a popular site for suicides. In 1921 the Westminster Coroner commented that there was 'something about the roar and rush of a tube train which was terribly fascinating to a person if he were alone on the platform'. Extrication of the body was a task of extreme difficulty, often delaying the train service for as much as an hour. Pits to facilitate extrication and reduce the risk of injury to persons throwing themselves in front of trains were first provided on the Morden extension in 1926. These were 1 foot 4 inches deep and ran along the whole length of the platform between the rails. During 1934–1935, all tube stations were provided with anti-suicide pits, and the number of attempts decreased noticeably.

[2] James, Bywaters and Pearce were associated with Holden for Bounds Green and Enfield West, and the Underground architectural staff for sub-stations, platforms, escalator tunnels and approaches.

The elements of the design comprised wide and inviting entrances from the street, often with concrete canopies, and high curtain-walls for the booking offices composed of alternate vertical masses of rectangular steel-framed cathedral-glass windows and unrelieved red-brown bricks. The walls were invariably capped by a wide concrete frieze and an overhanging cornice. A rectangular brick ventilation relief shaft was often incorporated in the structure. The brickwork was left plain inside most of the stations. Advertisement and notice panels, ancillary shops and doorways, telephone booths, stair-rails, clocks and signs were blended into a harmonious pattern. Bronze was freely used for booking office and shop window-frames, and for handrails, whilst the lower portions of the free-standing booking offices were faced with light-coloured sheet rubber of mottled pattern.

Much use was made of lamps mounted on columns 10-12 feet high, the light directed upwards by bowl-shaped reflectors immediately below the bulbs. In the design most commonly employed (known to LT staff as 'daffodils'), the fluted column and the bowl were circular in plan and finished in bronze, whilst in the centre of the bowl an object resembling a wide-mouthed, off-white vase served to diffuse the light and provide a decorative note. This type of lighting is most efficient where the light can be reflected down again by a low white ceiling, but in practice it was used both in the lower circulating areas and the lofty booking halls.

At Manor House booking hall, below the road surface, the lights were in larger circular opal shades attached to the ceiling. Surface platforms had opal cylindrical shades beneath the canopies, or, in the open, spherical shades mounted in iron rings on standards. 'Underground' bull's-eyes were embedded in the brickwork of the ventilation towers and spotlit at night.

Everything was planned to provide a smooth, easy journey from street to platform, with the ticket office or ticket-machine on direct line of route, if needed, but not obstructing season-ticket holders. On the tunnel sections, escalators invariably covered the full distance from booking hall to platform.

Station design was adapted to the layout of the site. Turnpike Lane and Enfield West booking halls were oblong in plan, Bounds Green octagonal, Arnos Grove and Southgate circular. Southgate, on an island site, had a much lower tower with sides composed wholly of glass in steel frame. It had a dumpy appearance, compared with the other stations. Wood Greeen had to be fitted into a row of shops at the High Road/Lordship Lane junction. It had a curved façade and was elliptical in plan.

Manor House had nine entrances at street level, with a 120-foot subway to the south-west junction of Green Lanes/Seven Sisters Road and a 100-foot subway to the west side of Green Lanes (north of the road junction), where the top of the stairs opened into a small brick shelter for Harringay-bound bus or tram passengers. The special feature at Manor House was the pair of tramway loading islands, 80 feet long, with seats, concrete canopies and stairs direct to the booking hall, faithfully carrying out the 1925 tribunal's recommendation. They fell into disuse when trolleybuses replaced trams in 1939, and were removed in 1951.

Turnpike Lane also had its subways to four points on the west side of Green Lanes, including two 80-foot tramway islands (unfinished when the line opened) which are still in use today by passengers for bus routes 217 and 231. Cockfosters had a low unostentatious street-level entrance, but the three-road train hall was an interesting exercise in angular reinforced-concrete piers and ribs, supporting a central clerestory.

Enfield West was one of the few tube stations to boast a buffet; this was perhaps intended to act as a *douceur* for those taken in by the station name. It was a good $2\frac{1}{3}$ miles from Enfield town, and Southgate Borough Council's protests resulted in the addition of 'Oakwood' to the nameboards early in 1934. The original name was dropped altogether from September 1, 1946.

NEW ESCALATORS

The escalators on the Cockfosters extension embodied the results of research by the Waygood-Otis company to achieve a faster, quieter, smoother-running machine, needing less maintenance and capable of a greater vertical rise than hitherto. In January 1930, experiments were made at Tottenham Court Road with an escalator travelling at 150 feet per minute instead of the usual 90 feet. The new escalators, of 'MH' type, first installed at Highgate (now Archway) in 1931, had a maximum linear speed of 180 feet per minute in the up direction with a maximum rise of 90 feet. A simpler 'M' type, for shorter rises and a maximum speed of 100 feet per minute, was also developed. The escalator equipment of the extension was as follows:

	Type	Vertical Rise (feet)	Number of Escalators
Manor House	MH	44·25	3
Turnpike Lane	M	35·88	3
Wood Green	MH	39·13	3
Bounds Green	MH	51·88	2
Southgate	M	34·75	2

To accustom the public to the speedier 'MH' type, these escalators were at first worked at 125 feet per minute, and later 165 feet. Escalator treads were painted in light and dark colours on alternate steps, as a safety measure.

NEW SIGNALLING

Signalling was of the standard, a.c.-track-circuit, colour-light, automatic-train-stop type, with electro-pneumatic operation of the latter, and of the points. The signalling on the original section of the Piccadilly line was also converted from d.c. to a.c. track circuits; additional signals and track circuits were installed, and the whole of the equipment modernized, so that a service of forty trains per hour could be handled.

Relay rooms containing the track-circuit apparatus and track-circuit feeds of the home and starting signals were provided in the station tunnel headwalls, reducing the amount of equipment in the tunnels themselves. This allowed inspection, maintenance and repairs to continue unhindered by the passage of trains.

To cope with the multifarious non-stopping possibilities on the Piccadilly line, a new type of train describer apparatus was evolved by the Underground signal staff, to supersede the drum-type describer, which had been adopted on the District in 1906 and on the Morden–Edgware in 1926. The salient feature was the use of holes punched in a paper ribbon to denote and store destinations and non-stop codes. It could handle 12 different destinations, and up to 12 different non-stop codes, embracing 15 non-stopped stations. The signalman set up the descriptions by pointer and plunger-switch, causing 35 impulses per description to be sent to each station along the line as far as the next signal-box. Positive impulses caused holes to be punched in the ribbon at each intermediate station. The stored ribbon hung down in a loop, and was automatically moved on with the passage of trains, by track-circuit control. Contacts through the holes in the paper caused appropriate displays to be made on the station indicators. This type of describer was later adopted on the Bakerloo and Central lines.

There were signal-boxes at Wood Green (controlling a 465-foot reversing siding between the running tunnels) with the pioneer relay-interlocking frame; Arnos Grove (59 levers, controlling the working of the three platforms and the fan of eight sidings south of the station); Enfield West (controlling two crossovers and access to the depot, 35 levers including spares); and Cockfosters (47 levers, controlling three terminal roads and five sidings).

Intermediate signal-boxes could be switched out and the train descriptions retransmitted; the Wood Green reversing siding could be worked automatically by the train describers at Hammersmith and Arnos Grove.

NEW SUBSTATIONS

Traction current was supplied by the associated North Metropolitan Electric Power Supply Co. from their substation at Watsons Road, Wood Green, to a large Underground substation and switch house at Jolly Butchers Hill, Wood Green. This had three BTH 1·5-MW rectifier units and transformers, and a control room for the remote control of substations at Cockfosters (two 1·5-MW rectifier units), Southgate (2 units), Arnos Grove (2 units), Manor House (3 units) and Holloway Road (2 units; a third was added in 1936). The input was at 11 kV, 50 c/s, and the output at 630 volts d.c.

The mercury-vapour continuously-evacuated steel-tank rectifier, first installed at Hendon Central in December 1930, offered many advantages over a rotary converter, including lightness, quietness, adaptability to remote control, greater efficiency at the industrial frequency of 50 c/s and the ability to take high momentary overloads without harm. At a meeting of the Institution of Electrical Engineers in November 1935, at which these rectifiers were described, Frank Pick said that the Underground group had taken its courage in both hands and ordered rectifiers for the Piccadilly line on the basis of the one installation at Hendon. The original rectifier buildings had large clearances because the engineers were not sure of the exact results of a breakdown—clearances could be smaller in future. These buildings were in the same general style as the stations.

Sixty thousand c.f.m. air-extraction plant was located underground at Finsbury Park tennis courts, and 70,000 c.f.m. plant in large fan houses at Colina Road and Nightingale Road, using the 12-foot working shafts in each case. All tunnel stations except Southgate had 20,000 c.f.m. pressure plants for local ventilation, feeding fresh air to grilles 7 feet 3 inches above each platform.

The car depot, between Enfield West and Cockfosters, was built of steel stanchions and stock bricks, the bricks used on the exterior being 'multi-reds' to harmonize with the stations and substations. The cleaning shed, measuring 450 by 100 feet, had three eight-car tracks, and the maintenance shed, measuring 450 by 190 feet, had nine eight-car tracks, all equipped with pits. North of the sheds were eleven stabling roads, with capacity for 250 cars. Equipment included floodlighting and loudspeakers.

TRAWLING FOR TRAFFIC

Liberal services were provided from the beginning. Wood Green had a $2\frac{3}{4}$-minute weekday service to town, and Arnos Grove $5\frac{1}{2}$ minutes until 8.00 p.m., then $8\frac{1}{4}$ minutes. On Sundays, Wood Green had a 5-minute service until 7.30 p.m., then 4-minute, and Arnos Grove half these. Six- or seven-car trains ran in the peaks, and three-car sets in the normal. Enfield West and Cockfosters began with 10-minute services. First trains to town left Arnos Grove at 5.23 a.m. (weekdays) and 7.50 a.m. (Sundays). Last trains from Piccadilly Circus to Arnos Grove were at the standard times of 12.30 a.m. (weekdays) and 11.30 p.m. (Sundays).

Running times from Piccadilly Circus were: Wood Green 21 minutes (a saving of 16 minutes on the combined tube and tram journey); Arnos Grove 25 minutes; Enfield West 30 minutes; Cockfosters 32 minutes.

Fares were low, particularly for longer journeys. Arnos Grove to King's Cross or Holborn was 6d, to Piccadilly Circus 7d; Enfield West to King's Cross was 7d, Leicester Square 8d, Hammersmith 11d. A shilling took one from Cockfosters to South Harrow or Hounslow West.

There were many new road-rail single bookings, and a network of road-rail season-ticket rates ensured that no opportunity was lost to siphon off the maximum amount of LNER traffic. Single fares and through seasons were available via Manor House to Kenninghall Road (Clapton), Edmonton Town Hall and Harringay; via Turnpike Lane to West Green station, Muswell Hill foot and Edmonton (The Cambridge); via Wood Green to Bruce Castle, Palmers Green, Winchmore Hill, Enfield Town; via Bounds Green to Grosvenor Road (Muswell Hill) and Hilton Avenue (Friern Barnet); via Arnos Grove to Russell Lane (Whetstone). To button up matters completely, not only were many of the stations laid out for easy interchange with road services, but new bus routes were started to fill gaps in the network.[1] A bell (later replaced by an illuminated sign) was installed outside Enfield West to tell bus crews when a northbound train was approaching,[2] and inside Wood Green station a sign informed passengers leaving the escalators '39A TRAM TO ENFIELD IS WAITING IN LORDSHIP LANE'. The Underground was trawling for traffic while the LNER was trying to catch it with rod and line.

The volume of traffic on the extension had been estimated at 36 million passengers a year. In February 1934 the estimated total traffic was at the rate of 25 million a year, of which 44% interchanged to or from the road services. In the *Report of Survey* for the County of Middlesex Development Plan, 1951, the daily users of stations in Middlesex were given; if these are multiplied up, and arbitrary ratios given for Saturday and Sunday traffic, it appears that the extension traffic had grown to over 70 million, whilst at the five LNER stations tested by Mr Pick in 1930, it had fallen from 12 million to $2\frac{1}{3}$ million.

PICCADILLY PUSHES WEST

The northern extension of the Piccadilly was balanced by a western extension which allowed tube trains to run beyond Hammersmith to Northfields (and later to Hounslow West) and South Harrow (later to Uxbridge). Between Studland Road Junction (just west of Hammer-

[1] On October 18, 1936, Walthamstow was linked to the Piccadilly line by the introduction of trolleybus route 623 to Manor House.
[2] This was the converse of the procedure adopted at Golders Green in 1907, where the ticket collector rang an electric bell when the Hendon bus hove in sight, so that train crews should wait for bus passengers.

smith station) and Turnham Green the extension made use of disused tracks which had a complicated history.

The first tracks on this alignment were opened by the London & South Western Railway on January 1, 1869, forming part of that company's line from the West London Railway (about half a mile north of Kensington, Addison Road) to Richmond. Through services ran between Richmond and Waterloo, but the routeing via Shepherd's Bush and Clapham Junction was a little circuitous.

On June 1, 1877, a short line was opened at Hammersmith to join the District Railway to this LSWR branch. An abrupt change from cutting to viaduct required a gradient that would do credit to a roller coaster (1 in 44 for 330 yards). District trains at once began to run through to Richmond. A further District extension was made from Turnham Green to Ealing Broadway on July 1, 1879, and a branch from Mill Hill Park (now Acton Town) was opened by the Hounslow and Metropolitan Railway Company to Hounslow Town on May 1, 1883, and to Hounslow Barracks [1] on July 21, 1884. Finally, on June 28, 1903, District trains ran north-west from Mill Hill Park to South Harrow, as related in Chapter 6.

From October 1, 1877, the Metropolitan had worked a service from Aldgate via the Hammersmith and City Railway and Grove Road (Hammersmith) over the LSWR line to Richmond, with GWR participation from January 1, 1894. From March 1878, Midland coal trains used the Studland Road/Hammersmith connection to reach the new depots at West Kensington and High Street, Kensington, but from January 1, 1907, the Aldgate–Richmond passenger service was withdrawn between Aldgate and Notting Hill & Ladbroke Grove, and worked by the GWR alone. Following the electrification of the District in 1905, the operation of LSWR, GWR and Midland steam trains over the Studland Road–Turnham Green section was a severe handicap to the full development of the District services, and in 1910 parliamentary powers were obtained for the construction of two additional tracks between these points. Mainly on the north side of the original tracks, these extra tracks were built by the LSWR and opened for traffic on December 3, 1911. The LSWR then used the northern tracks and the District and Midland the southern. The works included 120 brick arches of 20-foot span, eleven bridges over roads, and two flyover bridges west of Turnham Green to give non-conflicting paths to and from Richmond or Ealing. A new District station named Stamford Brook was built between Ravenscourt Park and Turnham Green and opened on

---

[1] Hounslow Barracks was renamed Hounslow West on December 1, 1925, and the intermediate stations at Hounslow Town (opened May 2, 1909) and Heston-Hounslow (April 1, 1886) were renamed Hounslow East and Hounslow Central on the same day. The original 1883 station and spur at Hounslow Town were finally closed on May 1, 1909.

February 1, 1912. The existing platforms at Ravenscourt Park and Turnham Green became islands.

These improvements enabled the District electric services to be intensified and accelerated, but the use made of the two northern tracks, now freed of District trains, was short-lived, as the GWR service from Notting Hill to Richmond was withdrawn after December 31, 1910, and the LSWR passenger service ran for the last time on June 3, 1916. After that these tracks were infrequently used and became quite grass-grown.

The Underground group doubtless foresaw the demise of the LSWR service when it obtained powers in the LER Act of 1913 to extend the Piccadilly line from Hammersmith to meet the northern tracks at Studland Road, in order to project Piccadilly trains to Richmond or Ealing. The scheme was revived in 1926 when the 1913 powers were again renewed in the London Electric and Metropolitan District Railways' Act of that year. Housing development in the area served by the District had been rapid (particularly at Hounslow, Ealing and Harrow) and the District service through Hammersmith was approaching the limit of line capacity. Factory development on the Great West Road, and at Park Royal and Alperton, had also brought more traffic. The 1926 powers included four-tracking the District from the flyover at Turnham Green to a point north of Acton Town.

A fresh agreement dated June 10, 1926, was made between the SR, MDR and LER. This gave the District and the LER the right to run over the LSWR lines in perpetuity, but the SR could run 'occasional and special' passenger and goods trains, and were to maintain earth-works and bridges. The MDR and LER assumed responsibility for other maintenance and for staffing and operating the SR section. The Underground were to keep all earnings from their own trains running over this section and to pay the SR a yearly rental for its use. Existing running powers were preserved.

As with the northern extension, the projection west had to await a solution of the financial problem, but as soon as this was achieved, work went ahead rapidly. By October 1930, work had begun on the demolition of property at Hammersmith, where the engineers had to replace two separate stations with a new joint station while both stations were handling heavy traffic, and excavate a new underpass beneath busy Hammersmith Broadway and Beadon Road. Part of the road surface was opened up, and at one stage trams on LCC routes 28 and 30 were running over a gaping hole, with the bared rails supported on piles and timbers. From May 31, 1931, the Piccadilly line had only one road in which to reverse, as the eastbound District line temporarily occupied the site of the westbound Piccadilly track.

West of Hammersmith, near Studland Road, the construction of the new northern track involved the demolition of five arches of the LSWR viaduct where it curved towards The Grove, but this gap was carefully replaced by a new reinforced-concrete bridge—it was rumoured that the military authorities were interested in preserving the connection to the Hammersmith and City. Balfour Beatty & Co. Ltd were the main contractors for the Hammersmith–Ravenscourt Park section.

On the four-track section, the working was to be down-down-up-up, with the Piccadilly trains in the middle, running non-stop between Hammersmith and Acton. Thus it was necessary (for the eastbound District stations) either to rehabilitate the old LSWR platforms (as at Ravenscourt Park and Turnham Green) or to build new platforms (at Stamford Brook, which was never served by the LSWR, and so needed a new eastbound platform; and at Chiswick Park, beyond LSWR territory, where the widening to four tracks swallowed up both platforms, so that two new platforms and a complete new station were built). At the west end of the platforms at Turnham Green, provision was made for a future connection between the Piccadilly and District tracks which would enable Piccadilly trains to run to Richmond.

On the District freehold territory from Turnham Green to Acton, retaining walls were built at the foot of the embankment, which was widened by depositing additional spoil. At road and rail crossings the two lines required the provision of extra bridges, including 85-foot girder bridges across Acton Lane and 72-foot across Bollo Lane.

At Acton Town the whole station was rebuilt to conform with the new track layout, and the existing flying junction north of the station was developed into a most impressive layout of flying and burrowing junctions constructed largely in concrete, giving non-conflicting paths between the Acton Town platform roads and the four Northfields or two Ealing Common tracks.

From Acton Town to Northfields the existing shallow cutting was widened. Northfields station was rebuilt on a new site east of North-field Avenue, to make room for a new depot between there and Boston Manor. Beyond Northfields the westbound running road burrowed beneath a depot approach road. There were no significant alterations between Boston Manor and Hounslow or between Ealing Common and Rayners Lane apart from the provision of car sidings at South Harrow on the site of the original electric car sheds erected by Yerkes in 1902.

Sir Robert McAlpine & Sons Ltd were the main contractors for the Turnham Green–Northfields widening.

East of Hammersmith the tracks had to be extensively rearranged. Between there and Barons Court there was now a long reversing siding between the two Piccadilly tracks, with access at each end. At Barons Court the roof of the northern platform was raised to give headroom to the taller District trains. East of Barons Court, the former layout of two loops (for Piccadilly stock workings to and from the Lillie Bridge depot) north of the Piccadilly tunnel mouth, and the two District lines south of it, was replaced by one loop and one District track on either side.

District passengers had their first experience of tube stock on District tracks on February 8, 1932, when new Piccadilly cars replaced vintage District units on some of the workings between South Acton, Acton Town, South Harrow and Hounslow West.

These cars gained direct access to their home ground on July 4, 1932, when half the 2¾-minute Piccadilly service from Finsbury Park to Hammersmith was extended the 8·77 miles to South Harrow. This gave a striking improvement in the service from South Harrow and points south; the journey time to Piccadilly Circus was reduced to 34 minutes (compared with 44) and there was now a through service to central London all day, whereas nearly all the District trains had been shuttles to Ealing Common, Acton Town or South Acton. Some of the peak-hour units were stabled at South Harrow. Sixty thousand free tickets were issued to 15,000 householders on the extension.

Following the opening of the extra tracks between Acton Town and Northfields on December 18, 1932, the Piccadilly service to Northfields (1.62 miles from Acton Town) began on January 9, 1933, with six trains an hour, and on March 13, 1933, the service was extended to Hounslow West (a further 4·08 miles) at the same frequency. The through Piccadilly trains non-stopped South Ealing and Boston Manor, except during the evening; the through District service to Hounslow now operated in the peaks only. For the Hounslow extension 40,000 free tickets were distributed.

It had always been intended that the Piccadilly trains should proceed over the Metropolitan tracks beyond South Harrow at least as far as Rayners Lane, but the assumption of control by the London Passenger Transport Board from July 1, 1933, brought the LER and the Met. into the same ownership, and facilitated the extension of the Piccadilly services to Uxbridge [1] (6·47 miles) on October 23, 1933. Previously, a sparse District shuttle service had run between Uxbridge and South Harrow, Ealing Common or Acton Town.

There was a basic twenty-minute service of Piccadilly trains to Uxbridge, more frequent in the peaks. These trains gave what was then the longest continuous electric-train journey on the London Transport system—31¾ miles from Cockfosters to Uxbridge in 82½ minutes.

Signalling for the new sections of track was on the standard electro-

---

[1] The Metropolitan Railway branch from Harrow to Uxbridge and the spur Rayners Lane–South Harrow were opened on July 4, 1904. Electric trains ran between Harrow and Uxbridge from January 1, 1905, and District electric trains between South Harrow and Uxbridge via Rayners Lane began to run on March 1, 1910. Until that date there had been no regular service between South Harrow and Rayners Lane junction.

pneumatic system, but in order to standardize with the existing District installation, pneumatic semaphore signals were installed for Piccadilly trains between Barons Court and a point about one-third of a mile west of Chiswick Park station. The District drivers had their signals on their left, and the Piccadilly drivers had theirs on their right. From Acton Town to Northfields the new signals were long-range colour lights.

Signal-boxes were built at Hammersmith (43 levers), Acton Town (119 levers) and Northfields (71 levers, including spares). These were equipped with Westinghouse electro-pneumatic frames, repeater lights for the controlled signals and points, and a separate ground-floor relay room.

The possibility of an assault on the tube tunnels at West Kensington by full-size trains (District or LMSR freight) was eliminated by a protective device installed over the eastbound tube track between Hammersmith and Barons Court. This consisted of a gantry holding three mercury-filled 'U'-shaped glass tubes which would be broken by any full-size train, thus releasing the mercury and opening the circuits for the eastbound Piccadilly home signals and train stops at Barons Court.

Substations were built at Barons Court, Chiswick Park, Northfields, North Ealing, Alperton (control station and switch house) and Sudbury Hill. The existing substations at Ravenscourt Park and Acton Town were augmented. Conversion from incoming a.c. was by GEC 1·5-MW mercury-arc steel-tank rectifiers, except at Acton Town where a BTH rotary converter was added to the existing plant of this type. Three rectifier units were installed at Barons Court, Chiswick Park, Northfields and North Ealing, and two each at Alperton and Sudbury Hill. The last four substations were remotely controlled from Alperton, whilst Chiswick Park was controlled from Ravenscourt Park. The new substations were fed from Lots Road, but those north and west of Acton were normally supplied from Neasden via a tie-line from Alperton. All the new substations on the northern and western extensions were designed to take 2·5-MW units eventually.

West of Northfields station, on the south side of the line, a new car depot was built as the western complement to Cockfosters. Of similar general design to the latter, it had accommodation for 304 cars. There was a 450-foot, 19-track car shed and a 430-foot length of 19 open sidings west of the shed. Of the 19 shed roads, two were for car cleaning and 17 with 4-foot deep pits for maintenance. There was also a lifting bay south of the main building, with two 300-foot tracks and an overhead travelling crane. Two vertical-spindle train washing machines were installed. With the opening of Northfields and Cockfosters depots, the original Piccadilly depot at Lillie Bridge was no longer used operationally, and became a permanent way depot. No doubt the operating and maintenance staff breathed a hearty sigh of relief when the last passenger train left Lillie Bridge—a derailment on the single access road could bottle up all the cars in the shed. This nightmare had become a reality on the morning of November 25, 1925, when only nine trains were out and available for the morning peak service.

Many of the stations served by the extended Piccadilly trains were completely rebuilt, but the work was spread over several years. Some of the earlier reconstructions were in the Morden extension style, with angular lines, Portland stone facings, and long vertical window panes, but most of the new stations were in the same style as those on the Cockfosters extension, and had tall booking-hall towers of red-brown Buckinghamshire bricks and wide steel-framed windows, surmounted by a concrete frieze and an overhanging cornice. These brick stations were either designed by Charles Holden or followed the general patterns and forms suggested by him.

At South Harrow a new station (opened on July 5, 1935) was built at the southern corner of the Northolt Road/South Hill Avenue road junction, a few hundred yards north of the Ealing and South Harrow Railway station building (which still exists). A reinforced-concrete bridge (south of the bridge over Northolt Road) was rolled into position on the night of September 15/16, 1934. The booking hall occupied the whole of the space beneath this bridge, and extended on each side in a single-storey, brick-concrete-and-glass structure with glass bricks in the roof and a low clerestory to give the maximum natural light. Two wide covered stairways (one with provision for future escalator installation) led up to the platforms. The platform buildings incorporated the usual cantilevered reinforced-concrete canopy, partly supported on pillars, and the station name bull's-eyes were sunk into concrete panels in the brickwork.

The new Sudbury Town was opened on July 19, 1931, replacing a corrugated-iron structure. A large rectangular booking-hall was impressively sited at the end of a long approach road, on the eastbound side of the line. Single-storey buildings on either side of the hall housed a refreshment room, bicycle store and staff quarters. The station name, flanked by the letters 'U-D', was displayed by large neon signs on the concrete friezes just below the booking-hall roof. The hall was lit internally by column-and-bowl type floodlights.

Sudbury Hill, completed in 1932, had a similar but smaller booking hall, made less impressive by a single-storey projection towards the roadway to house bookstall and tobacconist.

The rebuilt station at Alperton was also opened in 1932. Here the railway was on an embankment, and the booking hall was on the east side of the tracks, with a subway through the embankment to the westbound platform.

The original Park Royal station, sited just north of the GWR Birmingham main line, was closed on July 5, 1931, and a temporary station on Western Avenue, conveniently placed to serve the new factories, opened on the following day. On March 1, 1936, this was replaced by a permanent station with a large, square, brick tower flanked on the east by the booking hall and on the west by a curved parade of shops and two floors of flats. The circular booking hall had a concrete roof, supported by eight fluted concrete cantilevers and a brick wall. The band of windows with prominent closely spaced mullions set in a concrete border was a departure from the general style, in which the windows were set direct in the brick. The architects were Day, Welch and Lander, and the new station was named 'Park Royal (Hanger Hill)'.

North Ealing, the only station on the South Harrow branch to remain unaltered, still survives as an unspoilt example of District architecture at the turn of the century.

Ealing Common was rebuilt in 1931 in the Portland stone 'Morden' style. There was an unusual triangular passimeter booth, with bronze frame and glass above waist level. One side of the triangle was filled by a group of passenger-operated ticket machines which could be tended by the booking clerks from inside the office. The platforms were covered by a unique clerestory canopy.

The single-track section from just beyond Hounslow Central to Hounslow West was doubled on November 28, 1926, and the Hounslow West platforms rearranged to provide a three-road terminus. The old station buildings were later demolished and a Morden type heptagonal booking hall with shops, car park and bus forecourt was opened on July 5, 1931. The other Hounslow stations were not changed.

The old Osterley station at Thornbury Road was replaced by a new station on the Great West Road on March 25, 1934. The one-storey booking hall was surmounted by a plain brick 'sign tower' incorporating an illuminated bull's-eye. Above the tower was a small concrete lighting beacon. This carried a vertical row of opal glass lampshades on each face.

Boston Manor also had new street-level buildings, completed on the same date, with a further variation on this lighting tower theme, in brick and concrete with lamps inside a single vertical moulding. At rail level the original District wooden canopies remained.

Northfields was ready for the Piccadilly extension of January 1933. It was an example of the Holden brick style, with a large rectangular booking hall, and shops flanking a wide but low entrance. A separate exit at Weymouth Avenue, reached by a high-level footway, was finally closed in 1942.

South Ealing was only a quarter of a mile from Northfields, but 1·37 miles from Acton Town, and it was proposed to re-site it at Ascott Avenue (the Northfields Weymouth Avenue exit was part of this scheme). Meanwhile, at the old site, a temporary booking office was erected south of the line and the original platforms became islands. Subsequently the re-siting proposal was abandoned for various reasons, including the difficulties of providing bus connections and the long-established link between South Ealing station and the adjacent small shopping centre. A new 145-foot awning and glazed waiting room were provided on the eastbound platform in 1936, but the outbreak of war prevented complete rebuilding. The 'temporary' booking office of 1931 remains in use today—a sad little legacy of a 25-year-old change of plan.

Acton Town, rebuilt in the Holden brick style and completed on January 9, 1933, had a large oblong booking hall internally decorated with three bands of black tiles edged with red. The platform canopies were in rather stark concrete, with central glazing. Five platforms were provided, four for the Piccadilly and District through trains (giving cross-platform interchange between the Piccadilly trains on the inner pair of tracks and the District on the outer) and a short one for the District's South Acton shuttle service. There was a unique type of train indicator, consisting of a glass box housing enamelled plates which revolved to come into view.

The new Hammersmith station, ready in June 1932, had two island platforms, for Piccadilly–District cross-platform interchange, with canopies, columns, stairways and footbridges in uncompromising reinforced-concrete. Stairs led direct from the Broadway booking hall to the platforms, but the exit was by a separate set of stairs leading to a covered way at street level, which opened into the booking hall. A third footbridge and stairs served the Great Church Lane entrance. The Broadway booking hall was given a rather 'Morden' type exit to Queen Caroline Street, but Green's original façade to Broadway was retained. Barons Court underwent no substantial alteration.

A NEW HUB FOR THE EMPIRE

To cope with the extra traffic brought by the extensions and an increasingly lively West End, the policy of rebuilding and enlarging the central area stations was continued and expanded; wherever traffic justified, escalators were to replace lifts.

The most ambitious reconstruction of the period was at Piccadilly Circus, where traffic had risen from $1\frac{1}{2}$ million passengers in 1907 to 18 million in 1922. In Regent Street, schemes were in hand to replace the existing buildings by six-storey structures and there were other plans for redevelopment in the vicinity, all of which promised to increase traffic. The existing surface station, with its entrances from the Circus, Haymarket and Jermyn Street, had received extra booking offices in 1917, but the site would not allow further expansion. Another disadvantage of the position was the long distances between the lower lift landings and the platforms.

In these circumstances the only satisfactory solution was an entirely new booking hall beneath the Circus itself, and the necessary parliamentary powers for this were obtained in the LER Act of 1923. The original scheme included a domed booking hall immediately beneath the Shaftesbury Memorial, with the escalators for the Bakerloo and Piccadilly lines leaving the hall in opposite directions. After experiments at the Empress Hall, Earl's Court, with full-size wood-frame and millboard mock-ups of escalator heads and tunnels, a revised layout was adopted, with a flat ceiling to the booking hall and parallel banks of three and two escalators down to an intermediate landing whence a bank of three escalators would lead to each line.

The first part of this £500,000 work, begun in 1924, was to reinforce the concrete roadbed with steel rails. This had to be done piecemeal to avoid interference with road traffic. The only suitable working site was the road island occupied by Eros (the Shaftesbury Memorial). The statue was removed to a temporary home at Victoria Embankment Gardens, and in February 1925 construction of the 18-feet diameter, 92-foot deep working shaft was begun.

The ground beneath the Circus was riddled with service mains of various types, which were brought into a new 12-feet diameter pipe subway of cast-iron segments. This was 550 feet long and encircled the booking hall site although the depth varied from 18 to 30 feet. Gas mains and an LCC sewer also had to be diverted away from the station site.

The roadway above the booking hall had to be supported by a series of steel girders and joists. This spider's web of steel was assembled in Josiah Westwood & Co's yard at Millwall before being numbered and dismantled for piecemeal insertion.

In accordance with Clause 34 of the 1923 Act, mock-ups of the street-level stairwells were erected outside Swan & Edgar's store in February 1927 for inspection by Westminster City Council. The permanent stairwells were enclosed by bronze railings, and had the usual illuminated bull's-eyes above.

For the main opening ceremony on Monday, December 10, 1928, the walls of the booking hall were draped with coloured hangings, and a buffet was installed. The opening was performed by the Mayor of Westminster, Major Vivian B. Rogers, DSO, MC, JP, who started the escalators and bought a 2d ticket made of ivory from a machine. The public were admitted from 3 p.m.

The elliptical booking hall was 9 feet high and measured 155 and 144 feet along its major and minor axes. The peripheral wall was broken at five points to give access to the radial subways which connected with the stairways to the street. At the western end of the hall, the widest subway gave access to two stairways to the street outside Swan & Edgar's, and to the basement of the shop itself. This store displayed its wares in showcases lining the peripheral wall. Also in the wall were a row of telephone booths, and an auxiliary booking office to supplement the twenty-six automatic electric ticket machines. These machines, which delivered pre-printed tickets for denominations up to 6d, formed portals to the centre of the hall. The show cases, telephone booths, booking office and the five shops which backed on to the escalator headwall were finished in bronze with a frieze and skirt of Travertine marble, from Tivoli. A further touch of opulence was given by the red scagliola finish of the columns, with narrow bronze fillets at the angles. At the top of each column were twin lampshades. Near the five shops the 'world clock map' was later installed, showing the time throughout the world by means of a time-band moving across a static Mercator's projection.

On the wall above the escalator heads were oil paintings by Stephen Bone. The central feature was a map of the world, and the side panels, finished in March 1929, portrayed objectives of Underground travelling: Business and Commerce; Outdoor Pleasure; Shopping; Amusements of the Town.

Five Otis 100-feet-per-minute escalators (three in one shaft and two in another) connected with a 90 foot by 32 foot intermediate landing 42 feet below. Thence three escalators descended a further 21 feet to a chamber east of the southbound Bakerloo platform. From this point, stairs descended 10 feet to the platforms. Another three escalators spanned the 39 feet from the intermediate landing to a landing 12 feet above the north end of the eastbound Piccadilly line platform. The three-escalator shafts were 22 feet 4 inches in diameter, the two-escalator shaft 16 feet 4 inches. Bronze pedestal lamps shone upwards to the white matt distempered ceilings of the shafts.

The consulting engineer was H. H. Dalrymple Hay, and the architect Charles Holden. The main contractors were John Mowlem & Co. Ltd.

The old station was closed on July 21, 1929, and an arcade of shops built on the site, preserving the three street entrances. The arcade was connected to the new booking hall by a subway starting near the old Piccadilly Circus entrance

Much difficulty was experienced with water percolating from the roadway to the booking-hall ceiling and, after rainstorms, porters were always busy mopping up with pails and brooms. Remedial works began in June 1931 and were completed in the following year.

The Shaftesbury Memorial was returned to its former home on December 27, 1933, just in time to be traditionally damaged by a New Year's Eve reveller.

CENTRAL AREA IMPROVEMENTS

Most of the other central area stations of the Piccadilly line were rebuilt or modernized within a year or two of the extensions being opened. This reconstruction programme, together with the installation of new signalling and the provision of new rolling stock, transformed the Piccadilly into a modern tube railway.

The lifts between the Piccadilly line and the street at Earl's Court were converted to fully automatic operation from October 9, 1932, incorporating some equipment from a 1928 experiment at Warren Street. From October 1934, a new method was adopted to warn passengers that the gates would soon close—a beam of light shone through a moving strip of film on to a photo-electric cell; the film carried a sound track, and by the operation of an amplifier and control relays, loudspeakers gave the warning 'Stand clear of the gates, please', as they were about to close, replacing the buzzers used until then. 'Gates closing' signs were also illuminated.

At Knightsbridge, a new sub-surface booking hall with three separate street entrances was excavated beneath the Knightsbridge/Sloane Street road junction. The platform tunnels were re-tiled in what was described as the 'new biscuit colour' and the new station opened on February 18, 1934. A separate western entrance, with two escalators direct to the west end of the platforms, a booking hall beneath Brompton Road, and a subway approach from Hans Crescent, was added on July 30, 1934.

Brompton Road had long been lingering in the twilight of impending closure. So many trains passed through it without stopping that befuddled passengers trying to alight there felt constrained to write to *The Times*. Its fame spread to the theatre, and a Marie Tempest play was entitled *Passing Brompton Road*. It was hopefully closed during the 1926 coal strike, but reopened (on weekdays only) from October 4, 1926, after questions had been asked in the Commons. The construction of the Hans Crescent outpost of Knightsbridge station sealed its doom and the old station was shut for ever on July 29, 1934. The trains that had non-stopped Brompton Road now began to non-stop Barons Court instead, to the annoyance of local business people.

Hyde Park Corner was another example of rebuilding with a sub-surface booking hall and escalators. The new booking hall, opened on May 23, 1932, was beneath the extreme eastern end of Knightsbridge, with one stairwell on each side of the road. The escalators, with a fixed stairway between, fed the eastern ends of the platforms and the ground floor of the old surface station became a Lyons teashop. Showcases in the new booking hall housed a series of models illustrating the development of the LGOC bus, with appropriately painted street backgrounds.

The next station, Down Street, was closed after traffic on May 21, 1932, and a reversing siding built between the running roads on part of the site of the station tunnels. The new siding, available from May 30, 1933, was 836 feet long in a 14 foot 6 inch tunnel, and held two seven-car trains. This generous diameter gave room for a walkway and allowed a disabled car to be lifted from its bogies. A shallow pit and special lighting were available for car inspections. The siding was built for the dual purpose of providing a refuge for a disabled train and of reversing part of the service if the traffic from the northern section of the Piccadilly line should be heavier than that from the western. A seven-foot passage continued from the end of the siding tunnel to the Hyde Park Corner crossover tunnel. One of the Down Street lift shafts was retained for ventilation plant. The surface station, in Down Street some yards north of Piccadilly, was in a quiet locality, only 0·31 miles from Hyde Park Corner, and its very light traffic made it an obvious choice for closure.

Dover Street station was also hidden away north of Piccadilly on a second-best site. From the lower landing it was a long walk to the platforms under the main road. Improvement was achieved on September 18, 1933, when a subsurface booking hall was opened below Piccadilly at the eastern end of Green Park. On the south of Piccadilly, just west of the Ritz Hotel, two stairways led down from a small but dignified stone-clad pavilion erected on a strip of park land. A northern stairwell was accommodated in Devonshire House. Two escalators and a fixed stairway gave access to the western ends of the Dover Street platform tunnels. With the re-siting, the station was renamed 'Green Park' and the surface buildings put to other uses. They now have the distinction of housing *inter alia* the entrances to both a Lyons teashop and an Express Dairy teashop.

Although inaccurately named, Leicester Square station was well sited, but the booking hall was unsuitable for the growing traffic and did not lend itself to the installation of escalators. At the end of 1925, one of the lifts was experimentally speeded from 180 to 290 feet per minute, saving eleven seconds on the journey and providing a 30% increase in capacity. The other lifts at this station, and those at Holborn, were similarly converted in 1926–1927.

The financial assistance afforded by the Development (Loan Guarantees and Grants) Act made it feasible to proceed with a £400,000 scheme for a large circular booking hall partly beneath the site of the surface station and partly beneath the junction of Charing Cross Road and Cranbourn Street. Associated with the scheme was a new sub-station in Upper St Martins Lane with three 2-mW steel tank rectifiers and a switch house. The scheme (begun in October 1930) was generally similar to that at Piccadilly Circus, with the diversion of service mains to a 12-foot pipe subway, the construction of new sections of sewer, and the driving of headings to erect roof columns and girders. In some ways it was more complex than Piccadilly, as the existing lifts occupied part of the site. On the other hand, more working sites were available.

Before the booking hall could be built, the contractors had to underpin the surface station building, five columns of the Crown Hotel and two columns of the London Hippodrome. The old elephant run, dating from the time when circuses were held at the Hippodrome, disappeared during this work, and the new hall used its site. For the installation of the roof girders, the engineers had possession of half of Charing Cross Road and Cranbourn Street at a time.

One of the escalators was available on Saturday, April 27, 1935, for the Cup Final crowds, and the new station was opened on Saturday, May 4th, in time for the heavy Silver Jubilee traffic. Each 22 foot 9 inch shaft contained three 180-feet-per-minute escalators. The Hampstead escalators were 117 feet 6 inches long, with a rise of 58 feet 9 inches, and the Piccadilly set were 161 feet 6 inches long, with a rise of 80 feet 9 inches. These were the longest escalators in the world at that time, but have since been surpassed by others, including those at the Tyne pedestrian tunnel, and on the Moscow and Leningrad Metros.

In the centre of the new booking hall was a large circular booking office, supplemented by a crescent of ticket-and-change machines. The perimeter of the hall had shops, stalls and telephone booths. There were two street entrances on the west side of Charing Cross Road (one had direct doors into the Hippodrome theatre) and one on the east. A further Charing Cross Road entrance and another in Cranbourn Street were opened on June 8, 1936. The station tunnels of both lines were re-tiled. Seats and automatic machines were flush-fitted in the tunnel walls, and the signals and headway clocks in bronze panels in the headwalls. The main contractors were John Mowlem & Co. Ltd and John Cochrane & Sons Ltd, supervised by Sir Harley Hugh Dalrymple Hay, consultant engineer to the LPTB.

The 170-yard street interchange between Holborn (Piccadilly line) and British Museum (CLR) had long been a weak point of the London underground scene, and proposals for a low-level interchange subway had been mooted as early as 1907. In 1914 parliamentary powers were obtained for a costlier but more satisfactory scheme—moving British Museum station eastwards to Kingsway. With the outbreak of war the plan remained in abeyance. Fresh powers were obtained in 1930, and the contract was let in October of that year. The surface station, at the south-east corner of the Kingsway/High Holborn intersection, was enlarged and rebuilt to modern standards, with diffused lighting and rubber floor-covering. From the booking hall, a bank of four escalators in one shaft descended 76 feet to the intermediate landing. From here, passages and stairways led to the Central London platforms, which were housed in new 380-foot primrose-tiled tunnels built round the running tunnels. The platforms were, of course, on the outsides of the tracks as there was insufficient room between.

From the intermediate landing there was a further bank of three escalators descending 49 feet 8 inches to a point midway in height between the Piccadilly westbound and eastbound platforms, on the site of the lower lift landing. An easy flight of steps descended to the westbound main-line platform, but access to and from the eastbound and Aldwych platforms was awkward, as the long passages and steps between the two station tunnels were not altered.

Two of the upper escalators and the Piccadilly flight were ready on May 19, 1933, and the new station was completed (for the Piccadilly line) on May 22nd, being renamed 'Holborn (Kingsway)' from that date. The remaining escalators and the Central London platforms were in service from September 25th (British Museum station closed after traffic on the previous day). At British Museum, the surface buildings became shops and the platforms were demolished. A hump in the tracks remains as a reminder, together with the station tunnels and siding.

The new interchange facilities were immediately popular, and by 1938 the volume of interchange traffic had grown to ten times that previously exchanged between Holborn and British Museum.

North of Holborn the situation was less happy. At Russell Square a subsurface escalator station had been authorized by Parliament in 1930 and 1931, with street entrances from the corners of the Woburn Place/Bernard Street/Russell Square intersection. Unfortunately, the worsening financial situation compelled the Underground group to review its 1930 programme, and in February 1932 Lord Ashfield announced that, by agreement with the Treasury, they had cancelled the reconstruction of Russell Square, King's Cross, Edgware Road, Post Office (whose booking hall had been enlarged in 1929) and Sloane Square stations. This reduced the total capital expenditure to roundly £10,880,000 and the proportion ranking for Government grant to £8,672,000. These cancelled schemes were subsequently revived and completed by the LPTB (except Edgware Road and Russell Square). The powers for the latter have been kept alive, and the weight of present-day office, hospital and tourist traffic makes reconstruction more than ever justified.

North of King's Cross, York Road was closed after traffic on September 17, 1932, and a reversing siding built at platform level. In deference to the football club which had moved there from Plumstead in 1913, Gillespie Road was renamed 'Arsenal (Highbury Hill)' on October 31, 1932, and rebuilt in 1933 with a singularly ugly plain concrete façade, relieved only by an outsize bull's-eye.

Finsbury Park had been re-equipped in 1927 with a passimeter, improved lighting and tiling in the street-level subway, and a canopy with floodlit bull's-eye signs at Wells Terrace, but it was not otherwise modernized.

Associated with the extensions to Ealing Broadway, Edgware and Morden, and with the general growth in traffic, were various improvements to Hampstead and Central London Railway stations.

The reconstruction of Charing Cross lasted three years, and was completed in February 1929. With the construction of the new southbound Hampstead line platform, opened in September 1926, three new independent shafts were built between this platform and a greatly extended intermediate circulating area below the District tracks. Two of these shafts were equipped with reversible escalators and the centre one with fixed stairs. Two further escalators, making a total of eight, were available from December 4, 1928. These took passengers from the intermediate circulating area direct to street level. There was also an additional staircase from this level to the westbound District platform. Possible alternative routes were so numerous that the now-familiar Underground system of 'Follow the . . . light for . . .' was introduced experimentally. A green light led to Victoria, a blue one to Waterloo and, with a fine feeling for the appropriate, that for Piccadilly was red.

At street level a greatly enlarged booking hall was built, 112 feet long, 83 feet wide, 13 feet high and with an 18-foot centre clerestory supported by steel columns clad in glazed terracotta. As at Piccadilly Circus, the intention was that most tickets should be issued from automatic machines, some twenty machines and a change kiosk being available. There was also an auxiliary booking office, a season-ticket office, and an inquiry office. Shops were erected on the site of the old booking offices. The large space at the western end of the booking hall was used for small exhibitions. One early use was for a model underground railway, and on another occasion a nine-hole miniature golf course was laid down.

At Tottenham Court Road the 1925/6 reconstructions did nothing to relieve the congestion at the northern end of the Hampstead platforms, where the exit and interchange passages were concentrated. A new passageway was opened on April 27, 1933, between the Hampstead platforms at their south ends, leading to a single 'up' escalator. At the top of this, a long subway brought the passenger to the foot of the upper bank of escalators.

Warren Street was rebuilt as a two-flight escalator station. The four escalators were in service from September 27, 1933, and the new, resited booking hall in 1934. Above the station were three floors of flats, semicircular in plan.

Camden Town acquired two escalators from October 7, 1929, and was also equipped with passimeters. Kentish Town had two escalators from November 21, 1932, a new booking hall with tiling in biscuit, blue and black, and a wooden passimeter.

The reconstruction of Highgate (now Archway), promised in 1927, was completed in 1931. Two escalators replaced the lifts, and there was a new octagonal booking hall, with a central passimeter, and shops

Confusion at Finsbury Park—rush hour passengers from the Piccadilly and GNCR tubes struggling to board trams and buses for Tottenham, Stamford Hill, Wood Green and Enfield in April 1930.—*London Transport.*

Order at Manor House—the special tramcar loading islands provided at the new Manor House tube station in 1932 (LCC tram 1613 on route 53 and MET tram 249 on route 27).—*London Transport.*

Sentinel steam wagons removing excavated clay from the site of Turnpike Lane station, Piccadilly line, March 26, 1931.—
London Transport.

Sudbury Town—the first and perhaps the finest of Dr Charles Holden's 'brick-style' stations; photographed in April 1934.
—*London Transport.*

PICCADILLY PROGRESS

and telephone booths in the perimeter. Arcades with stalls led to the original street entrance in Junction Road and the 1912 entrance in Highgate Hill. At the latter entrance there was a tall façade of metal-framed windows set in concrete, in the Morden style.

In 1928 Golders Green acquired the inevitable passimeter and an additional covered exit from an extended northbound platform, leading down the south side of the embankment at a point halfway between the booking office and the Finchley Road; this new exit was designed to speed the interchange from northbound trains to trams.

Monument and Bank stations were linked on September 18, 1933. Two escalators, with fixed stairs between, rose 60 feet from the south end of the Morden–Edgware platforms at Bank to a point beneath King William Street, whence subways and stairways led to the western ends of the District platforms at Monument. Passengers for the Morden–Edgware line could book at Monument.

On the Central London, Marble Arch booking hall was rebuilt beneath the road junction, with street entrances from the north side of Oxford Street, and from the south side at the corner of Park Lane and near the Hyde Park gates. Two escalators and a staircase replaced the lifts, and the new station was opened on August 15, 1932.

Chancery Lane was similarly reconstructed, and the street access was moved about 100 yards east, away from the thoroughfare from which the station took its name, to the junction of Holborn and Grays Inn Road, with two street stairwells south of Holborn and one on each northern corner. Three escalators descended 65 feet to the eastbound platform level and from here two more escalators, flanking a fixed staircase, covered the remaining 15 feet to the westbound platform. The eastbound platform tunnel was lengthened slightly, and the new station was opened on June 25, 1934. In the next month it was renamed Chancery Lane (Grays Inn).

At Oxford Circus a third Bakerloo escalator, in its own shaft, was placed in service from October 2, 1928.

NEW AND IMPROVED ROLLING STOCK

Whilst the civil engineers were implementing this intensive programme of station improvements, their mechanical engineering colleagues were carrying out an equally energetic scheme of rolling-stock modernization.

Service experience with air-operated doors on the Cammell-Laird and standard stock had emphasized their marked superiority over gates or hinged doors, the principal advantages being shorter station stops and substantial economies in labour costs. It was evident that the whole fleet would have to be equipped with air doors before long, and attention was first directed to the possibilities of converting existing stock.

As mentioned in Chapter 9, twenty of the 1906 motor cars had been 'air-doored' to run with the 1920 Cammell-Laird cars, and towards the end of 1925 a similar conversion programme was begun for the whole of the CLR stock. The gated platforms were enclosed and bulkheads demolished, thus incorporating the platform space in the main saloon. Two 3 foot 6 inch door openings were made in each side of the trailers, with a single-leaf sliding door for each opening, or four per car. The motor cars also had the platforms enclosed, but with the guard's compartment separated by a bulkhead and with its own hinged door in the car side. Further along the motor car was a larger opening with double-leaf sliding doors.

The door control system was interesting. Probably owing to the wooden construction of the trailers, control was pure-pneumatic instead of electro-pneumatic. A high-pressure air line along the train was filled or exhausted by the guard's control. This worked a pneumatic relay in each car which operated the valves for the low-pressure door engines. The closing of the doors proved to be very slow, and electrically operated valves were fitted to each car to speed up the exhaustion of the control line.

Internally the cars were transformed into very good copies of the contemporary standard stock, with the usual grey and brown lozenge-patterned moquette upholstery, standard lighting in white opal hemispherical shades, and improved ventilation. One car was experimentally converted in 1925, and the rebuilding of the whole 259 cars was completed by March 1928. Two guards shared the task of controlling the doors on a six- or seven-car train. Eight of the 1903 motor cars were re-equipped with 240-h.p. motors, new motor bogies and automatic control equipment, and eight further trailers were converted to control trailers (in addition to the sixty-four already converted) to provide four additional trains for the Ealing service.

The price of more doors had to be paid in fewer seats. From 48 or more on the trailers, 42 on the 1903 motor cars and 32 on the Ealing motor cars, the number of seats declined to 40 on the trailers, 36 on the control trailers, and 30 on the motors, or, for the whole fleet, from 11,716 to 9,192 seats (1920 compared with 1928—259 cars).

From a financial viewpoint, the conversion was an outstanding success, since what was effectively a new fleet had been obtained for the cost of reconstruction. These cars continued in passenger service until 1938/9, by which time most of the trailers were thirty-nine years old.

With this success, thoughts turned towards a similar conversion programme for the 1906/7 stock of the LER lines, but the steel bodies were less amenable to alteration than the wooden ones of most of the Central London stock, and it was decided to scrap the gate stock completely and replace it.

Following the 1923–1925 orders for standard stock described in Chapter 9 new orders were placed with manufacturers as follows:

Year Ordered	Year Delivered	Manufacturer	Motor Cars	Trailers	Control Trailers	Total
1926	1928	Metropolitan Carriage, Wagon and Finance	64	48	—	112
1927	1928/9	Ditto	110	160	36	306
1928	1929/30	Union Construction Co.	77	37	68	182
1929	1930/31	Ditto	18	17	18	53
1930	1931	Metropolitan-Cammell Carriage, Wagon and Finance	22	20	20	62

The 1926–1929 cars were intended to replace the 1906/7 stock on the Bakerloo, Morden–Edgware and Piccadilly, and for general augmentation. The 1930 order was to replace the existing joint stock on the Bakerloo service beyond Queen's Park.

The Union Construction Co., Ltd, manufacturer of the 1928 and 1929 batches of cars, was an associate company of the Underground group, and was older than the UERL. The construction company had been founded in 1901, presumably with the intention that it should act as subcontractor to the UERL for constructing the lines or rolling stock, on the pattern of the construction companies in the American cities with which Yerkes was associated. The name 'Union' appears to be a direct Yerkes importation from Chicago, with its 'Union Traction Co.', 'Union Elevated Loop', etc. As events turned out, Yerkes found it more expedient to use the UERL itself as the 'contractor', and the UCC lay dormant for many years. Its entire issued capital consisted of 157 £20 shares, all held by the UERL. The name was changed to 'Union Construction & Finance Co., Ltd' early in 1929.

In November 1927, Frank Pick told reporters about a programme to replace the gate stock, mentioning that Acton Works had been enlarged (as described in Chapter 9) and that a temporary works had been provided at Feltham, Middlesex. The buildings, at Victoria Road, Feltham, had been constructed during the 1914–1918 war by the Whitehead Aircraft Company. After the end of the war they had lain derelict until part was taken over by the Army for an Ordnance Depot, and the other part by the Union Construction Co. about 1925, initially for the rehabilitation of the Central London cars. The Feltham works produced a total of 241 tube cars, 45 District Railway Cars, 103 tramcars (the famous 'Felthams') and 60 trolleybus bodies (for the 'Diddlers')—a very creditable achievement for a paid-up capital of only £980! The Feltham trams had several features in common with contemporary tube stock, including metal body framing and sheeting, aluminium-alloy doors (with an air-operated sliding door at the front), a smooth external finish, air brakes, weak field control notches and electric heating. The company vacated the works on March 25, 1932, and was voluntarily liquidated on the formation of London Transport, which had no power to manufacture railway rolling stock (except experimental cars).

The Morden–Edgware line was the first to benefit from the delivery of new stock, and its last gate-stock train was withdrawn after close of traffic on January 31, 1929. Next, the Piccadilly gate-stock was replaced with standard stock between March and July 7, 1929.

Deliveries of standard stock also released the 1920 Cammell-Laird stock for use on the Bakerloo after conversion to modern standards. One car had been experimentally converted in 1925, with cross-seats and moquette upholstery, and the rest were done in 1929/30, with extra lights above the windows, standard lampshades and colour schemes of green and cream internally, red and cream externally. Control and communication equipment, and the end doors of the control trailers, were also modernized. The 1906 motor cars that had been converted to run with the 1920 trailers were either scrapped, demoted to works cars, or converted for use on the Aldwych branch. The latter were two of the French cars, converted to air-doors and double-ended for driving from either end, to run singly in the off-peak and coupled in a pair in the peak; they replaced two double-ended Hungarian cars with capstan-operated doors and were scrapped in 1956. The converted 1906 motor cars were replaced by 20 of the 1928 batch of UCC motor cars, and these 20 three-car sets of 1928 motor cars and 1920 trailers ran between Elephant & Castle and Queen's Park until 1939, when the 1920 cars were withdrawn from passenger service.

Thus the last gate-stock car of the Underground group ran on the Bakerloo line, leaving the Elephant & Castle at 12.15 a.m. on January 1, 1930. At that time it was claimed that 1,219 new tube and surface cars had been put into service at a cost of nearly £5 million. The Brush and Leeds Forge cars of 1914, and the 1906 trailers converted to motors, were also scrapped in 1930.

The savings in staff costs were material. A six-car train of gate-stock needed a driver, guard and four gatemen. With air-doors, one guard controlled a maximum of four cars up to April 1927, but after the successful development of 'intercom' telephone equipment between driver and guard, trains of up to seven cars were controlled by one guard. Guards' wages were increased by 4s 10d per week, and the displaced gatemen were retained as porters to give additional platform supervision at their old rate of pay.[1] For a six-car gate-stock train, trainmen's wage costs per car-mile were just under 1·2d, but for a similar air-door train only ½d. Total train running costs per car-mile fell from 10d in 1925 to 8½d in 1929, and were further reduced when the surplus staff had been absorbed.

[1] Some were allowed to transfer voluntarily to the LGOC platform staff. In 1931 the financial position put an end to the employment of surplus staff.

With air-doors, station stops were substantially cut (from 50 seconds to 25 at busy stations), enabling more frequent services to be run with the same number of trains.

The stock ordered from 1926 to 1930 was of generally similar design to the 1923–1925 stock described in Chapter 9, but detailed improvements were made as deliveries progressed. On the driving bogies of the motor cars, the wheel diameter when new was reduced to 3 feet from the 1926 order onwards and the wheelbase to 6 feet 6 inches. Roller armature bearings were fitted from the 1928 order onwards. The diameter of the wheels on the trailer bogies was reduced to 2 feet 6 inches with the cars ordered in 1927. From the 1926 order onwards, the moulded external waistbands disappeared, and the windows were set flush with the body sides. The roof-level roller blinds were replaced by destination plates in the lower part of the right-hand window of the driver's cab; the train running number was also transferred to this position some years later. The sliding doors of the later deliveries were externally smooth instead of ribbed, and the motor cars after the 1927 Metropolitan and 1929 UCC batches were without the fixed post between the doors.

The drop windows of the 1923 stock gave place in the 1924 batch to inward-tilting quarter-lights, and the old heavy metal frames to the draught screens were replaced from the 1926 batch onwards by a rimless design, with the grab rail extending from seat level to the lower edge of the clerestory. Lampshades had shorter and neater stems.

A variation affecting all tube cars was the distinctive marking of non-smoking cars, instead of marking those reserved for smokers as hitherto. From November 22, 1926, such cars, which were motor cars, carried a blue star with 'no-smoking' in white, superimposed on a green disc. Concurrently with this change, smokers were allowed a greater proportion of cars than before; they were also given the freedom of the lifts from February 1926.

The Watford replacement stock ordered in 1930 embodied two technical advances to make these cars suitable for longer high-speed runs in the suburbs—electro-pneumatic brakes and weak field control. The last-mentioned addition to the control system allowed higher speeds at the expense of increased current consumption. At first it came into operation automatically, after the 'full parallel' notch, but a 'flag-switch' was later interposed in the circuit. This was closed by the driver on reaching certain fixed points on the open sections, and switched out when leaving the open sections for the tube.

It was possible to use stock with standard floor heights because, on the Queen's Park–Watford section, platforms were at 'compromise height' (2 feet 9 inches). On this stock, the wheel diameters reverted to 3 feet 4 inches on motored axles and 2 feet 8 inches on the trailers.

The 1920 Watford Joint stock had defied attempts at air-door conversion, and was withdrawn after the very short life of 10/11 years, except for six motor cars and three trailers rebuilt by the LMSR with modified doors and used in three three-car sets (M-T-M) for the Watford Junction–Rickmansworth and Croxley Green shuttle services until 1939. They were scrapped early in 1949.

The Piccadilly extensions of 1932 and 1933 called for cars capable of fast loading and unloading and of high speeds on the long runs between stations, particularly on the $2\frac{3}{4}$-mile stretch between Hammersmith and Acton Town. Longer distances between tunnel stations required more attention to car ventilation.

In its final fling as a manufacturer of tube stock, the Union Construction & Finance Co. built a six-car experimental train, which ran on the Piccadilly line from mid-December 1930 and on the Morden–Edgware in 1930–1931. The two motor cars, 50 feet $9\frac{1}{2}$ inches over headstocks, were about 1 foot longer than standard stock cars. The extra length was at the trailing end of the car, where the sides were tapered in. The guard's compartment had the usual metal bulkhead in one car, glazed screens in the other. The access doors to the guard's vestibule were of the single-leaf air-operated pattern, and could be used by passengers in the absence of the guard. The trailers, at 51 feet $9\frac{1}{2}$ inches, were 2 feet longer than their predecessors. On two of them the extra length was mainly devoted to increasing the opening left by the two double doors from 4 feet 6 inches to 5 feet 2 inches, and the number of seats was kept at forty-eight. On the other two, there were four openings on each side of the car, two with double doors with a 4 foot 6 inch opening, and, at the car ends, two with single doors with a 2 foot 5 inch opening. The extra end doors gave speedier loading and unloading at the expense of eight fixed seats per car. These cars had improved tipping quarter-lights, and a continuous outside wind scoop between the doors, connected to grilles inside. The 48-seat trailers had a double layer of coir anti-noise screening beneath the floor, and twelve electric radiators.[1] All cars had additional lighting, brighter upholstery, electro-pneumatic brakes and some aluminium fittings.

When sufficient experience had been obtained with the experimental cars, orders were placed early in 1931 for the Piccadilly extension fleet: 145 motor cars from Metropolitan-Cammell, 90 trailers from Birmingham and 40 trailers from Gloucester—275 cars in all. These were delivered in 1932 and 1933. The main features of the experimental cars were retained, including the increased length (51 feet $5\frac{1}{8}$ inches over

---

[1] The 1923 experimental cars and the 1923–1925 batches of standard stock were equipped with heaters when delivered, but as these early heaters trapped dust and became a potential fire risk, they were taken out of use. An improved type was fitted to the 1930 Watford Replacement stock and to the 1931 and 1934 stock described below. Older stock was equipped with the new heaters from 1930 (Piccadilly) and 1935 (Morden-Edgware and Bakerloo).

headstocks) and the adoption of four door openings per side on trailers. The loss of eight seats per trailer was partly mitigated by the installation of four tip-up seats, two in each end-wall, and two similiar seats in the motor cars. On each motor car, a 10 foot $7\frac{1}{2}$ inch length was still sterilized by the control and compressor gear so that the number of fixed seats remained at thirty.

Single-leaf doors gave a 2 foot 3 inch opening, and the double-leaf gave openings of 5 feet 11 inches on the motors and 4 feet 6 inches on the trailers. The guard's controls were incorporated in neat panels in the end walls of the motor cars, at the ends remote from the cabs, and glass screens replaced the metal bulkhead separating the guard from the passenger compartment. Door engines and controls were by G. D. Peters, the electric traction control equipment and air compressors by BTH, and the 240-h.p. traction motors by GEC. Weak field control was fitted, allowing a balancing speed of 50 m.p.h. on level track.

Westinghouse automatic electro-pneumatic brakes gave quicker and smoother operation, and synchronized braking action throughout the train. The Westinghouse quick-acting pneumatic brake was retained for emergencies. To reduce the unladen weight, doors, seat frames and many minor fittings were made of aluminium alloy; fittings designed to be handled by passengers were covered in black plastic material. The power of the lamps in the passenger compartments was raised from 40 to 60 watts, but they were still fed direct from the conductor rail; emergency lights were, as usual, supplied from batteries.

Seat moquette of various colours and designs was used, and light-oak woodwork instead of mahogany. The most usual colour scheme was flowery blue moquette, and blue and cream internal paint-work. Both internally and externally, the cars had a bright smooth appearance, the result of careful attention to design details and the use of flush panelling—the smoothed exterior also allowed the cars to pass easily through the new washing machines.

These cars were joined by a final batch of twenty-six Metropolitan-Cammell motor cars ordered in 1934, of very similar general design (but with roller bearing axles) and the last examples of standard stock to be built. Their delivery enabled the remaining six-car trains on the Piccadilly to be made up to seven cars.

FIGHTING ICE AND SNOW

As soon as the tube railways were extended for long distances above ground they became liable to disruption when frost or frozen sleet formed an insulating layer over the conductor rails. At first, wire brushes were fixed to the shoe-beams of ordinary motor cars and would be lowered to clean the conductor rails during traffic hours.

In the early 'thirties, special ice-wagons were built. These had serrated rollers and steel brushes, and carried tanks of calcium chloride solution which was electrically heated and blown on to the conductor rails by compressed air, through sprayers. These wagons, with a motor car at each end to supply traction and air, ran at 10–15 m.p.h. between service trains, or after traffic hours. The use of 'Kilfrost' anti-freeze solution, which came later, simplified matters as it could be fed by gravity to the steel brushes. If an ice-film formed suddenly or the ice-train stalled, wooden-handled scrapers were pushed over the rails by staff, the train following slowly behind.

SERVICE DEVELOPMENTS

Many aspects of the Piccadilly line replanning gave evidence of the pursuit of higher speeds—more and wider car doors, and weak field control; wider station spacing on the extensions; the closure of little-used central area stations and the non-stopping of others; the four-track section from Barons Court to Northfields; wider platforms, re-signalling. These factors made the Piccadilly appreciably faster than the other lines. The average speed for all the Underground group railways was 18 m.p.h., but from Piccadilly Circus to Rayners Lane or Hounslow West it was now 23·7 m.p.h., to Northfields 24, to Arnos Grove 21 and to Cockfosters 22. From Finsbury Park to Cockfosters it was 25 m.p.h. (all these speeds include the time taken for station stops).

During most of the era under review, new timetables were issued twice a year, for summer and winter operations. The services were reduced in the summer and increased in the winter, to cater for the heavier demands brought by bad weather and by theatre traffic. Thus, each autumn, the energetic Underground press relations office was able to make the most of what was to a large extent a normal annual event. Throughout this period, the Bakerloo peak service hovered round the 26 trains per hour level, the Piccadilly 24 tph and the Central London 30 tph,[1] compared with 1910 frequencies of 34, 26 and 27 respectively. Nevertheless, there was some real progress. With the gradual development of the open country beyond Golders Green, the Hampstead service increased from 24 tph in 1925 to 28 (29 in the maximum hour) in 1929, although this was but a pale shadow of the 44 lighter-loaded trains per hour of 1912. A phenomenal traffic increase on the Morden line resulted in the operation of 36 tph by 1928.

Perhaps the most striking feature of this era was the development of weekday normal hour services. As early as 1925, most lines had a midday normal frequency roughly equal to that of the peak hours, although with shorter trains, usually of three cars, sometimes four. In

---

[1] In July 1924 it was claimed that improved signalling on the Central London was allowing 40 tph, but this had reverted to 29 tph by 1926.

March 1930 the Central London midday normal service was increased to 37 tph.

The Hampstead line was again in the forefront of non-stop developments. Theatre trains were resumed in the early 1920s, leaving Charing Cross between 11 and 11.20 p.m., calling at all stations to Tottenham Court Road, then non-stop to Hampstead or Highgate. An interesting non-stop service for business traffic was the 8.58 a.m. train from Edgware introduced on June 13, 1927. This ran non-stop to Golders Green, overtaking the previous train at Brent by a loop outside the platform roads. It also passed Mornington Crescent and Warren Street, arriving at Charing Cross at 9.28, giving an average speed of 22 m.p.h. There was later a corresponding evening return working, using a northbound overtaking loop at Brent. The loops, which were used continuously by the 8.58 a.m. train from Edgware and the 5.12 p.m. from Tooting, were taken out of service on the night of August 22, 1936.

Previously, from September 1925, alternate morning peak trains from Edgware to Charing Cross had run non-stop to Hendon, whilst the non-stopping of various central and inner suburban stations had been in force for several years. During 1926 and 1927, steady progress was made with the development of the LCC Watling Estate, and private enterprise housing also began to contribute traffic to the Edgware extension. The 1927 traffic was 45% greater than the 1926 figure of 8½ million passengers. By 1929 the figure was 17 million, and a deputation from Hampstead Garden Suburb called at 55 Broadway to discuss complaints. The Underground group replied that they were aware of the problems and were seeking solutions. Between 8 and 9 a.m., 15 trains left Golders Green for the West End and 14 for the City. Additional seven-car trains were promised. At this time, 51 of the 90 trains on the Morden–Edgware line were still six-car, and this was the standard peak formation on the Bakerloo and Piccadilly.

The other classic non-stop line, the Piccadilly, continued mainly on pre-war lines, but in 1930 four morning trains from Finsbury Park ran non-stop to King's Cross, arriving at Leicester Square in 10½ minutes, after missing Russell Square or Covent Garden.

THE MONSTER MODERNIZED

More frequent services, longer trains and new extensions all increased the strain on Lots Road. At first, additional plant was installed to work alongside the old, and when this had met immediate needs, the old was progressively replaced.

In 1925 there were 8 old 6-MW sets, 3 new 15-MW sets, 64 old boilers and 8 new. The million-pound replacement programme involved the provision by 1932 of 6 more 15-MW turbo-generators and 24 50,000-lb.-per-hour boilers, thus replacing all the 1905 generators and boilers.

The turbines of the new sets worked at higher steam temperatures than hitherto—275 lb./in.² and 650° F. The 1922 and 1925 boilers, of Babcock & Wilcox CTM type, produced 39,000 lb. per hour at 290 lb. in.² and 700° F. To maintain the steam output during the changeover, four CTM boilers were erected in 1927 outside the west end of the main boiler house.

Other Lots Road improvements at this time included the replacement of all the high-tension switchgear, the construction of a new control room at the east end of the engine-room building, the installation of a water-sluicing system for ash disposal, and driving a new 9-foot tunnel to the river for condensing water. In 1935 a new 15,000-ton reinforced-concrete coal-storage bunker was built on the neck of land south of Chelsea Creek. Plant was provided to unload both rail- and river-borne coal, and the new storage site was connected to the main power station by a bridge and conveyor belt.

FINANCIAL RECONSTRUCTION

In consonance with the thorough reconstruction and renovation of the Underground's physical assets, the UERL's financial structure was altered substantially during the later 1920s. One of the main objectives was the elimination of two anomalous and burdensome classes of capital.

Reconstruction had to await two events—market parity between the dollar and the pound sterling, and the likelihood of some dividends being paid on the UERL ordinary shares. The need for parity arose from interest on the $4\frac{1}{2}\%$ and 6% bonds being payable, at the choice of the holder, at a fixed rate of exchange in either New York, Amsterdam or Frankfurt-am-Main, as alternatives to London. As long as the dollar was worth more in terms of sterling than the fixed rate, it paid European holders to cash their dividend coupons in New York and then exchange the dollars for pounds. The loss on foreign exchange had cost the UERL several tens of thousands of pounds in each post-war year, but the leak was stopped by Britain's return to the Gold Standard in 1925.

Secondly, the suburban extensions of the tube railways were building up a solid longer-distance traffic, and the 1924 London Traffic Act stopped the growth of bus competition, so that in March 1926 Lord Ashfield was looking forward to a dividend on the UERL ordinary shares.

The first step was taken in June 1926 when the UERL directors made a proposal to consolidate the £10 ordinary, and the 1s 'A' shares (issued on the take-over of the LGOC and the New Central Omnibus Co.). Each £10 ordinary share would be replaced by 7·1 new £1 shares, and ten 'A' shares would be replaced by eleven £1 shares. This was approved, and the ordinary capital then consisted of £5,068,878 in £1 shares.

In 1926 $1\frac{1}{2}\%$ was paid on the ordinary shares, the first dividend on the ordinary capital since the company was formed in 1902.

The next, and more difficult, step was to lift the millstone of the £6,330,050 nominal value of 6% tax-free Income Bonds issued mainly in 1908.

The 'lever' held by the UERL was its right to redeem the bonds at par at any time. In July 1927, after payment of a 2% interim ordinary dividend, the UERL announced its proposals for extinguishing the bonds: the interest on the bonds would be taxed, bonds would be exchangeable for ordinary shares, priced at 22s, for two years, and, during this period, bonds would not be compulsorily redeemed, and redemption thereafter up to July 1937 would be at 3% premium. In response to requests from bondholders, the option period was extended to three years and the conversion value of ordinary shares reduced to 21s.

The scheme was approved at an Extraordinary General Meeting of the company on August 26, 1927. Tax liability began from July 1, 1927, and the conversion period extended until June 30, 1930. Only £414,270 of bonds were unconverted when the option expired and these were redeemed on March 2, 1931, at 103%, leaving the UERL with an ordinary share capital of £10,700,990 in £1 shares.

Meanwhile, the dividends on the ordinary shares had been increasing steadily. During the 1920s and early 1930s the ordinary dividends of the UERL and a typical subsidiary were:

UERL			LER		
No dividend before 1926			1919	1⅛%	
1926	1½%		1920	1⅛%	
1927	5%		1921	3¼%	
1928	7%		1922	4%	
1929	8%		1923	4%	
1930	8%		1924	4%	
1931	7%		1925	3¼%	
1932	4½%				
			1926	3½%	
			1927	4%	
			1928	5%	
			1929	5%	
			1930	5%	
			1931	4½%	
			1932	3%	

The proportional distribution of the Common Fund was altered in 1921 and 1928. The latter alteration introduced the principle of dividing the fund in the ratio of the ordinary capital of each participant, with the LGOC's capital counted as double. The proportions were:

	CLR	CSLR	LER	LGOC	MDR
From 1.7.21	16%	6%	41%	25%	12%
In 1929	13·31%	6·56%	41·38%	24·4%	14·35%

One result of the fact that tube extensions could be financed only by Government-guaranteed debenture issues to the public was that the UERL's share of the total capital of its subsidiary companies fell from 35% in 1921 to 25% at the end of 1932. Their capital had increased from roundly £48 million in 1919 to £80 million in 1932. £26 million of this increase had been spent on railway extensions and improvements, and passengers carried on all services rose from 1,376 million to 2,311 million. The UERL had every reason to be proud of this record of its stewardship of London's passenger services.

NEW OFFICES

Since 1898 the District Railway had had offices at St James's Park station and the UERL also found a home there. The Bakerloo offices were at London Road depot and the Piccadilly's at Piccadilly Circus station. From April 1907, the three Yerkes tubes had a combined office at London Road, but in October all moved in with the District, the ensemble becoming known, from 1909, as 'Electric Railway House'. This soon threatened to split its sides, and in August 1922 work began on a new six-storey block for the Underground group on an adjoining site over the MDR tracks. Completed in 1924, these offices were called simply '55 Broadway'.

With the development of traffic, and the growing scale of the Underground group's operations, further office accommodation was needed, and demolition to make way for a new building began in May 1927. Seven hundred reinforced-concrete piles were driven into the ground, and nineteen steel girders positioned over the District tracks and platforms. On these foundations was erected a 35,000-ton superstructure with a steel framework and Portland stone facing. Designed by Charles Holden, this new headquarters building was cruciform in plan, with nine floors of offices. The external walls were set back at the seventh and ninth floor levels, and above that level was a central three-storey tower stepped back at each floor. The tower was 175 feet high—in fact, so high that under the London Buildings Act the ninth floor and above could not be used for office purposes. At ground- and first-floor levels, the offices filled the space between the arms of the cross. The circulating area on each floor was faced with Travertine marble. The building was heated in winter by water brought to 300° F by twelve 112-kW immersion heaters, fed from Lots Road during the off-peak.

Externally, the features attracting most attention were the sculptural groups by Jacob Epstein—'Night' on the north facade at first-floor level, and 'Day' over the east façade, symbolizing the day and night service given by the Group's vehicles. Between the sixth and seventh floors were eight bas-reliefs depicting 'The Winds', including three by Eric Gill and one by Henry Moore.

The building's plain Portland stone facings, metal-framed windows and strictly rectilinear motif had much in common with Holden's Morden extension stations. The new headquarters were officially opened on December 1, 1929. E. V. Lucas thought the building 'Babylonian', and lesser lights daubed the Epstein sculpture with tar. The partners, Adams, Holden and Pearson, were awarded the Royal Institute of British Architects medal and diploma of 1929 for this work.

SCHEMES UNFULFILLED

During this period three important schemes for new tubes were publicized and investigated. Parliamentary powers were obtained for two of them but none was built.

The local authorities of south-east London had been pressing for a tube railway for many years. Deptford Borough Council had approached the Minister of Transport in 1920, and deputations from the south-east London boroughs waited on Lord Ashfield in May 1924 and on the Minister of Transport in February 1925. The question was again raised at the South East London traffic inquiry in October 1926. Giving evidence, Frank Pick was careful to state that the Underground group would not consider additional tubes in south-east London until the Southern had had full opportunity to test the results of electrification; he added that his company and the Southern had an agreement for mutual consultation before promoting railways that would be in competition. The Bakerloo could with advantage be extended to Camberwell and he agreed with an answer that Ashfield gave to the 1924 deputation that the area should have underground facilities—'if not today, then tomorrow'. The inquiry report recommended that the LER, in conjunction with the Southern, should be invited to consider the practicability of extending the Bakerloo in a south-easterly direction from the Elephant & Castle, in order to reduce street congestion there.

No further progress was made, despite pressure from Camberwell Council in 1929, but after the programme of new works qualifying for Government assistance had been agreed in 1930, the Underground group made a discovery that must surely be unprecedented in the history of London underground railways. Owing to lower costs, skilful planning and improved techniques, the agreed works would, they calculated, cost less than anticipated. In order to expend the agreed sum on the relief of unemployment it was thought better to concentrate on one large work rather than many small ones, and as there was an impending LCC development scheme for the Elephant & Castle area which would expunge the CSLR surface station, the occasion was opportune to combine the work of rebuilding the two Elephant stations with a Bakerloo extension to Camberwell. With such an extension, the traffic at the Bakerloo's Elephant station would drop, so that a combined station would comfortably serve both lines. Parliamentary powers for a $1\frac{3}{4}$-mile extension, with a terminus at Denmark Hill and an intermediate station at Albany Road, Walworth, were obtained in 1931, but the Act had scarcely received the Royal Assent when there were second thoughts on the finance of the extension. In September 1931 the UERL announced that, with the other works in hand, the Camberwell extension would exhaust for the moment the entire resources of the Underground group. By 1932 the financial situation had deteriorated so much that the work was postponed indefinitely, although parliamentary powers were kept alive.

In another part of London, the LNER pondered long and deeply over the overcrowding of its suburban trains between Liverpool Street and stations on the Ilford line, and in January 1930 announced that Messrs Mott, Hay & Anderson had been engaged to prepare plans and estimates of costs for a Liverpool Street–Ilford tube railway. The scheme, and variations, were examined, but the cost (£7½ million) was reputedly too great for the LNER to bear and the proposal died from the combined effects of the 1931 financial crisis and uncertainty about the London Passenger Transport Bill.

Meanwhile, the Ilford and District Railway Users' Association was campaigning for better Ilford facilities, including a Central London extension direct from Liverpool Street to Redbridge Lane/Eastern Avenue, then via Eastern Avenue to Romford with a branch to Barkingside and Claybury. In March 1933 it petitioned the Minister of Transport, and in December sent a deputation to Lord Ashfield, who pointed out that LNER electrification or a Central London extension could not possibly be self-supporting. However, this was not the end of the story, as we shall see in the ensuing chapter.

In the north-west, the Metropolitan Railway had developed its residential traffic with such success that serious congestion threatened at the northern approaches to Baker Street. The section between Wembley Park and Finchley Road had been four-tracked in 1914–1915, but south of Finchley Road, the addition of extra tracks parallel to the two existing ones in tunnel would have involved a very complex and expensive engineering operation. In the autumn of 1925 the Metropolitan deposited a Bill for a £2·05 million avoiding line 3 miles 7·9 chains in length. This would leave the existing line immediately south of Willesden Green, then at Mapesbury Park, just north of Kilburn and Brondesbury station, burrow under the existing tracks and enter deep-level tube tunnels running under Kilburn High Road, Maida Vale and Edgware Road, finally joining the Circle by a 200-yard cut-and-cover section ending a few yards west of Edgware Road (Metropolitan) station. The tubes would be 15 feet 6 inches internal diameter to accommodate the Metropolitan rolling stock and locomotives. Three intermediate stations were proposed, at Quex Road, Kilburn Priory and near the Regent's Canal. Edgware Road station was to be rebuilt with four eight-car platforms at which some of the trains from the tube section would terminate. As a further measure to relieve congestion, the east-bound track of the 'Widened Lines' was to be electrified. In the same Bill, the Metropolitan also sought further extension of time for the completion of the old scheme to extend the GNCR to Lothbury.

Royal Assent was given on August 4, 1926, but apart from the reconstruction of Edgware Road station (complete with train indicators which included 'Aylesbury' and 'Verney Junction'), nothing more was done.

Congestion on the Finchley Road–Baker Street section grew worse. As a palliative, St John's Wood and Marlborough Road stations were closed during both peak periods from October 1, 1929. The closings were said to allow the operation of twenty additional trains in the morning peak. The service at Swiss Cottage was improved at the same time, but there were protests from local residents, and the other two stations were reopened in the evening rush hour.

METROPOLITAN TO EDGWARE

The Metropolitan Railway was quick to jump on the Development (Loan Guarantees and Grants) bandwagon, and by November 1929 it had secured approval for a £300,000 scheme of station enlargement, platform lengthening and an extension of automatic signalling. More frequent and longer trains were envisaged. With a further look at the prospects of Treasury grants, it applied for powers to build a 4-mile branch from half a mile north of Wembley Park to Stanmore, and this also qualified for assistance.

At the annual general meeting in February 1930, Lord Aberconway explained that the Stanmore catchment area was being rapidly developed for housing, and that he looked upon it as the legitimate territory of the Metropolitan. To the very pertinent question put by a shareholder—how would they handle the additional traffic south of Finchley Road?—his Lordship gave the soothing answer that they were building many additional coaches of an improved type and lengthening trains.

This part of the Metropolitan Bill was 'certified' for quick passage, being for relief of unemployment, and was (wisely) not opposed by the Underground group. It gained Royal Assent on June 4, 1930, and construction began soon afterwards. The branch was officially opened by the Minister of Transport, Mr P. J. Pybus, on December 9, 1932. Public traffic started the next day, with 144 trains daily. The fastest train completed the 11½ miles from Stanmore to Baker Street in twenty-two minutes. In the normal hours there was a Stanmore–Wembley Park shuttle.

The double-track branch left the main line by a 15-chain curve and was at first on embankment. At Kingsbury there was a deep cutting, but the line was on embankment again for most of the remaining distance to Stanmore, which was at the summit of an almost continuous rise from Wembley Park. There were stations in a pleasing domestic style of architecture by C. W. Clark, at Kingsbury (in cutting, 1·77 miles from Wembley Park), Canons Park–Edgware (on embankment, 3·66) and Stanmore (4·49), where stabling sidings were provided. A fourth station, Queensbury (2·61 miles from Wembley Park), was opened on December 16, 1934, well sited for the new housing development on the former Stag Lane aerodrome and the small factory area in Honeypot Lane.

The signals were of the usual Metropolitan three-aspect colour-light type, fitted with train stops, but the whole branch was controlled from the Wembley Park box, under a novel centralized traffic control system installed by Westinghouse, the first such installation in Great Britain, and the first in the world for purely suburban traffic.

Power was supplied from Neasden through two new substations at Preston Road and Canons Park, each with transformers and rotary converters of 1,500-kW capacity (three at Preston Road, two at Canons Park).

Traffic on the new branch developed slowly. The Metropolitan fares were on the main-line scale; the Edgware tube, with its substandard fares, was only about a mile east of Queensbury or Canons Park, and was fed by numerous cross-country bus services.

The opening of the Stanmore branch served to push still more traffic into the Finchley Road–Baker Street bottleneck, and opening out became essential. This problem was bequeathed to the LPTB, and the next chapter will tell how it was solved.

TOWARDS FULL CO-ORDINATION

This story of the Stanmore branch, serving territory which did not really need an additional railway, and exacerbating the congestion on the main line, was symptomatic of the lack of co-ordination between the agencies of passenger transport in London which had been manifest throughout the 1920s.

Innumerable commissions and committees had recommended greater co-ordination between the various passenger transport operators, but little was done until a serious tram strike in March 1924 made action imperative. Independent bus competition had a disastrous effect on tramway finances, restricting the ability to award wage increases, and the Labour Government introduced a Bill to allow the Minister to designate streets as 'restricted' on which bus services could be statutorily controlled. He was advised by a new London and Home Counties Traffic Advisory Committee, representing local authorities, Government departments, labour and transport interests.

The London Traffic Bill received Royal Assent on August 7, 1924, and the first restriction orders were issued early in 1925. The LGOC was able voluntarily to acquire independents without the fear that newcomers would replace them, and the number of independent buses fell from 662 in February 1926 to about 200 in December 1930.

In 1925 and 1926 the Advisory Committee held inquiries into transport facilities in north/north-east, east and south-east London, from which it concluded that no lasting solution of the passenger transport problem was possible unless wasteful competition was replaced by unified management.

At the North London inquiry, Pick had suggested a common fund and common management for all London passenger transport operators, and in April 1926 the Minister of Transport asked the Committee to look into this suggestion. The consequent 'Blue Report' of October 1927 strongly supported the idea. The Underground group, LCC and the main-line railways approved the common management/common fund scheme, but the Metropolitan was unenthusiastic, and the Labour party strongly opposed it, on the grounds that public ownership was the proper solution.

As the Government could not find time for a public Bill to implement these proposals, the Underground group and the LCC deposited in the autumn of 1928 a London Electric Railways (Co-ordination of Passenger Traffic) Bill and a London County Council (Co-ordination of Passenger Traffic) Bill, giving the parties powers to establish a common fund and common management, for their own undertakings.

The majority of the Traffic Advisory Committee members supported the Bills (the 'White Report'), but Liberal and Labour MPs opposed them. After the 1929 dissolution, they were carried forward to the new Labour Government, but at the third reading on July 17, 1929, Herbert Morrison (now Lord Morrison of Lambeth), Minister of Transport, who was determined on public ownership, advised the Commons to reject them, and with Liberal support they were thrown out.

In October 1930 Morrison issued a memorandum setting out his plan for a public board—a statutory authority taking over all London's public passenger transport except the suburban services of the main-line railways, which would be co-ordinated. The board would be small in size, running a self-supporting enterprise by commercial methods.

Morrison's London Passenger Transport Bill was deposited in autumn 1930, and provided for a Board of five members appointed by the Minister. Within the London Traffic Area, the Board would have a monopoly of road services. There would be a standing Joint Committee with the main-line railways, whose tasks would include examining railway co-ordination and development schemes.

The Bill was before a Joint Select Committee for thirty-five sitting days, starting on April 28, 1931. On May 12th, Counsel for the promoters was able to announce agreement with the Underground group and the main-line railways. Negotiations with the provincial bus companies resulted in the Board's area becoming roughly egg-shaped with the monopoly section confined to a rough circle within. The Metropolitan Railway fought the Bill tooth and nail.

On July 20th the Committee (which had only three Labour members) voted 5 to 4 in favour of the Bill proceeding, but a new Government was formed on August 24th and the Bill had to be considered by the new Parliament. In October 1932 it was carried over to the next session, and was supported by the National Government. The chief Government amendments were the introduction of Appointing Trustees to appoint the Board (the chairman of the LCC, a representative of the Traffic Advisory Committee, the chairman of the Committee of London Clearing Bankers, the President of the Law Society, and the President of the Institute of Chartered Accountants of England and Wales) and an increase in the size of the Board to seven members.

Debate was animated and prolonged, but the committee stage was concluded on December 13, 1932, and the report stage on February 13, 1933. Finally, Royal Assent was given on April 13, 1933.

The Appointing Trustees met on May 18th and chose members for the new Board: Lord Ashfield (chairman) and Frank Pick [1]—full-time members for seven years; John Cliff and Patrick Ashley Cooper—part-time members for five years; Sir John William Gilbert, Sir Edward John Holland and Sir Henry Maybury—part-time members for three years.

The 'Appointed Day' for the London Passenger Transport Board to commence operations was Saturday, July 1, 1933, and on June 27th, Ashfield issued a message saying that the Board would seek to carry on the best traditions of its predecessors, hoping for continued public support.

Thus ended the independent existence of many companies with famous names and long and honourable careers—the Metropolitan District Railway, the London Electric Railway, the City & South London Railway, the Central London Railway,[2] the Metropolitan Railway, the London General Omnibus Company, the Metropolitan Electric Tramways and the London United Tramways. The municipal tramways and the London bus business of Tilling and British Automobile Traction Ltd were also transferred on the appointed day.

Thus, on July 1, 1933, the London Passenger Transport Board became responsible for passenger transport in an area of 1,986 square miles, estimated to contain 9,400,000 people.

A challenging era of co-ordination, rationalization and development lay ahead.

[1] Pick was subsequently appointed vice-chairman and chief executive officer (he had been managing director of the Underground group since 1928).
[2] I.e. all the tube railways except the Waterloo & City, which remained part of the Southern Railway until nationalization on January 1, 1948, when it came into the Southern Region of British Railways.

CHAPTER 11
London Transport
1933–1939

OFF WITH THE OLD

ALTHOUGH the Board had taken over the operation of all the larger constituent companies from July 1, 1933, it was some time before obsequies were completed. The London & Suburban Traction Co., the LUT, MET, LGOC and SMET were dissolved on March 18, 1935, but the Metropolitan Railway, Metropolitan District Railway, LER, CSLR, CLR and the Lots Road Power House Joint Committee lingered until March 10, 1939. Long familiar names disappeared from the door plates at 55 Broadway, to be replaced by one which embraced them all.

The final meeting of the UERL was held on June 27, 1935, and the company was dissolved on September 28th. Liquidation had involved the preparation and issue of stock certificates to 26,500 stock and shareholders, and the distribution of cash in three payments to 18,500 shareholders. A holder of £100 of UERL ordinary shares received £117 of 'C' stock, $4\frac{6}{11}$ Northmet £1 ordinary shares, $13\frac{1}{2}$ AEC £1 ordinary shares and £3 9s 3d in cash. Lord Ashfield delivered the valediction—'We need have no regrets at the passing of the Underground Company. It lived an honourable and useful life, and the work which it began will be continued upon a larger and more complete scale.'

Towards the end of 1933, Sir William Beveridge arranged a series of lectures at the London School of Economics on the London Passenger Transport Act. Frank Pick's lecture on the 'Practical Aspects of the Board' was a model of acute discernment and informed prognosis. He recounted the official list of Appointing Trustees, and confessed that when he first saw it, he had laughed: 'They are, with one exception, possibly entirely ignorant of the problems and needs of London Traffic. Yet to them is allotted the task of maintaining a competent and efficient Board . . . it is a typical English invention. . . . It smacks of compromise. . . .' After describing the meagre rights of stockholders, he went to the heart of the matter: 'Under the Board, capital has lost its power. It cannot appoint the management or interfere with it. It cannot use its investment to serve any other end or aim. It has no rights except the right to receive a specified return or reward. The power has been transferred, let me say it with bated breath, to a bureaucracy. In the escape from capitalist control, in the escape from political control, we have almost fallen into a dictatorship.'

FULL CO-ORDINATION

As provided by the 1933 Act, a Standing Joint Committee with the main-line railways was established in July 1933. The London Transport members were Ashfield, Pick, Ashley Cooper and Maybury, whilst the main lines were represented by Sir James Milne (GWR), Sir Josiah Stamp (LMSR), Sir Ralph Wedgwood (LNER) and Sir Herbert Walker (SR). To assist with its duties the SJC appointed four informal advisory committees composed of officers of the constituent undertakings, to advise on traffic (including new works), accounts, fares and charges, and engineering.

An early task was the preparation of the statutory Pooling Scheme. The basis of the scheme was that each party should pay its passenger receipts into a pool after deducting 'Operating Allowances' (operating expenses varying with car mileage or directly with fleet size) and 'Additional Allowances' (interest on approved additional capital expenditure, and associated additional expenses). The pool would then be divided between the five participants in 'Standard Proportions' based on the actual receipts of the main lines and the predecessors of the LPTB in a standard year (the Committee chose 1932), adjusted for items such as capital expenditure which had not then fully fructified. The Standard Proportions could be revised if there were material alterations in circumstances or development not covered by an Additional Allowance. The agreed proportions were approximately:

	%
LPTB	62·0
Main Lines	
GWR	1·3
LMSR	5·1
LNER	6·0
SR	25·6
Total Main Line	38·0

(from July 1, 1934, the LPTB proportion increased by about 0·1%, the SR decreased by nearly that amount, and there were fractional alterations for the other railways).

The Scheme applied to the whole of the Board's receipts, and to the main-line receipts for journeys local to the Board's area or to a few stations beyond the border where the Board provided road services.

It was formally confirmed by Order of the London Passenger Transport Arbitration Tribunal on June 18, 1935, and applied retrospectively from July 1, 1933.

A NEW WORKS PROGRAMME

The way was now clear for railway improvements to meet the traffic needs of Greater London as a whole, improvements which could be planned in an atmosphere free from the inhibitions of the competitive era and without regard for the effects on the separate revenues of the five undertakings that operated London's rail transport.

The preparation of a scheme took about two years. At the request of the Minister of Transport, particular attention was paid to east and north-east London, where the Ilford and District Railway Users' Association was still actively canvassing its proposal for a Central London line extension from Liverpool Street via Eastern Avenue. In Leyton, too, there was discontent with the LNER services. North London authorities were pressing for the electrification of the ex-GNR suburban lines, and the extension of the Hampstead tube to Highgate (LNER).

Unemployment was still causing concern. In March 1935, J. H. Thomas said that the Government would welcome, in the prevailing conditions of good Government credit and cheap money, any proposals for new tube railways and London railway electrification if they could be proved necessary. Finally, on June 5, 1935, the Chancellor of the Exchequer, Neville Chamberlain, announced to the House of Commons a bold scheme for east, north and west London, financed by Government-guaranteed loans—12 miles of new tube railway, the electrification of 44 miles of suburban railway and the doubling and electrification of a further $12\frac{1}{2}$ miles. In *The Times* the following day this was the main news item, and the newspaper commented that the House was 'jubilant' at the announcement, which came as a 'welcome surprise'.

The main points were:
(i) The electrification of the LNER from Liverpool Street to Shenfield.
(ii) The extension of the Central London line in tube from Liverpool Street to Leyton via Stratford, and from Leytonstone to Newbury Park, and the operation of Central London trains over the electrified Loughton branch and Grange Hill loop-line; the lengthening and improvement of central area Central London stations.
(iii) The extension of the Morden–Edgware line from Highgate to East Finchley, and a connection of the Great Northern & City line and the LNER at Finsbury Park, to enable tube trains to run on electrified tracks to Edgware, High Barnet and Alexandra Palace. The Finchley (Church End) to Edgware single-track section to be doubled.
(iv) The construction and electrification of two additional tracks

alongside the GWR Birmingham line from about half a mile west of North Acton to Ruislip, for use by Central London trains diverging from the Ealing line.

(v) The construction of a 2·14-mile tube railway from Baker Street to Finchley Road to enable Bakerloo trains to be extended to Stanmore; and associated improvements on the Bakerloo and Metropolitan lines. The realignment of tracks and station improvements between Finchley Road and Harrow.

(vi) The reconstruction of Aldgate East station and junction.

(vii) The replacement of trams by trolleybuses on 148 route-miles.

(viii) The reconstruction of King's Cross, Post Office and other central area stations.

(ix) Improvements on the Uxbridge branch.

(x) Power supply improvements and certain ancillary works.

The plan was hailed by Herbert Morrison as 'the first big fruits of the London Passenger Transport Act', but it brought protests from the South London local authorities and faint praise from the *Transport World*, which wanted to see *all* the ex-GER suburban lines electrified.

On June 20, 1935, a formal agreement was made between the Treasury, the LPTB, the GWR and the LNER. The main provisions were that the works should be completed by September 30, 1940, that the Treasury would guarantee the principal and interest of a sum not exceeding £40 million in cash, to be raised by a new finance company, which the Treasury would 'cause to be formed', and lent to the LPTB (70%), GWR (5%) and LNER (25%). These three undertakings would pay interest to the finance company at the same rate as the latter would borrow money. The Treasury guarantee was confirmed by a Government measure—the London Passenger Transport (Agreement) Act 1935—and the Board obtained powers to borrow from the finance company in the London Passenger Transport (Finance) Act 1935, to which was scheduled the agreement of June 20th.

The London Electric Transport Finance Corporation Ltd was incorporated on July 10, 1935, and seven days later made a public issue of £32 million Guaranteed Debenture stock at the price of 97% and the low interest rate of $2\frac{1}{2}$%. Lists closed within sixty-five minutes, and the issue was oversubscribed. In January 1937 the Corporation issued a further £9·65 million of the same stock; this time at $92\frac{1}{2}$%, so that the total cash yield for the nominal value of £41·65 million was £39,966,250.

The major powers for the New Works Programme were obtained in 1936, but there were certain other applications, including some for works additional to the main programme. The following summary gives the principal tube railway powers granted during the years 1934–1939:

1934

London Passenger Transport Act
Reconstruction of King's Cross tube station (abandoning certain 1930 powers) and Moorgate station.

1935

London Passenger Transport Act
New tube railway between Baker Street and Finchley Road, and the associated improvements at Baker Street station, improved terminal facilities at Elephant & Castle, and the rearrangement of the Metropolitan tracks between Finchley Road and Harrow.
A new terminus at High Street, Uxbridge.

1936

Great Western Railway (Ealing and Shepherd's Bush Railway Extention) Act
Two additional tracks from North Acton to West Ruislip.
London and North Eastern Railway (London Transport) Act
Track improvements associated with the Shenfield, Loughton and Grange Hill loop electrifications.
Track improvements associated with the Edgware, High Barnet and Alexandra Palace electrifications, including the doubling of the Finchley/Edgware section.
Revival of the powers obtained in the LNER 1924 Act to link the GNCR and LNER at Finsbury Park.
London Passenger Transport Act
New tube railway between Highgate and East Finchley (LNER), with an intermediate tube station beneath Highgate (LNER).
New connecting line between Edgware LNER branch and Edgware LT station.
Lengthening of Bakerloo station tunnels (to take eight-car trains) at all stations between Kilburn Park and Elephant & Castle inclusive except Baker Street (covered by 1935 Act).
Escalator tunnel at Edgware Road (Bakerloo).
Reconstruction of South Kensington and Gloucester Road stations, with escalators.
New tube railways between Liverpool Street and Leyton via Stratford, and between Leytonstone and Newbury Park.
Lengthening of all existing Central London station tunnels from Shepherd's Bush to Bank (to take eight-car trains). (Liverpool Street was under private property.)
New subway at Earl's Court, with escalators, to new exhibition hall.
LPTB running powers to Ongar, also over the Grange Hill loop, the GNCR link at Finsbury Park, and the Edgware, High Barnet and Alexandra Palace branches.

Repeal of part of the CLR Act of 1909 restricting the eastward extension of the CLR from Liverpool Street.

1937

Great Western Railway Act
Two additional tracks from West Ruislip to Denham.
London Passenger Transport Act
New railway from Edgware to Bushey Heath.
Reconstruction of Highbury, Marylebone, Monument, Liverpool Street and Notting Hill Gate stations to allow installation of escalators.
Improvements at Gloucester Road, Wood Lane and Ickenham stations.

1938

London Passenger Transport Act
New station and tracks at Wood Lane (Central London).
Additional interchange subway, Tottenham Court Road.
Additional ventilation shafts.
Works at Charing Cross station, including an electric control room.
Powers to take electricity from the Central Electricity Board in emergencies.

1939

London Passenger Transport Act
Additional interchange subway at Leicester Square.
Additional ventilation shafts.
Lengthening of Northern line station tunnels for nine-car trains at Camden Town, Euston and Kennington (to enable drivers to change over).
Powers to run electric trains over certain level crossings (subject to Ministry of Transport consent) between Leyton and Theydon Bois, notwithstanding provisions in LNER and LPT Acts of 1936 prohibiting electric traction until level crossings had been replaced by road bridges.

TROUBLES ON THE MORDEN–EDGWARE

With one exception, there was little opposition to these proposals, although MPs sometimes took the opportunity to raise extraneous issues when the Bills came before the Commons.

The exception was the 1937 Bill, which was considered at a time when the Morden–Edgware line was in bad odour with the travelling public.

Peak-hour congestion on both the Morden and Edgware sections had grown steadily more severe as the rows of LCC cottage dwellings and the 'desirable residences' of private speculators spawned new legions of commuters. We have seen that public discontent with the Edgware service were manifest as early as 1929. Over the years the grumbles and complaints seethed and bubbled to the surface in the form of Questions in the House or letters to *The Times*. A series of letters in November 1935 complained of up to eighty-eight standing passengers in a car, and the need to fight to board a southbound train at Hampstead in the morning peak.

The year 1937 began inauspiciously with a signal failure at Tooting on January 28th, followed on February 4th by a week's 'go slow' protest by the drivers (which also affected the Bakerloo) about the observance of speed limits.

A further 'go-slow' occurred on the Morden–Edgware at the beginning of March against the new reduced running times which followed the equipment of all trains with electro-pneumatic brakes and weakfield control. Then on April 5th the Edgware section made headlines with a passengers' 'stay-in strike'.[1] On that evening a defective spring on a maintenance wagon broke off all the train stops from Edgware to Golders Green. Station-to-station working was introduced, and during the re-forming of the service an Edgware-bound train was curtailed at Colindale at 11.25 p.m. to maintain the southbound intervals. The 'all change' cries of the porters touched off a revolt amongst the occupants of the first car, who refused to alight and sent two of their number along the train to enlist more support. After some delay, the doors were closed and the protesting passengers hauled into the siding to cool down. Subsequently the train returned to the southbound platform and the tired passengers got out. This demonstration was a culmination of the feeling aroused by many similar curtailments of trains, and was followed by announcements from the Board explaining the need for turning trains short of their advertised destination, and regretting the inconvenience. But, it was added, the Board 'could not tolerate' interference.

Meanwhile the Morden–Edgware line overcrowding continued, to be well aired in Parliament and in newspaper correspondence columns. There was another stay-in strike on the evening of April 6th when a Morden train was turned at Tooting Broadway. Two days later a new Morden–Edgware timetable was put into operation, giving ten more trains from Colindale in the morning peak (divided equally between the West End and City branches), and seventeen in the evening peak (two West End, fifteen City). There were further additional trains from Golders Green, and corresponding increases on the Morden section. On April 16th the Board claimed that 95% of the trains were now running to time.

[1] 'Stay-in strikes' were not new. As early as 1909, Golders Green passengers had refused to leave trains which were being reversed at Hampstead to put the timetable right after late running; similar incidents occurred up to 1914.

THE BATTLE OF BUSHEY HEATH

This row about Edgware line overcrowding came at a time when there was increased awareness of the dangers of unrestricted metropolitan growth. The Board's proposal to extend the Edgware line through virgin territory to Bushey Heath (otherwise called 'Aldenham') brought opposition from those who deplored the constant expansion of London, and arguments and counter-arguments were deployed in *The Times*. On March 16, 1937, the headmaster of Aldenham School wrote to complain of the prospective development of Aldenham as a dormitory suburb and the consequent worsening of the Edgware line overcrowding. He asked whether the Board was to be allowed to create artificial centres of population in the few precious and unspoiled areas of London under the pretence of 'serving' communities which either had no desire to be served or did not exist. Further correspondents complained of the 'intolerable' overcrowding and sympathized with the stay-in strikers.

Frank Pick entered the lists on April 12th with a letter which began: 'It is somewhat unfortunate that a matter which is before Parliament and will be considered judicially by a Parliamentary Committee, should have become the subject of outside controversy.' He thought that the Board's intentions had been misunderstood, and claimed that the electrification and doubling of the LNER line to Edgware would divert about a quarter of the traffic at present using the Golders Green line. New rolling stock and improved services would give the area 40% more seating capacity. The Bushey Heath extension was primarily needed to provide a new depot as the existing and planned provision at Golders Green, Edgware and Wellington sidings (Highgate) would be wholly inadequate. Without a new depot it would be impossible to work the northern railways efficiently. The built-up area already extended much beyond Edgware, so that the extension would deal with the existing traffic more conveniently. The extent of any housing development beyond Aldenham was in the hands of local and town-planning authorities, but the Board would be glad to see a green belt completed across Bushey Heath station, as this would afford an off-peak traffic.

A letter from Mr A. J. Child of Bushey Heath took up Pick's, point by point. Mr Child regretted that Pick was misinformed about housing development beyond Edgware. The last house was a mile from Edgware, or in other words, two miles short of Bushey Heath. He was surprised that the Board welcomed the green belt at Bushey Heath and wondered why they had made provision for shops adjacent to the station. The higher price of land would make it more difficult for local authorities to buy it for retention as green belt. Why had the Board rejected an alternative depot site just outside Edgware? The reasons must have been cogent, since at a time when other projected works were being postponed on the score of economy,[1] they proposed to extend a line 1¾ miles further, cut through a 100-foot contour, fill in a lake, and erect a station and shops where they did not desire to create a new community.

Further letters on similar themes followed, and MPs were well briefed when the Board's 1937 Bill came before the Commons on April 26th.

Mr W. J. Kelly, Labour MP for Rochdale, moved its rejection on the grounds that the scheme would do nothing to ease the 'discreditable and disgraceful' overcrowding on the Edgware line. He brought out the time-honoured 'cattle travel better' comparison, but added a spicy touch of originality—'Young girls and men are crowded in such a way that the question of decency even comes up.' He asked MPs to sample the Hampstead line between 8.30 and 9.00 a.m., but to leave their womenfolk at home.

Other speakers voiced the arguments that had appeared in *The Times*. Mr P. J. Noel-Baker foresaw 'ribbon development of an absolutely atrocious kind' along the Watford By-Pass, and, stating that land near Aldenham had increased enormously in value when the Board's plans were divulged, he declared that 'a rise in land values is always the surest sign of the jerrybuilder's wrath to come'.

The Parliamentary Secretary to the Minister of Transport answered by quoting the London Transport argument that the passengers who would use the extension were largely those using Edgware station already, and the Bill was passed by 134 votes to 55.

When introducing the Bill before the Lords Committee on June 8th, Counsel pointed out that a similar extension had been authorized in 1903 (The Watford and Edgware Railway) and disclosed that a return of 6½% was expected on the expenditure of £815,000. Much land had already been acquired. The Bill was passed by the Committee, but it met strong opposition at the third reading on July 14th. Lord Brocket moved an amendment to delete the Aldenham extension, suggesting an alternative site for the depot which did not impinge on the green belt. Lord Ashfield made one of his rare appearances to defend the scheme, saying that Lord Brocket's site had been surveyed but was wholly inadequate. After discussion, and some lecturing on the enormity of the crime of upsetting a Select Committee's decision, the amendment was rejected by 52 to 24. The Bill received the Royal Assent on July 20, 1937.

MUCH ADO AT MORDEN

The southern section of the line was also congested at peak periods. A diagram appearing in the Board's report for 1935/1936 showed that

[1] This was a reference to the Board's announcement on April 9, 1937, that, owing to the rise in prices of engineering materials, they had postponed consideration of the Bakerloo extension to Camberwell, and that the 1935–1940 programme itself might have to be curtailed.

at 8.00 a.m. the northbound traffic at Oval station was at the rate of 12,250 passengers per half-hour, but the corresponding seating capacity was 4,500. In the report the maximum figure of 13,000 per half-hour was quoted for the Morden–Edgware line, exceeded only by the District line at 13,500. The Board advocated the staggering of working hours as a possible solution to the problem of rush-hour overcrowding, but did not adopt the principle for its own office staff until May 1939.

The idea of staggered hours also found favour with the Minister of Transport, Dr Leslie Burgin, who advocated it in the Commons during June and July 1937. On the latter occasion he gave the information that, with additional signalling, the Morden–Edgware line was now carrying forty trains per hour, but it was hoped to exceed this figure with the new rolling stock.

The Morden line figured prominently in the 1937 report of the London and Home Counties Traffic Advisory Committee. A group of seven local authorities in North Surrey had suggested that it be extended to Epsom, and that the overcrowding be relieved by running longer trains (with extended station platforms) or by building express tunnels from Morden to Kennington. The Board's reply, accepted by the Committee, was that the morning peak seating accommodation had increased by 55·4% since 1927, and the service was now up to the limit of line capacity. Longer trains would involve platform lengthening at forty-two stations, prohibitively expensive for trouble occurring in only short periods of the peak. The line would be resignalled to take 42 trains per hour (instead of 38) and new rolling stock would give a 25% increase in capacity, but not enough for the extra traffic from an extension beyond Morden.

A new underground railway in developed or developing areas cost £750,000 per mile, fully equipped, and the Board calculated that this required 45,000 persons per mile of line, living within about one mile of the line, to make it self-supporting. A new 12-mile line from Kennington to Epsom, express tube to Morden and surface beyond, would need 300,000 to 400,000 completely new passengers, and there was no hope of attracting anything like that number.

The Committee decided that there were insufficient grounds for it to make official representations to the Board under Section 59 of the 1933 Act, and recommended greater use of the Southern Railway services and the staggering of working hours. It pointed out that the problem arose from building development taking place without regard for transport facilities, and recommend closer consultation between the railways and the developers, with some means of attracting houseseekers to areas where railways had spare capacity or where additional facilities were in prospect.

AND SO THERE WERE NINE . . .

The Board's railway operating department took up the north Surrey authorities' suggestion of longer trains, but at the Edgware end of the line. The general manager (Railways) J. P. Thomas and chief mechanical engineer W. S. Graff-Baker devised an ingenious scheme. In October 1937 the *Evening News* carried accounts of platform-lengthening from Golders Green to Colindale and of secret tests with nine or ten-car trains at dead of night. The first nine-car train ran in public service on November 8, 1937, and was considered to be such a success that three further trains of this length were put into operation.

The trains consisted of two motor-trailer-motor sets with three trailers in the middle, and all the cars were freshly painted. The guard was on the seventh car. On the open section, the station platforms were extended at Burnt Oak (north end), Colindale (south, where the trains terminated until Burnt Oak had been altered in February 1938), Hendon Central (south), Brent (north) and Golders Green (south). At Colindale the reversing siding was extended about 100 feet north, and at Golders Green the scissors crossover at the south end of the platforms had to be removed. At Edgware, No. 1 platform (the longest) was used, but even so the southernmost car and a half were off the platform, and passengers had to detrain through the end doors via car No. 7. The scheme also involved extensive alterations to signals and track circuits.

On southbound journeys, the two rear cars were reserved for passengers for stations to Golders Green or for Tottenham Court Road. After leaving Golders Green these two cars stopped in the tunnel at all stations to Goodge Street inclusive. At Tottenham Court Road the front cars were in the tunnel and the fortunate occupants of the last two could alight at the platform, having had a much less crowded journey than the passengers on the rest of the train. From Leicester Square to Kennington the two rear cars again stopped in the tunnel, and were out of passenger use.

On northbound journeys from Kennington, the two leading cars were stopped in the tunnel as far as Leicester Square, and the two rear cars were reserved for traffic to Leicester Square or to Golders Green or beyond. At Tottenham Court Road, the two rear cars were in the tunnel and the (hitherto empty) two front cars were at the platform; a similar stop was made at all stations to Hampstead.

There was full signposting on the platforms and cars to make sure that passengers did not board the wrong car, but if they did go wrong they could use the end doors to reach the correct position and travelling ticket inspectors were available to help them. Each train carried a notice in black on yellow, '9-car train' in the right-hand car window. At stations where end cars were booked to stop in the running tunnel, the tunnel segments were painted white, and red handrails were installed.

One of the objects of the scheme was to clear the congestion at the north end of Tottenham Court Road northbound platform, caused by the location of the entrance at this end. It was claimed that these trains could maintain practically the same service speed as a seven-car train, and that they could hold 1,000 passengers; they undoubtedly had traffic value, but probably their most useful contribution was in the field of public relations. Their presence showed the highly-compressed and oft-delayed Edgware line passengers that 'They' were 'doing something'.

Real relief for the Northern line [1] came with the introduction of the 1938 stock, described later in this chapter. Included in the order for new cars were ten nine-car trains, consisting of two driving motor cars, five non-driving motor cars and two trailers. The new nine-car block trains came into service from June 1939, but were withdrawn upon the outbreak of war, and the trailers removed, leaving seven-car, all-motor-car trains. The non-standard formation of these trains later resulted in the unbalanced rolling-stock orders of 1949 and some extensive re-marshalling, as described in Chapter 13.

NORTH FROM HIGHGATE

Charles Brand & Son Ltd began to construct the new tunnels from Highgate (LPTB) to East Finchley in November 1936. Six 12-foot Greathead shields, one 12 foot 3 inch rotary excavator and one 21 foot 2½ inch station tunnel shield were used from four working sites. The rotary excavator was claimed to advance 170 feet a week, or about twice as fast as a Greathead shield. The works were inspected by journalists in February 1938, and at the subsequent luncheon Mr J. P. Thomas disclosed that 9½ million passengers used Highgate (LPTB) station annually of whom 4·6 million transferred to or from the trams or buses running along the route of the new tube extension. The new Highgate (LNER) station would have 35 peak trains per hour, 21 from the tube level and 14 from the surface platforms. Northern line traffic was expected to increase from 150 to 180 million passengers per year when the scheme was complete.

At East Finchley, the tunnel portals were on either side of the LNER tracks, and, just beyond, at the Great North Road crossing, new single-track 145-foot plate girder bridges were erected on each side of the existing 100-foot bridge, which was replaced by one of 145 foot span. The work included the diversion of Brompton Grove and demolition and re-erection of the White Lion.

In December 1937 the LNER placed a £25,000 contract for doubling the single line between Church End (Finchley) and Edgware. This

---

[1] Thus renamed from the 'Morden–Edgware' on August 28, 1937; the Central London was renamed 'Central Line' at the same time.

work necessitated the closure of the line on some Sundays and during the midday normal period on some weekdays. When the line was closed, a replacement bus service was provided, and this first ran on Sunday, April 10, 1938. To accelerate the widening and electrification work, the whole line was closed after the last train on September 10, 1939, and a substitute bus service began the next day. This was operated with single-deck buses from Cricklewood Garage, and ran hourly between the forecourts of Finchley (Church End) and Edgware LNER stations, calling intermediately at Mill Hill East and Mill Hill LMSR stations. In the event, the Edgware–Mill Hill East section never again had a passenger train service.

At Edgware, where a third platform had been added on November 20, 1932, the rail layout was to be completely remodelled, with the tracks from Burnt Oak terminating in a central two-road bay, while those from Finchley to Bushey Heath would serve the outer faces of the same platforms. Connecting tracks would permit the operation of through trains to Bushey Heath from either Mill Hill or Golders Green. There was also provision for a third terminating track on the west side of the station, with its own island platform giving cross-platform interchange to northbound Bushey Heath trains. The new track layout cut across the site of the existing bus garage, and the contract for a new garage, further west, was signed in July 1938. At surface level there was to be a station of new design, 'different from any existing station of the Board', with covered road-rail interchange. Some work here began in October 1937, and there was further activity from September 1938, with the demolition of the east wing of the surface buildings, thus wrecking the symmetry of the 1924 design. On the Bushey Heath section, construction began in June 1939.

The No. 1, 1938 edition of the Underground map gave 1939 as the opening date for the whole of the Northern line works, but in the House of Commons on February 2, 1938, the Minister of Transport forecast that tube trains would be running to High Barnet in summer 1939, to East Finchley 'some months earlier', and to Alexandra Palace and Barnet from the Northern City line [1] early in 1940.

To avoid having two stations with the same name, Highgate (LPTB) station was renamed 'Archway (Highgate)' from June 11, 1939.

On June 28, 1939 Lord Ashfield and Sir Ronald Matthews (Chairman of the LNER) acted as hosts to a party of press men and railway officers. Sir Ronald drove a special train from Archway to East Finchley, and the incomplete Highgate station was inspected. Lunch was taken at the re-sited White Lion hotel at East Finchley, and Lord Ashfield mentioned in his speech that the original £40 million New Works Programme would now cost £45 million.

The public opening of the 2·01-mile extension to East Finchley fol-

[1] Thus redesignated from 'Great Northern & City' in October 1934.

lowed on July 3rd, publicized by an advertising campaign with the slogan 'New in London'. Neat folder maps were issued showing the extension against a geographical background of roads and road services. There were 200 tube trains a day to the West End, plus 60 steam trains to King's Cross and the City. In peak hours there was a four-minute tube service. The reconstruction of East Finchley station and its bridges was not complete at the opening date, and tube and steam trains at first shared the outside tracks; trains passed the unfinished Highgate station.

Work continued on the electrification to High Barnet, and in the early morning of Sunday, March 3, 1940, in the blackout, two steam locomotives dragged 250 tons of trackwork for a distance of 120 feet north of East Finchley in preparation for the through running of tube trains. In the same month a contract was placed for the reconstruction of Highgate station at surface level.

The 5·38-mile section to High Barnet was electrified from April 1, 1940, for crew-training purposes, and opened to traffic from April 14th, with intermediate stations at Finchley Central (thus renamed from 'Finchley (Church End)' on April 1), West Finchley, Woodside Park and Totteridge & Whetstone. There was a ten-minute peak service from Barnet and a five-minute service from Finchley Central, with all trains running via Charing Cross. The normal-hour headways were twelve and six minutes respectively. Each day 212 tube trains served High Barnet, and 396 Finchley Central and stations south; this compared with 114 steam trains in the pre-war timetable. The 32-minute journey from High Barnet to Leicester Square cost 11d (all fares on the High Barnet branch were brought down to the London Transport scale from April 14th). Steam train services on the Alexandra Palace branch were unchanged, and LNER and LMSR steam trains ran in weekday peak hours and Monday–Friday evenings between East Finchley and Finsbury Park, King's Cross, Moorgate or Broad Street.

This extension also had its folder map and guide, and an advertising campaign with the slogan 'More for You'. (After that, apart from the short Mill Hill East extension, the Northern line theme became 'No More for You'(but more of that later)). The special bus service between Edgware and Finchley Central for railway-ticket holders now started from Edgware LT station instead of Edgware LNER; the booking office at the latter station was closed and LNER tickets were issued at the LT station from April 14, 1940. The bus journey took 20–30 minutes compared with ten minutes by steam train.

After being used by shelterers from September 1940, Highgate station was opened on January 19, 1941. Steam trains were withdrawn from East Finchley on March 3, 1941.

Two months after the outbreak of war, the various partly-completed sections of the New Works Programme were reviewed, and, as far as the Northern line scheme was concerned, it was decided to postpone the Finsbury Park–Alexandra Palace electrification, the link with the Northern City line at Finsbury Park, the doubling and electrification of the Finchley Central–Edgware section, and the extension to Bushey Heath. A further review in June 1940 resulted in the suspension of nearly all work, but owing to wartime traffic needs of Mill Hill barracks, the ·89-mile section from Finchley Central to Mill Hill East was electrified on one track only, and served by Northern line trains from May 18, 1941. Trains ran from Mill Hill East to Morden until 7.00 p.m. on weekdays; on weekday evenings and on Sundays, from Mill Hill East to Finchley Central only.

The existing single-deck bus service 240 between Edgware and Mill Hill Broadway was simultaneously extended to Mill Hill East station (via Hammers Lane and the Ridgeway), and the railway bus service calling at stations between Edgware and Finchley Central was discontinued. Railway tickets issued at Edgware LT and Mill Hill LMSR stations were available on bus 240 for journeys to or via Mill Hill East, and similar tickets were issued from underground stations in the reverse direction. This facility, though little publicized, is still in force at the time of writing, but season tickets via this route have been restricted to existing holders since August 31, 1951.

A limited through service had been introduced between High Barnet and the City section of the Northern in January 1941, and from May 19, 1941 this was increased to a regular ten-minute service in each peak.

HIGH BARNET LINE DESCRIBED

At Archway the running tunnels had continued for a short distance beyond the station to provide two dead-end sidings. The northbound siding was extended to form the northbound running tunnel to East Finchley, and the southbound was retained as a nine-car reversing siding between the two running roads, with the new southbound running tunnel joining the old from the east side, just north of the station. This tunnel came in with a 15-chain curve at the foot of a long 1 in 50 descent from Highgate, and speed-control signalling was installed to ensure that southbound trains kept within the 30 m.p.h. speed limit. Two successive signals would clear only if the train passed through two timing sections in rear at less than 40 and 30 m.p.h. respectively. If the train had exceeded these limits, and was held at the signal as a result, the signal would clear automatically after the train had been brought to a standstill.

Both running tunnels were driven beneath the Archway Road to a point just before Highgate. Here the platforms were 490 feet long, to take nine-car trains. Cross-passages between the station tunnels led to the bank of two escalators and one flight of fixed stairs which rose 60 feet from platform level to the new booking hall which had been constructed beneath the LNER station. From this there were stairway entrances from Priory Gardens and Wood Lane on the east side, and from Shepherd's Hill and Archway Road at The Woodman on the west. At the latter point the street was 60 feet above the booking hall, and there were plans for a twin flight of escalators ending in a circular brick shelter which would be surmounted by a weathercock and a statue of Dick Whittington and cat. Wartime conditions precluded construction and for more than sixteen years passengers grew weary climbing the steps up to the Archway Road; relief came at last on August 26, 1957, when a single escalator was opened to span this final section of the long ascent from train to street.

The old LNER station buildings were demolished and work started on a new island platform with reinforced-concrete canopies and shelters, and steps leading down to the new booking hall. This was nearly complete by the time the deep-level station was opened to traffic in 1941.

From Highgate the line continued beneath the LNER in 12-foot iron-segment tunnels, at first rising at 1 in 61 and then at 1 in 100. Near the tunnel mouths there were short stretches of 1 in 50 northbound and 1 in 40 southbound. There was no room for a bell-mouthed tunnel entrance to minimize the effect of air pressure as a southbound train entered the tunnel, and pressure relief was obtained by constructing a series of fifteen air vents in the top of the first 100 yards of the tunnel. Mercury-filled glass tubes of the type first used at Barons Court guarded the tunnel mouths from full-size trains.

At East Finchley a new building in concrete and red-brown bricks was designed by Adams, Holden and Pearson, in a development of the 1932 style. Staff offices spanned the two new island platforms, which were connected to the ground-level booking office by stairs and subway. The latter joined the main booking hall, on the east of the station, to an auxiliary entrance on the west side. Above the northern bridge abutment was placed the statue of an archer, pointing towards London. Designed by Eric Aumonier, and about 10 feet high, this figure was formed of 6 cwt of beech built round a steel support and covered with 5 cwt of sheet lead. The bow was of English ash covered with copper and gilt.

Further north, the steam railway stations were little changed, except for minor structural alterations, platform height adjustments and new bull's-eye name signs. Finchley Central had a new island platform, but war conditions precluded the construction of the planned 88-foot-wide road overbridge, with booking hall beneath.

From Finsbury Park to Alexandra Palace, High Barnet and Mill Hill East, the LNER lines were resignalled to accord with the Board's a.c. track-circuit, e.p., colour-light system, with stop signals, repeaters and fog repeaters. To handle the steam-hauled goods trains, distant signals were installed in rear of the two-aspect signals to provide a minimum braking distance of 500 yards. These were 20-inch yellow discs with a black fishtail band. The signal overlaps at the two-aspect signals were based on the braking performance of tube trains, and goods trains were accordingly subject to a 20 m.p.h. speed limit. Steam locomotives working over this section had to be equipped with tripcocks. Mechanical ground frames were provided for shunting at Woodside Park and Totteridge. Beyond Mill Hill East, there were no signals, and the line was operated on the 'one engine in steam' principle, the driver obtaining possession of the key to the Edgware ground frame before proceeding. New power signal-boxes were built at Park Junction (74 levers), East Finchley (28), Finchley Central (53), and High Barnet (26). Some of the old signal-boxes were retained for night working of steam-hauled coal trains. Signalling current at $33\frac{1}{3}$ cycles and the compressed-air supply for train stops and point operation, came from the substations.

Substations were built and equipped at Crouch Hill, Highgate, Muswell Hill, East Finchley, Finchley Central, Woodside Park, High Barnet, Mill Hill and Edgware. (Later, following the curtailment of the Northern line scheme, the equipment was removed from Crouch Hill, Muswell Hill and Mill Hill). East Finchley was the main substation and control point, receiving the 11 kV, 50 c/s current from the Northmet at Watsons Road, Wood Green, via three new high-tension cables laid beneath the streets, through the Alexandra Palace grounds and along the Alexandra Palace railway branch, and distributing it to the other substations by parallel feeders. East Finchley had three English Electric water-cooled, steel-tank 1,500-kW rectifiers, the others two each. The main transformers were oil-immersed and were situated in the open, adjacent to the substation building.

The Wellington carriage depot of the LNER, on the east side of the Highgate–East Finchley line just north of the Alexandra Palace branch, was altered to take tube stock, and a new depot was built for the LNER at Wood Green. Further tube car stabling was provided in sidings parallel to the Alexandra Palace branch.

NORTHERN CITY COMES INTO LINE

Before the former Great Northern and City line could be brought into the new scheme, it was necessary to convert it to normal standards. Its positive and negative conductor rails were outside the running rails, and the signalling had to be altered. By May 1939, there was enough spare standard stock displaced by the arrival of 1938 stock to equip the Northern City, and the line was closed between 3 p.m. on Saturday, May 13th, and start of traffic on the Monday. The negative centre rail had been installed some time before, but during this week-end twenty miles of old current rail were removed and replaced by ten miles of new positive conductor rail in the standard outside position. In the stations the original platforms were now about ten inches above the car floors, and the line was resignalled to London Transport standards, with a new box at Drayton Park controlling the scissors crossover at Finsbury Park.

The surviving fifty-four gate-stock cars were despatched to Neasden depot for the removal of electric motors and other equipment before being sent on to Chesterfield for scrapping.

By August 1939, work was well in hand on the tunnels and ramps between Drayton Park and Finsbury Park, and at the latter station the site for the new platforms had been cleared and the laying of foundations begun.

WASTED WORK ON THE NORTHERN HEIGHTS

But for the war, tube trains would be running today between Moorgate and Alexandra Palace and between Bushey Heath and Kennington via Edgware, Finchley and Charing Cross, and no one would give them a second thought. In the summer of 1939 the opening dates were announced as autumn 1940 and spring 1941, but, as already mentioned, wartime requirements snuffed out all activity. When this happened, some of the work on these projects was well advanced.

At Finsbury Park, the old GNCR terminus was to be retained for a peak-hour shuttle service to Moorgate. The new service was to diverge at Drayton Park and climb through tunnels on either side of the existing tube mouths, then converging and continuing in cutting and later on embankment to a new bridge over Seven Sisters Road which would bring them into the LNER station at Finsbury Park. The ramps from Drayton Park were to have 1 in 45 gradients northbound and 1 in 50 southbound; a new connecting line for steam trains proceeding to the up Canonbury line was included in the scheme. At Finsbury Park LNER station the new lines were to run on either side of an island platform supported on steel girders, forming an eastward extension of the existing station. Improvements included a new booking hall beneath this girderwork and an imposing new façade to Station Road. North of Finsbury Park the up flyover bridge would be modified to carry both tube tracks on to the Edgware line. The existing LNER stations between Finsbury Park and Alexandra Palace, with the exception of Highgate, were not to be substantially rebuilt.

The tunnels and ramps from Drayton Park were completed and the earthworks beyond to the Seven Sisters Road bridge were almost finished. The bridge itself was delivered and placed on site ready for installation. At Finsbury Park the bridge abutments were almost finished and the girder work for the new platforms erected. Between Finsbury Park and Drayton Park, rails were dumped ready for laying, and the new signal box at Drayton Park was completed and opened. The southbound ramp at Drayton Park was later connected to the LNER goods yard and has since been used regularly to transfer empty tube

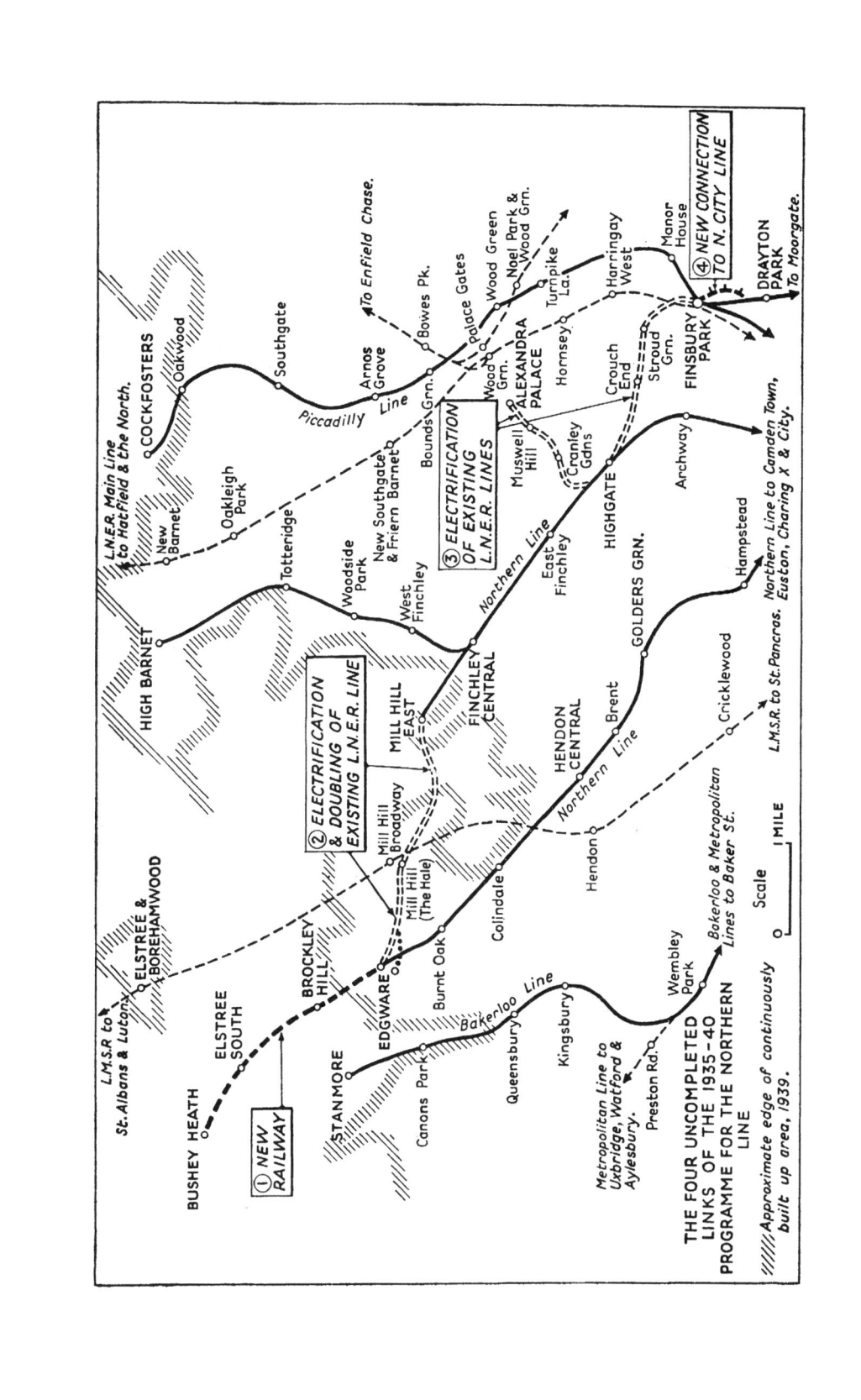

THE FOUR UNCOMPLETED LINKS OF THE 1935-40 PROGRAMME FOR THE NORTHERN LINE

stock to and from the Northern City and Acton Works via Highgate and King's Cross.

Between Finsbury Park and Alexandra Palace lineside cabling was completed and conductor rails laid all the way. New subsidiary signal boxes at Alexandra Palace and Cranley Gardens were well begun and substations completed at Crouch Hill and Muswell Hill. At Crouch End station the platforms were reconstructed.

The single line between Finchley Central and Edgware was to be doubled, with two-platform stations at Mill Hill East and Mill Hill (The Hale), the latter with a footbridge to a new ticket hall serving both the LNER and LMSR stations. At Edgware, the new lines were to curve into the reconstructed LT station which has already been described. Beyond, the 2·86-mile extension would burrow under Station Road and then run for 500 feet between retaining walls. After this it would then be in cutting and on embankment until it reached a 530-yard brick and plate girder viaduct which would carry it over Watford Way and through the proposed Brockley Hill station. The Brockley Hill booking hall was to be in an arch of this viaduct, and at the far end another plate girder bridge would span the site of a proposed new lateral road. North of the viaduct, embankment would give way to cutting, then to twin tube tunnels of 12-foot diameter [1] which would debouch into deep cutting containing Elstree South station. The platforms of this station would be reached by steps from a booking hall in Elstree Hill. Skirting the north-east side of the proposed Aldenham depot, the two running lines would continue to a three-road, two-platform terminus at Bushey Heath, situated at the junction of the Watford by-pass and the Bushey–Elstree road.

Cabling between Finchley and Edgware was completed, and conductor rail was laid and the second track placed in position as far as Mill Hill (The Hale). New platforms at the latter station were partly finished, and new substations were completed at Mill Hill (Page Street) and Edgware. The engineering works for the junction with the Golders Green line at Edgware were almost ready. At Edgware itself, the bus garage was demolished and re-erected on a new site to allow for widening of the station area. New platforms, signal-box and retaining wall were constructed and the original station building at road level was partly demolished to make way for the new. The strip of land between Edgware and Brockley Hill was a relic of the Watford and Edgware scheme. It had long been owned by the Underground company and housing development had left it untouched. Most of the route to Bushey Heath was fenced and levelled, and earthworks and drainage

---

[1] These tunnels were to be of usual tube construction, cast-iron lined, but some use was to be made of segments removed when the Bakerloo and Central line station tunnels were extended. Packings of iron and pitch pine would bring the diameter from 11 feet 8¼ inches to 12 feet.

were well advanced. The underpass beneath Station Road at Edgware had been started, and the foundations of Brockley Hill station and arches of the viaduct were constructed. The tunnels near Elstree Hill were partly bored and segments dumped on site. Near Elstree a sub-station was half built. The Aldenham depot was completed and overhead travelling cranes installed.[1]

The proposed pattern of service on weekdays was as follows:
Bushey Heath Seven peak, six normal-hour trains per hour to Charing Cross section via Finchley Central and Archway (this replaced an earlier plan for a Bushey Heath-Edgware shuttle).
High Barnet Seven peak, six normal-hour trains per hour to Charing Cross section, plus seven peak-only trains per hour to Moorgate via Northern City line.
Finchley Central Through trains from Bushey Heath and High Barnet as described above, plus short-working of seven peak-only trains per hour to City section via Archway and Euston.
East Finchley Through trains as above (28 tph peaks, 12 normal) plus six normal-hour-only trains per hour to Moorgate via Northern City.
Alexandra Palace Seven peak, six off-peak trains per hour to Moorgate via Northern City.
Drayton Park Through trains as above (14 tph peak, 12 normal) plus 14 peak-only trains per hour from the old Northern City terminus at Finsbury Park to Moorgate.
Archway Through trains as above (21 tph peak, 12 normal) plus six normal-hour-only trains per hour to City section via Euston.

It was intended that all open-air stations on the Northern line extensions should take nine-car trains, except on the Alexandra Palace-Drayton Park section, where eight-car trains were proposed.

A FARE ROW

One of the standard answers to complaints of overcrowding on the Edgware line was that, once the Stanmore branch was linked to the Bakerloo, the through tube facilities to the West End would attract passengers from the Northern line.

Those who lived near the Stanmore line preferred to walk or take a bus to Edgware, Burnt Oak or Colindale and use the tube service. There were two reasons for this—the obvious one of the inconvenience of changing at Baker Street (also at Wembley Park on many slack-hour trains) with a longer journey to the City or West End, and secondly

---

[1] Used as an aircraft factory during the war, the Aldenham depot was afterwards converted to a temporary bus maintenance works to relieve the pressure on Chiswick Works. Two of the 15-ton gantry cranes installed for railway use were adapted for handling buses. In 1952 work began on permanent conversion to a bus overhaul works; this was completed in 1956.

the powerful attraction of lower fares on the tube line. In accordance with the expansionist Underground policy of the 1923–1932 decade, the Edgware line fares, as well as being basically lower than main-line fares (approximately 1d a mile compared with 1½d) were on a tapering scale, or in today's parlance, 'sub-standard'. On the other hand, the Stanmore branch had been part of the Metropolitan Railway, and therefore had fares on the main-line scale. Thus, Edgware to Charing Cross was 7d, but Canons Park to Charing Cross was 1s 2d.

Local authorities and Members of Parliament pressed long and hard for the Stanmore fares to be brought down to the Edgware level. There were difficulties in the process of equation because the Metropolitan fares had been governed by the Railways Act of 1921 and the ex-Underground group fares by the various Acts authorizing the constituent railways, but on June 11, 1939, the Board raised fares to meet rising costs, and the opportunity was taken to bring down the Stanmore branch fares to the standard LPTB scale of 1d a mile.

BAKERLOO TO STANMORE

As we have seen, the two-track section between Finchley Road and Baker Street had long been a major obstacle to the efficient handling of the Metropolitan's extension line traffic. Straightforward widening was out of the question for both physical and financial reasons, and deep-level tunnels were the only alternative. The Board adopted the imaginative solution of linking them with the Bakerloo line. This had the great advantage of giving a direct service to the West End from many suburban Metropolitan stations, and cross-platform interchange at Finchley Road from the remainder. The Baker Street–Queen's Park section of the Bakerloo was relatively lightly trafficked, and could stand some reduction in service if trains were lengthened from six cars to seven. There was therefore little difficulty in injecting more trains on to the Bakerloo south of Baker Street.

From Finchley Road to Wembley Park the track arrangement was down slow, up slow, down fast, up fast, but at Wembley Park the Stanmore branch left by a flat junction, and on the main line to Harrow the arrangement became—down fast, up fast, down slow, up slow. Under the new scheme, the two inner tracks and intermediate stations between Finchley Road and Wembley Park would be used exclusively by Bakerloo trains. North of Wembley Park a new burrowing junction would allow the Bakerloo trains to continue on to the Stanmore branch without fouling the southbound Metropolitan, whilst on the main line the Metropolitan tracks would be rearranged: down fast, down slow, up slow, up fast. Between Baker Street and Finchley Road the intermediate Metropolitan stations would be closed and replaced by two stations on the new tube section. Metropolitan trains would call at Baker Street, Finchley Road and Wembley Park only.

The scheme also included the complete reconstruction of Neasden Works as a running depot for the Metropolitan and Bakerloo lines, with a flyunder connection between the north end of the depot and the northbound lines south of Wembley Park.

Charles Brand & Son Ltd began work on the Baker Street–Finchley Road line in April 1936, and shafts were sunk at Baker Street station, Marylebone goods yard (LNER), Acacia Road and Adelaide Road.

By November 1937 the two 12-foot diameter tunnels had been driven through the clay for the whole 2¼ miles from Baker Street to Finchley Road. North of Swiss Cottage, where the Bakerloo tracks were to rise to subsurface level, some extremely intricate engineering work was involved. A new southbound Metropolitan tunnel was built, mainly beneath the shops on the east side of Finchley Road. The northbound Bakerloo tunnel broke into the old southbound Metropolitan tunnel, which was temporarily filled with concrete and clay to facilitate boring. The Finchley Road sewer and service mains had to be diverted, and the property underpinned. Some of the tunnelling was done from the basements of these shops, and some by opening up Finchley Road itself, although sufficient width of roadway for two lines of traffic had to be left undisturbed at any one time. The tube tunnel mouth was protected with mercury tubes.

At Baker Street a second southbound platform was provided slightly to the north of the original one, so that trains from Stanmore could be held at a platform if the junction was not clear instead of waiting in tunnel outside the existing platform. The step-plate junctions with the existing Bakerloo running tunnels were therefore east of the station southbound, but west of it northbound. They consisted of iron-segment tunnels, reducing from 27-feet internal diameter at the widest part of the junction to 14 feet at the apex, and were built round the running tunnels, which were kept in place with timbers and wedges until they could be dismantled. Down to the 19-foot ring, they were constructed while trains were running, but between 19 and 14 feet they had to be built in non-traffic hours.

Another difficult piece of work at Baker Street was the enlargement of the circulating area below the Metropolitan extension line to house the upper ends of the new escalators coming from the southbound tube platforms. The retaining wall east of the Metropolitan No. 4 platform had to be re-sited, and the tracks to the two bay platforms (Nos. 1 and 4) carried on new girder bridges.

North of Finchley Road the work consisted mainly of putting in new connections between the four tracks, building new centre platforms and signal-boxes, and installing new signalling. From September 18, 1938, the 'fast-slow-slow-fast' track arrangement came into use between Finchley Road and just north of Dollis Hill, with new island platforms at Finchley Road, West Hampstead and Dollis Hill stations. The new southbound Metropolitan tunnel south of Finchley Road was taken into use on the same date. From November 6th, the new track layout was extended to beyond Preston Road, and the Wembley Park flyunder opened. The Neasden flyunder came into use on March 26, 1939, and on the following day standard tube stock took over the Stanmore branch shuttle service, to allow the staff to gain experience with tube trains.

The Bakerloo scheme was delayed by the diversion of men to ARP work in September 1938 and August 1939, and the planned opening dates (first spring, then June, then October 1939) could not be maintained. Finally, the through service began on November 20, 1939, when a tube train left Stanmore for Elephant & Castle at 5.9 a.m., and 11·18 miles were added to the existing Bakerloo line mileage of 20·82. Intermediate stations served by through tube trains for the first time were Canons Park, Queensbury, Kingsbury, Wembley Park, Neasden, Dollis Hill, Willesden Green, Kilburn, West Hampstead, Finchley Road, Swiss Cottage and St John's Wood. There were 7 trains per peak-hour from Stanmore and 14 from Wembley Park. Trains were six cars long all day, and of the 20 needed for the Elephant–Stanmore service, 17 were composed of standard stock and 3 of 1938 stock. (By this time all 24 trains on the Queen's Park and Watford service were 1938 stock). A folder was issued bearing the slogan 'New for You', with maps of the central area and the Stanmore branch, and details of services, fares and season ticket rates.

From Stanmore to Piccadilly Circus the single fare was 10d, with the 11d fare extending to Elephant & Castle. Marlborough Road and Lords [1] (Metropolitan) stations closed after traffic on November 19th. It had been intended to reopen Lords for cricket matches, but these were not held because of the war and the station did not reopen. The LPTB had continued the Metropolitan's practice of opening these stations at 9.45 a.m. on weekdays.

Many of the works associated with the Stanmore extension were incomplete on November 20th. In particular, the lack of describers for northbound trains caused confusion both to the public and the staff. The manufacturers were engaged on more urgent tasks and as a stop-gap measure, from December 12th, trains for Wembley Park or Stanmore carried below their front centre windows a sign with a large 'M' (for Metropolitan) in a diamond. When eventually installed, the platform indicators showed the destinations of the first and second trains.

At Baker Street the new southbound station tunnel was 377 feet long, comfortably accommodating a seven-car train. Between the old and new southbound tunnels, and linked to them by cross-passages, was a new circulating area at platform level, whence two new escalators

---

[1] Thus renamed from St John's Wood on June 11, 1939.

rose to the enlarged concourse beneath the Metropolitan extension platforms. From here, two further escalators led up to a new booking hall at street level, in the south-west corner of the main station, but these works were not ready until November 24, 1940.

At St John's Wood the surface buildings, by S. A. Heaps, were a variation of the 1932 style. Surmounted by a low circular tower, the booking hall was 17 feet high and 35 feet wide, with entrance corners rounded off. The oblong window with its concrete mullions struck a new note, but the red-brown bricks and bold low canopy of the earlier style were retained.

Two escalators, with a fixed stairway between, descended 60 feet to platform level between the two station tunnels, which were 435 feet long and 21 feet 2½ inches internal diameter, giving an 11-foot wide platform.

At Swiss Cottage, the other intermediate tube station, the escalator and station tunnel arrangements were similar, but the booking hall was below the surface of Eton Avenue and Avenue Road, with two stairwells from street level and a subway connection from the Metropolitan booking office at the south-west corner of Belsize Road and Finchley Road.[1] A brown and yellow colour scheme was adopted at St John's Wood and brown and green at Swiss Cottage. Following an experiment with a paper name strip at Tottenham Court Road in 1937, the station name was constantly repeated in a strip of embossed tiles at the 'picture rail' level of the station tunnels. The basic tiling was of the usual biscuit-coloured faience, but some tiles were embossed with armorial designs representing the various counties that the Board served.[2] Seats for waiting passengers were recessed into alcoves. Both stations had passimeter booking offices.

Finchley Road station was rebuilt with a modern booking office, new platforms and a roof across the centre tracks. At West Hampstead and Dollis Hill new island platforms were placed between the local roads, but at Kilburn the former up local platform was converted to an island, and the old down local platform abandoned. These islands were given neat shelters and waiting rooms in brick, steel-framed glass and reinforced concrete. Canopies were steel framed but boarding was used for the filling, covered on top by bituminous roofing material, below by wallboard. This method, first tried at South Ealing, was also used at Finchley Road. The platforms had flower-beds bordered by low brick

---

[1] The Metropolitan station remained open, with eight trains per hour in the rush hour, but was finally closed on August 17, 1940, the Belsize Road entrance being retained for the Bakerloo station.

[2] These tiles were also used at rebuilt Central area stations and on the Central line eastern extension. Designs also included Thomas Lord (at St John's Wood), the Houses of Parliament, St Paul's Cathedral, as well as the London Transport symbols—the bull's eye and the griffin.

walls, and old booking halls were refurbished with passimeters and biscuit-coloured tiling.

Beyond Neasden, a new single-span steel bridge of impressive proportions carried the North Circular Road over all tracks. The old terminal platform for Stadium traffic at Wembley Park was replaced by a southbound platform on the track from the car sheds. Platforms served by tube trains from Finchley Road to Stanmore inclusive were built or reconstructed to the compromise height of 2 feet 9 inches above rail level.

The line between Finchley Road and Wembley Park was completely resignalled to the Board's standard system, with new signal-boxes at Finchley Road (59 levers, with 24 route-setting levers), Willesden Green (59 levers), Neasden depot (south) (83 levers), Neasden depot (north) (47 levers), Wembley Park (existing box and 95-lever power frame modified) and Stanmore (47 levers). The Stanmore box was installed primarily to signal the fan of six sidings built on the site of the goods yard, and was all-electric, as compressed air was not then available at Stanmore; its installation meant the end of the pioneer CTC system. The intermediate automatic signals were the starters at Kingsbury, Queensbury and Canons Park. A seven-lever power frame was installed in a new cabin at Baker Street to control the tube junctions. A new feature introduced in the tube tunnels to Finchley Road was the excavation of recesses near the signals, so that signal staff could work safely while trains were running.

NEASDEN DEPOT

The new depot at Neasden, on the site of the old Metropolitan works, was completed in 1938. Tall metal-framed windows in the red-brown brick walls, and lavish provision of roof glass provided ample natural lighting. The roofs were supported by light steel trusses which rested on the walls or on a steel framework supported by girders standing between the tracks. At each end there was a deep rectangular glass frieze, bearing the serial numbers of the roads beneath.

To the north of the depot, there was an extensive fan of sidings, and altogether Neasden could accommodate 728 cars. The main car shed contained eight eight-car roads for examination and cleaning, and three sixteen-car roads for repairs and three for inspection. (Inspection roads had pits at side and centre.) The cleaning shed, at the south-east end of the main shed, had four eight-car roads. Near the depot entrance was a car-washing machine, a small brick structure with eight motor-driven vertical rollers carrying wiping cloths. Water was sprayed on to the roof and sides of the cars, and on to the rollers as a train was driven slowly through. Trains normally entered the depot at the north end, and could be successively examined, cleaned and washed on their journey through.

The adjacent Neasden power station remained as a reminder of Metropolitan Railway architecture. To the north of the main depot was a small steam locomotive shed, for ballast and service trains, and a shed for three road vans carrying railway breakdown equipment. Neasden became the main depot for the Metropolitan and Bakerloo lines, but London Road, Croxley Green and Queen's Park were retained for stabling.

Acton Works was now overhauling Metropolitan stock, and new buildings there (including paint shop, motor shop and an enlarged trimming shop) were finished before the outbreak of war, but the rearrangement of work was not completed until 1949. The intended new paint shop was used for car body work, and painting done in the older building. The office block, built in 1932, gained an additional storey at the rear in 1959–1960.

BAKERLOO IMPROVEMENTS

At all Bakerloo stations from Kilburn Park to Elephant & Castle inclusive, the tunnels were lengthened from 291 to 377 feet so that the extended platforms could take seven-car trains. In some cases there were curves at either end of the original station, and the extra section of platform had to be on a curve. These works were completed by 1940, except for finishings.

The Elephant to Queen's Park section was resignalled to modern standards to handle a more intensive service, and new signal frames were installed at Elephant & Castle, Lambeth North, Piccadilly Circus and Paddington, as well as at Baker Street, as mentioned above.[1] Speed-control signalling was installed at two points, to allow trains to draw up close together at the approaches to busy stations.

At Lambeth North a new scissors crossover was installed to improve the access to London Road depot (replacing a trailing crossover) and at Elephant & Castle works were carried out to improve the terminal facilities. The platforms were extended at each end to give a new length of 377 feet, and beyond the southern ends of the platforms two 515-foot tunnel sidings were driven beneath the Walworth Road and connected by a scissors crossover. These cut partly across the existing sidings beneath the New Kent Road, which were abandoned.

The new sidings followed the route of the intended extension to Camberwell and were constructed under the powers obtained for that line in 1931. The south London authorities had seen to it that southern extensions of the Bakerloo line had been kept very much in the Board's eye. In 1935 they protested loudly at the north London bias of the New Works programme.

---

[1] On the LMSR section between Queen's Park and Watford, colour light signalling had been installed in 1932/3, with train stops. A new station was opened at South Kenton on July 3, 1933 by the LMSR.

Yet another deputation, from Camberwell and Southwark, called on Lord Ashfield in October 1936, to be told that the extension would cost £2 million and would not pay. In February 1937 London Transport announced that an extension from Elephant & Castle was being considered by the Standing Joint Committee of the Board and the main lines, but in April of that year they dashed any hopes by stating that any new works were out of the question owing to increased costs.

UXBRIDGE BRANCH RENEWED

The extension of Piccadilly line trains to Uxbridge gave a further spurt to housing development in the meadows west of Harrow, and the extra traffic called for new stations to replace the little shack halts at Rayners Lane, Eastcote and Ruislip Manor. Before tube trains came, the Metropolitan Railway had done much to promote speculative building, and also pursued its own housing projects through its Surplus Lands Committee and Metropolitan Railway Country Estates Ltd.

One of the most ambitious Metropolitan schemes was 'Harrow Garden Village' near Rayners Lane station. Traffic at this station had increased from 22,000 passengers a year in 1930 to 4 millions in 1937. The halt opened in 1906 had remained isolated in open country until about 1930, when building began in earnest. It consisted merely of two wooden platforms, with tiny corrugated-iron shelters and, at street level, a timber booking hut. As the first step, a temporary wooden booking hall (including newspaper and tobacco kiosks) was opened on March 14, 1935, and the construction of a new station was then begun. A new bridge to carry Rayners Lane over the railway was provided at an early stage, and the new station was opened on August 8, 1938. Similar in style to the Piccadilly line stations of 1932, it had a tall, boxy booking hall of brick and glass capped by a wide frieze and overhanging cornice, both in concrete. It drew attention to its presence by jutting 25 feet into the pavement from the bulding line, providing an easy natural entry from north or south side straight into the booking hall. The hall, with its coffered ceiling, was 32 feet wide by 34 feet long by 32 feet high. Internal walls were finished with a black brick dado 7 feet high, surmounted by a 3 foot white coping; above this, the Buckinghamshire facing bricks were exposed. A covered footbridge and stairways gave access to the reinforced-concrete platforms, which had brick shelters and wide concrete canopies.

The rebuilt Eastcote station, bearing some architectural affinity to Northfields, was opened early in 1939, and Ruislip Manor on June 26, 1938. At the latter station the booking hall was in the bridge abutment, and stairs led up to the platforms 25 feet above. No radical alterations were made to the Metropolitan station at Ruislip, which had been built on a fairly generous scale in 1904 in the expectation of traffic development.

Few good words could be said for the existing terminus at Uxbridge. Tucked away in the quiet backwater of Belmont Road, it had but two tracks, and the station buildings were concentrated on the southern platform, the northern having merely a diminutive shelter.

In 1935 powers were obtained to build a new three-road terminus in Uxbridge High Street, connected to the existing tracks by a new line 836 yards long. Work began on September 14, 1936. The new line, in a cutting with concrete retaining walls from 6 to 30 feet high, led to a terminus at street level.

On December 4, 1938, the new works were opened for traffic. Six days before, the peak-hour Piccadilly line service from Uxbridge was increased from four to eight trains per hour, giving a total of 176 Piccadilly and 138 Metropolitan trains per day.

The semicircular High Street façade had two storeys with the upper one set back and the station entrance in the centre was flanked by parades of shops. These shops, which looked on to a bus layby, housed businesses from older shops in and behind the High Street which had been demolished to make way for the station. The booking hall was distinguished by an impressive stained-glass window at the High Street end and contained the usual kiosks, lavatories and staff accommodation, also a buffet and bicycle store. Tickets could be obtained from machines or at the passimeter. The overall roof, with its stout concrete ribs and centre clerestory was remarkably like that at Cockfosters, but it had to be higher, to accommodate Metropolitan trains. There were two wide platforms between the three terminal roads, giving four platform faces. North-west of the station a new bus station was laid out, with direct access to the booking hall.

Associated works included the provision of sidings on the site of the old goods yard, the construction of a new bridge to carry York Road over the railway, and the building of a new signal-box equipped with the Board's standard electro-pneumatic system. The increased service required two new substations, sited at Uxbridge and Rayners Lane.

The Uxbridge branch had been resignalled with automatic three-aspect Westinghouse colour lights in 1930, but at Rayners Lane the original mechanical-frame signal-box remained in the wedge of land between the lines to West and South Harrow. This box was wrecked by a runaway ballast train in the early morning of November 22, 1934, and its place was taken by a new box at the Uxbridge end of the eastbound platform which came into full use on November 17, 1935. This had a 35-lever power locking frame with sixteen push-pull route-setting levers. A new reversing siding was provided west of the platforms between the running lines, controlled from the new box.

CENTRAL LINE MODERNIZATION

During the 1930s the Central London line was worked quite efficiently with its reconstructed cars, but it was obvious that they would have to be replaced by stock of standard type when the eastern and western extensions came into use. Two alterations were essential before standard stock could be used—the installation of a fourth current rail, with the positive rail outside the running rails, to replace the original three-rail arrangement, and realignment of the tunnels at points where clearance was limited. A further improvement the Board wished to make was the replacement of the remaining ten miles of bridge-type running rail on longitudinal sleepers by bull-head rail on cross-sleepers, completing a task begun in 1912. It was also desired to lengthen the station tunnels from 325 feet to 427 feet to take eight-car trains, and resignal the line to give increased capacity.

The original Central London line had been built when tube tunnelling technique was in its infancy; boring was necessarily imprecise, and segments were up to 8 inches out of true alignment. A better alignment was needed to accommodate standard-size stock, to allow higher running speeds, and to give enough clearance for the fourth rail at certain points. Even so, clearances were still tight after realignment; the positive rail had to be fitted $4\frac{1}{2}$ inches above the running rails instead of the usual 3 inches and some positive rail is of inverted L-section to make an easier fit.

The tunnels were surveyed to see where segments had to be adjusted, and an experimental realignment was made between Bond Street and Marble Arch in the autumn of 1936. The work was done in the short $3\frac{1}{2}$-hour non-traffic periods, and the main scheme took two years. Each night a train of service wagons left Wood Lane yard behind a battery locomotive, stopping to drop a wagon and its crew at each working site. The upper segments were the ones usually altered. After the signalling and current cables and tunnel telephone wires had been moved out of the way, the key segment at the top of the tunnel was removed by a special screw-jack and the adjacent segment unbolted and removed for return to Wood Lane for grinding. The concrete grout and clay behind it were cut away to the required depth with pneumatic picks and shovels, and a newly-ground segment from Wood Lane was inserted with packing pieces to produce the required diameter. Any space behind the newly-installed segment was filled with new cement grout, injected through holes in the segment at a pressure of 100 lb. per square inch. Gauging and the restoration of cables, etc., completed the night's work, in time for the first train. Some 10,000 segments were thus repositioned. One mild surprise was the discovery of a tunnelling shield which had been left in position from the original construction near Post Office station. When this impinged on the planned realignment it was removed with oxy-acetylene torches.

The siting of the stations on 'humps', and the Ministry of Transport recommendation that platforms should not be on a gradient steeper than 1 in 260, caused problems in platform lengthening. Usually, the greater part of the extension was made at the approach end, where the decelerating gradient was 1 in 60 uphill, rather than the departure end with its 1 in 30 descent for acceleration. Normally the answer was to make a slight gradient in the platform extension itself, raising the adjacent running line and roof of the running tunnel for about 80 feet. Adjoining the extended station, running tunnel enlargement was done by methods similar to those used for realignment, but with new and longer side segments inserted. At Shepherd's Bush, the exit stairways and escalators precluded the lengthening of the tunnels at the western end, whilst at the east end the lines converged towards a crossover tunnel. This tunnel was enlarged, the tracks moved further apart, and the platform extended between them, as an island.[1] At Liverpool Street the dead-end siding tunnels were extended to serve as reversing sidings between the running tunnels, and the platforms were extended from their existing length of 400 feet.

The tunnel realignment was completed by September 1938 and standard stock was gradually introduced, temporarily modified with the centre collector shoes used for positive. The last of the old CLR stock was withdrawn from passenger service on the night of July 12/13, 1939, and the sole surviving Central London electric locomotive (the Wood Lane shunter) was taken out of use at the same time. The installation of standard track, and the station tunnel extensions (except the finishings), were completed by April 1940, and the four-rail system came into operation on the 21st of that month. Much of the original conductor rail survived as the negative rail, recognizable by its shallow, inverted-channel cross-section.

CENTRAL LINE EXTENSIONS BEGUN

The first contract for the tunnelling on the eastern extension of the Central line, for the difficult Mile End–Leyton section, was placed with John Mowlem & Co. Ltd in October 1936[2] and work began in the same month. This section was chosen for an early start because the tunnelling, mostly through water-bearing strata, would take longer to complete than the other sections. The presence of water also boosted the price per mile by £250,000 to roundly £1 million.

In March 1937, Charles Brand & Son Ltd. were given the Liverpool Street–Mile End section, and in July the contract was placed for the Hainault depot; in August, contracts were completed and work begun on the extra tracks from North Acton to Greenford, and Ruislip depot. By then, engineers were in possession of a working site at Gants Hill, and in February 1938 a Parliamentary question brought forth the in-

---
[1] See Filor, C. G. H., J.Inst.C.E., February 1940.
[2] Except the approach ramps to Stratford station which were awarded to John Cochrane & Sons Ltd in February 1938.

formation that the tube trains would reach Stratford by summer 1940, and Newbury Park by December. In May 1938 Charles Brand & Son secured the contract for the Wanstead–Gants Hill tunnels, and in August the last remaining section of new tunnel, between Gants Hill and Newbury Park, was given to Edmund Nuttall Sons & Co. (London) Ltd.

When press representatives were shown the work between Liverpool Street and Leyton in July 1938, work was proceeding from fourteen sites. From Liverpool Street to a point west of Mile End station the rotary excavators and shields cut through the stiff blue clay without difficulty. Further east the subsoil was waterlogged, and much of the work had to be done in compressed air. From Old Ford Road to Stratford the tube tunnels were beneath the LNER embankment, and conditions were particularly difficult where they had to pass below the various branches of the River Lee and the adjacent abutments of railway bridges. On this section, and also near Mile End and between Stratford and Leyton, the Joosten chemical soil consolidation process was employed. This process had been previously used at Monument–Bank and Knightsbridge.

At Mile End station, provision had to be made for tube tracks outside the existing District line platforms. On the south side this was relatively simple, as there was already an open cutting behind the platforms, but on the north the existing retaining wall behind the platform had to be demolished and replaced by a new mass-concrete wall beyond the planned eastbound tube track, underneath the Mile End Road.

During 1937, the rearmament programme caused a shortage of iron and steel, and experiments were made at Wood Lane with reinforced-concrete segments for lining the running tunnels. One test tunnel section was half iron, half concrete, with its top 2 feet 6 inches below the surface of the ground. The concrete successfully withstood the weight of 175 tons of iron, and a 12-foot, all-concrete tunnel likewise resisted 228 tons. Another test which was passed successfully was in a tunnel under construction, where the hydraulic rams pushing the shield forward bore against the newly-erected segments with a force of up to 560 tons. The concrete segments, which were used on the $2\frac{3}{4}$ miles of tunnels east of Redbridge Lane/Eastern Avenue, made a tunnel of 12 feet 3 inches internal diameter, compared with 12 feet for iron segments. The segments were 2 inches thick, compared with $\frac{7}{8}$ inch for cast iron. They were bolted together in the same way as iron segments, but the bolt-holes were lined with steel ferrules and the surrounding concrete reinforced with steel.

At Redbridge the original plan was for the line to come to the surface to cross the River Roding by bridge. This plan was changed so that the line remained underground, but it was so shallow at Redbridge station that cut-and-cover construction was used, with reinforced-concrete tunnels of rectangular cross-section. It was necessary to tunnel in compressed air in the water-bearing soil of the Roding Valley.

The construction of the shallow tunnels from Stratford to Leyton was also very difficult, as the ground was water-bearing and the route lay athwart the LNER main lines.

As late as June 1940, the London Transport staff magazine *Pennyfare* predicted that tube trains would be running to Loughton and Hainault in November of that year. But this was not to be. Early in July came the announcement of the Minister of Transport's decision that work should cease until hostilities were over, apart from the completion of certain bridges and revetments for safety reasons. By then the tunnels east of Liverpool Street were complete and many miles of the Loughton branch and the Hainault loop had been equipped with current rails. Some progress had been made in installing London Transport signalling, both in the tunnels and on the surface. A new station at Loughton was opened on April 28, 1940, and the rolling-stock depots at Hainault and Ruislip were virtually completed. The road-bed of the western extension to Ruislip was finished, and some work had been done on station reconstruction. The GWR had completed two additional tracks between Wood Lane and North Acton in October 1937.

IMPROVEMENTS IN THE CENTRE

Between 1933 and 1939, the Board carried on the Underground's programme of improving central area stations to cope with new traffic brought by the extensions. The work was concentrated on the most important of the unreconstructed Northern and Central line stations.

Following the experiments at Warren Street described in Chapter 10, and the installation of automatic lifts at Earl's Court in 1932, three of the lifts at Strand were modified for automatic working in spring 1935. An improvement on the Earl's Court installation was the adoption of automatic control by telephone-type relays, to give a regularly-spaced combined service (every thirty-seven seconds from each landing when all three lifts were working). The audible warning, 'Stand clear of the gates, please' (first tried at Earl's Court in 1934), was recorded on film sound-track attached to a drum revolving at constant speed, and additional warnings were given by illuminated signs.

On March 4, 1937, three high-speed automatic lifts were brought into service at Goodge Street. The lifts were small—they held only seventeen people—but were the fastest on the London Underground, giving a service every thirty-four seconds in the peak, with a speed of 600 feet per minute in the 85-foot shafts. The doors were deliberately designed to prevent passengers from observing the progress of the lift, thus ensuring that the more timid ones remained comfortable. At the landings, the door surrounds were neatly tiled, and indicators showed the next lift to leave. As at Strand, control gear automatically ensured even time spacing between cars, whether two or three were in use. Manual press-button control was available.

Tottenham Court Road was a chronic patient, always in the operating theatre. On May 8, 1937, four years after the addition of a sixth escalator, a relief passenger subway was opened alongside the westbound Central London platform, and on November 15, 1939, a new 210-foot deep-level interchange subway was available between the eastbound Central and southbound Northern lines.

On the City section of the Northern line, Moorgate station was extensively reconstructed. The CSLR subsurface booking hall was enlarged to serve the Metropolitan, Northern and Northern City lines, with a new subway connection from the Metropolitan. The CSLR already had escalators, which came up on the west side of the new booking hall. These were converted from shunt to comb, and new escalators installed for the Northern/Northern City interchange (in single shafts) on June 19, 1936. The new booking hall was in full use on October 2, 1936, and the GNCR booking hall was closed; on the same day, two escalators in one shaft were opened between the new top station and the Northern City line.

King's Cross was another place where there was a group of separate booking halls and lifts built at different times. Here a joint booking hall for the Piccadilly and Northern lines was constructed beneath the LNER forecourt and approach road, with passenger subways into the main-line station and access to the existing subway to St Pancras station. Other subways led to the south side of Euston Road and to a new Metropolitan booking hall and station beneath Euston Road. The new tube booking hall incorporated the site of the old CSLR hall, and replaced the LER surface station. Work began in the early months of 1936 and the tube booking hall and escalators were opened on June 18, 1939. Three escalators with a 56 foot 6 inch rise and a speed range of 125–180 feet per minute led to a circulating area at Piccadilly platform level. Two further escalators connected with the Northern line platforms, 18 feet 6 inches below. The existing low-level interchange subway between the Northern and Piccadilly lines was retained for traffic interchanging in that direction only, the Piccadilly/Northern change being via the lower escalators. The construction of the new Metropolitan station continued after other works of the 1935 plan had been suspended, and it was opened on March 14, 1941.

On the Central line, work was started in October 1935 on a new booking hall for Post Office, beneath the junction of Newgate Street, Cheapside and St Martins-le-Grand. This station was re-named 'St Paul's' on February 1, 1937, and on January 1, 1939 a new booking hall came into use with two flights of escalators (three escalators between booking hall and westbound platform, one of which has since been taken out; two escalators and fixed stairs between westbound and eastbound platforms). Five lifts were replaced and the street booking office at King Edward Street/Newgate Street closed.

To serve the new exhibition hall at Earl's Court, a new subway was driven from the circulating area beneath the District platforms, ending at a pair of escalators to a new booking hall beneath the exhibition entrance. The existing street-level gallery from the District station to Warwick Road, and the Warwick Road booking hall were rebuilt, mainly in reinforced concrete. The old footbridge across the Lillie Bridge sheds, and the part of the sheds south of it, were demolished to make room for a car park. The new facilities came into full use on October 14, 1937.

Good progress was made in replacing the shunt type, non-reversible 'A' escalators by faster, reversible comb types ('M' or 'MH'), and during 1936, this work was carried out on eighteen installations at Charing Cross, Earl's Court, Kilburn Park, Liverpool Street, Maida Vale, Paddington and Warwick Avenue. Another development was 'speedray control', tried at Manor House in 1934. A light ray and a photo-electric cell at the lower end detected the movement of passengers on to the escalator. Breaking the ray caused relays to be operated which speeded up the escalator, and also set a time-switch in motion. If no further passenger had broken the ray after an interval long enough for the initial passenger to reach the upper end, the time-switch completed the contacts which put the escalator back to idling speed, thus saving wear and tear. By 1960, thirty-seven machines were thus converted, always on the 'up' escalators because passenger movement on them is less even than on the 'down' ('up' passengers come in bunches from trains).

Another use of photo-electric cells is to detect the arrivals and departures of trains, and thus control operation of 'gap lights', which illuminate platform walls and part of the roadbed at sharply curved stations where there is a gap between the middle doors of the cars and the platform edges. This system was adopted at Waterloo (Bakerloo northbound) and Charing Cross (Hampstead northbound) in June 1934.

A striking illustration of the traffic-promotion value of modernized stations was given in the Board's report for 1937–1938. Since 1931 the originating traffic at Marble Arch had increased by 48·4% and at Green Park 41·0% compared with an average increase of 5% at unmodernized stations. At Moorgate, the annual Northern City/Northern interchange traffic grew from 1·3 million to 1·7 million by the end of 1937.

In 1936 the platforms at the main central area interchange stations were numbered, odd numbers for southbound and westbound, even numbers for northbound and eastbound. Staff were trained to memorize the numbers and give directions for platforms at other stations.

BETTER VENTILATION

With the programme of extensions and consequent increase in the length and frequency of trains on the central sections came the need for additional ventilation. Apart from this, some sections of line had become unpleasantly warm, and better ventilation was needed.

The primary purpose of the ventilation system on tube railways is to remove heat, and it has been estimated that 20% of the heat released when coal is burnt at Lots Road is eventually emitted in the tube tunnels from the motors, resistances and brakes. Heat from passengers' bodies adds a little more. The subsoil surrounding the tube tunnels has been gradually warming up ever since the lines were opened and as the zone of warmed soil is ever-widening, an increasing burden is thrown on to the ventilation equipment.

The Board secured powers in 1938–1939 for additional ventilating shafts at various points on the Bakerloo and Northern lines, and in August 1938 announced preliminary plans for a £500,000 programme of ventilation improvements, with thirty new plants, bringing the total to over one hundred.

On the tube extensions the standard arrangements included high-capacity axial-flow fans, situated between stations and extracting the warmed air from the tunnels, and lower-capacity centrifugal fans at the stations, blowing in air from rooftop level, which was dispersed through grids in the station tunnel walls. Working shafts were used as ducts wherever possible, and if not so used were often retained as relief shafts, thus reducing draughts at stations.

Some interesting experiments were made. At Oxford Circus (Central line) ducts were installed in the station tunnel wall opposite the platforms, and level with the car windows, with the intention of blowing fresh air into the cars, but only about one-sixth of the air was changed during the necessarily short station stops. Early in 1938, at Trafalgar Square northbound, atomized water spray was employed to cool the air, whilst at Tottenham Court Road, near the Northern line southbound platform, a refrigerator and fans were installed in a disused lift shaft. Air was blown over pipes containing water just above freezing point, giving an effect equivalent to melting 50 tons of ice per day.

THE CAMPAIGN AGAINST NOISE

Another aspect of the drive to make tube travel more attractive was the effort to reduce the noise reaching passengers when the trains were running in tunnel. Tube travel is fundamentally noisy because the tunnel walls, within a foot or so of the train, reflect noise from the running wheels, gears, motors, etc., back into the cars through the window and ventilation openings.

In the experimental rolling stock of 1923, much of the noise was confined to its source by shrouding the bogies, but the arrangement was inconvenient in service, and the feature was not incorporated in the standard stock. Some noise reduction was obtained in that stock by asbestos lining in the body sides.

Subsequently, the Underground group made another approach to the problem by trying to trap the noise in the tunnel lining. At the end of 1931 an asbestos mixture was experimentally sprayed on to Piccadilly line tunnel walls, and two further experiments were made early in 1932. Five hundred feet of the eastbound Piccadilly tunnel between York Road and Caledonian Road were lined with $1\frac{1}{2}$-inch asbestos mattresses, held on rods between the segments. In the other experiment, in the southbound tunnel north of Hampstead, a $\frac{1}{2}$-inch coating of asbestos fibre was sprayed on the tunnel walls and asbestos mattresses placed between the segment flanges.

The asbestos, which was expensive, formed a perfect trap for the brake-block dust which is constantly settling in tube tunnels, so in 1934 a third approach was tried—to prevent the noise penetrating the car bodies. A standard car was modified at Acton Works with $\frac{1}{2}$-inch asbestos lining on the inside of all metal panels (ceiling, sides and floor), and fixed windows with double glazing and air conditioning. This experiment also ended in a blind alley, and attention was redirected to the tunnels and track. On the Cockfosters extension, some reduction in noise had been obtained by ballasting the 4-foot way, using 90-foot rails and welding the conductor rails.

Experiments continued, and in July 1936 Lord Ashfield optimistically told a meeting of Conservative MPs that 'the day of the quiet tube has now arrived. It will soon be possible to carry on conversation in the tube trains in a normal voice.'

One gate-car had been equipped for rail-grinding in 1931, and a second gate-car was adapted in 1936. These cars had specially shaped carborundum blocks which could be pressed down against the track surface. In September 1938 a new car was introduced, with two $49\frac{1}{2}$-inch electrically-driven grinding wheels.

In 1936 and 1937, experiments were made with 90-foot rails welded together. Rail ends were fashioned into Brogden (or 'scarfed') joints. A vertical cut was made along the centre line of the rail for 9 inches, and half the rail cut away. The other half overlapped the corresponding half of the adjoining rail, thus preserving a continuous bearing surface for the wheel tread. Each rail end was supported by a 'joint chair'. A large-scale programme of track relaying, with rail welded into 300-foot lengths, and Brogden joints, began on the Northern line in September 1937, after the installation of a flash butt welding plant at Lillie Bridge permanent way depot.

In February 1938 a new mobile welding plant was demonstrated, mounted on two bogie well wagons to tube loading gauge. It was intended for use at rail dumps on the extensions, and was used for a time at Wellington sidings for the High Barnet extension.

Trials of tunnel screens were continued, and eventually a satisfactory design was evolved. In December 1937 a contract was awarded for the soundproofing of 6½ miles of Northern line tunnels between Camden Town and Golders Green. The standard system used foam-slag filling at wheel level on each side of the train. The slag filled a space of roughly triangular cross-section, bounded by the tunnel wall, a reinforced-concrete top cover, and a perforated asbestos facing sheet, the whole assembly being supported by steel brackets attached to the tunnel flanges.

The Northern line extension to East Finchley, and the Bakerloo to Finchley Road incorporated the new standard features of welded rails and a simpler type of noise-absorbing screen, consisting of a horizontal fin at floor car level, formed of concrete slabs lined underneath with sound-absorbent material.

1936 STOCK

The adoption of electro-pneumatic brakes and weak field control on the Watford replacement stock of 1930 and the Piccadilly stock of 1931 proved successful in service conditions, and contracts were placed for the similar modification of the earlier standard stock, most of which was working on the Morden–Edgware line. This work was completed by the beginning of 1937, except for twenty 1928 stock motor cars working with the 1920 stock on the Bakerloo, which were not modified until 1939. Other developments with existing stock were the substitution of 'London Transport' for 'Underground' as a fleet name from the end of May 1934, and the display of the line name on the side panels of many cars, in gold-leaf letters, from about December 1934 to 1937. From June 1932, Bakerloo trains running beyond Queen's Park were distinguished by a blue stripe on the cream upper panels to compensate for a lack of platform describers on that line. This device was removed by November 1937.

As modified, the standard stock was reasonably fast and reliable, but the motor cars suffered from the same disadvantage as all the earlier stock—the ten feet or so of potential passenger space occupied by the resistances, contactors, switches, compressors and other equipment. The only possible solution to this problem was to mount the control and other gear beneath the floor, and the first of four experimental tube trains with underfloor equipment was shown to press representatives on November 17, 1936, after trial runs from November 9th.

These trains incorporated so many new features that they represented an immense technical advance on the final standard stock—to use a London bus analogy, it was as if the 'RT' had directly followed the 'NS'.

Unfortunately, the Board temporarily forsook the tradition of good functional design, and fell victim to the contemporary fashion of streamlining. All twenty-four cars were driving motors, and eighteen of them had streamlined cabs, semicircular in plan, with the body panels and windscreens sloping inwards from floor to cantrail level. The driver sat in the middle of the cab, in a swivelling leather armchair, with a 'joystick' lever in each hand—one for the controller on his right and one for the e.p. brakes on his left. This arrangement of semicircular cab wasted some of the valuable extra body space that the whole design had been planned to achieve, and the presence of streamlined 'ends' in the middle of the trains (each formed of three two-car sets) added a ludicrous touch.

The fourth train showed a return to common sense—its car ends were virtually flat. Flat, but not ugly, for the ends of the roof were smoothed down to meet the vertical body ends, and the body corners were rounded.

The cars were 52 feet 6 inches over headstocks, and all twenty-four bodies were built by Metro-Cammell. The underframes were of welded steel sections, with the centre longitudinals forming ducts which were used on two cars of the first train for an experimental ventilation system by Stone of Deptford. These two cars had double windows permanently closed. Fans blew filtered air (warmed in winter) into the cars. The other cars had normal resistance heaters.

A flush external finish was achieved by using rebated glass in the windows of all cars, whilst window pillars of triangular cross-section afforded the least possible obstruction to vision. Body panels were sprayed with asbestos composition to reduce noise. Each car had two double-leaf and one single-leaf air-operated door per side; the single-leaf doors were at the end remote from the cab, which had separate hinged doors for the driver's use. There were forty fixed passenger seats per car, in bays of 6 longitudinal, 8 transverse (4×2), and 6 longitudinal per car side plus 2 tip-up seats at the non-driving end. The seats (experimentally Dunlopillo-padded on one unit) were covered in light-green moquette, and were similar to those of contemporary Green Line coaches. The driver's seat could be reversed for use by the guard.

The control and lighting current came from a 50-volt 5-kW motor-generator, thus avoiding the two bugbears inseparable from lighting direct from the current rail—interruption at crossings and reduced intensity when the train is accelerating. The lamps were wired in parallel, and their light was diffused by fluted oblong shades in chromium-plated frames. These were mounted in two rows, one beneath each side-ceiling (above the front edges of the longitudinal seats). Between the lampshades were supports for standing passengers consisting of a long rubber stem and a plastic bulb.

The inner ends of the two-car sets were semi-permanently coupled, but the outer ends had Messrs G. D. Peters' automatic 'Wedglock' centre coupler. All mechanical, electrical and pneumatic connections were made simultaneously, under electro-pneumatic control by push button from the driver's cab.

Bogies were of all-welded construction, with extended bolsters to control side oscillation. There were various combinations of leaf and coil springs so that different theories could be tried, and as the presence of the electrical equipment beneath the floor precluded the use of ordinary brake rigging, a new design was evolved. This used Westinghouse independent self-lapping electro-pneumatic wheel brakes, with two brake shoes per wheel and one cylinder for each shoe, and automatic retardation control.

The floor maintained the normal level of 2 feet 1 inch from the rail at doorways but rose slightly between them. This low level necessitated the adoption of 2 foot 7 inch wheels and specially designed compact 138-h.p. traction motors by Crompton Parkinson. There was one motor per bogie, or two per car, and roller motor-suspension bearings were used. The motored axles were 2 foot 9 inches from the bogie centre, and the (outer) trailing axles 3 foot 6 inches from the centre. A 'maximum traction' effect was thereby obtained, and 58% of the train weight was available for adhesion. This factor, combined with multi-notch control and reduced train weight, gave an acceleration rate of 2 m.p.h. per second, whilst a maximum brake application gave deceleration of 3 m.p.h. per second. The gears between the motors and the driving axles were specially designed for silent operation. Two stages of weak field running were provided.

Four electrical manufacturers were given a free hand 'to demonstrate their interpretation of the multi-notch principle', to quote *Pennyfare*. The systems they installed were

(i) *Crompton Parkinson* On two sets, Allen West face-plate controllers, giving 19 notches. On one set, Allen West motor-driven camshaft control.

(ii) *G.E.C.* Contactor control by means of a camshaft driven by a servo motor, with 56 notches.

(iii) *B.T.H.* The American P.C.M. (Pneumatic Camshaft Mechanism) with air-operated units controlled by magnet valves.

(iv) *Metropolitan-Vickers* 45-notch oil-driven power drum for resistance switching (on the flat-ended trains).

These trains entered service on the Piccadilly line from April 8, 1937, perhaps in response to complaints of overcrowding on the South Harrow and Uxbridge branch. They remained at work until 1939 when they went into store at Cockfosters. Ten years later, they were adapted for further use, as we shall describe.

1938 STOCK

Early in April 1937, after comparatively short experience with the 1936 stock, £3 million of orders were placed for 750 new tube cars. The

Metropolitan-Cammell share of 500 cars was publicized as the 'largest order ever placed with any British rolling-stock builder'. Later supplementary orders brought the Metropolitan-Cammell contract up to 751 cars and that of the Birmingham Railway Carriage & Wagon Co. to 370 cars.

In the same month, the British Thomson-Houston Company received the £1 million order for the whole of the control apparatus for the new trains, and in June 1937, GEC and Crompton Parkinson each received an order for 800 traction motors, of the same design, at a total cost of £600,000.

The first deliveries of the new stock arrived at Golders Green towards the end of May 1938, and the first train went into regular service on June 30th.

Many features of the experimental cars were incorporated in the new stock and it was very similar in appearance to the non-streamlined fourth train of 1936, with its elliptical roofs, rounded down over the cabs. A new departure for London urban railway stock was the 'non-driving motor car' (a motor car without a cab, having the same body shell as the trailer). On the Northern (and subsequently on the Bakerloo) the standard train formation, from the north end was: DM-T-NDM-DM + DM-T-DM, giving five motor cars per train compared with three in the standard stock, and allowing the division of a seven-car train into four-car and three-car units for off-peak operation.

Driving motor cars were 52 feet $3\frac{3}{4}$ inches over body, and trailers and non-driving motor cars 51 feet $2\frac{3}{4}$ inches. All cars were 9 feet $5\frac{1}{2}$ inches high, from rail to roof top, and 8 feet $6\frac{3}{16}$ inches wide over body panels. All the 644 driving motor cars and 107 of the non-driving motor cars were by Metropolitan-Cammell; the remaining 99 of the last-mentioned type, and all the 271 trailers, were by Birmingham. The LNER owned 172 of the driving motor cars, 41 of the non-driving motor cars and 76 trailers, ownership being indicated by a plate on the solebar. This was regarded as that company's share for operating the Edgware, Barnet and Alexandra Palace branches.

The arch-roofed steel bodies, and steel underframes, were of combined welded and riveted construction, and the body and underframe formed a combined stress-bearing structure.

The trailers and non-driving motor cars had four sets of air-operated doors per side, two single-leaf of 2 feet 3 inches opening, and two double-leaf of 4 feet 6 inches opening. The driving motor cars had one single- and two double-leaf doors per side, and 42 fixed, plus 2 tip-up seats. The other types had 40 fixed and 4 tip-up seats, all the fixed seats having deep cushions and bucket-type backs, arranged in the same way as on the 1936 stock. Provision was made for the passengers to open any door by individual push buttons located in each door-post, under the guard's overriding control.[1] This feature made the door-control mechanism much more costly and complicated as there had to be a separate electro-pneumatic control valve at each doorway, instead of one for each car side. However, these separate door valves, combined with re-design of the door engine, did enable door-opening time to be reduced from four seconds to two.

The windows, window pillars and artificial lighting closely followed the 1936 pattern. Windows adjacent to the sliding doors were double glazed, so that the door leaf slid between them when open. Consequently there were only six tilting quarter-lights per side on a non-driving car, compared with eight on a 1931 trailer. Alternative sources of ventilation comprised the drop windows in the end doors and openings in the car ends, which discharged fresh air through grilles in the ends of the passenger compartments. In cold weather, electric heaters beneath the seats could be switched in to the line voltage.

The 50-volt motor-generators for the control and lighting circuits were located on the driving motor cars; they supplied current for the traction control, door control, coupling and brake control circuits. A 56-ampere-hour lead-acid battery floated across the generator terminals; this would provide partial lighting and feed all control circuits in emergency.

Wedglock couplers were fitted at the cab ends of driving motor cars, but elsewhere the cars were linked by semi-permanent bar couplings, and 40-wire jumpers. The Wedglocks gave instantaneous automatic coupling of 32 electrical and all mechanical and pneumatic connections. An air-driven electro-pneumatically controlled camshaft operated the various valves in correct sequence; it ensured that the pneumatic connections were made without loss of pressure in the train line and that each separate unit had one operative tripcock. Control was from buttons in the cab.

The bogies had many features in common with those of the 1936 stock, including all-welded construction, and extended bolster, and maximum-traction axle positions. The inner axles were 2 feet 9 inches from the bogie centre, and the outer axles 3 feet 6 inches, even on the trailers, which could later be converted to non-driving motor cars if required, as the underframes were also standard. Laminated side-bearing springs and auxiliary helical springs gave a much softer suspension than that of the standard stock. On each motor bogie was a self-ventilated nose-suspended motor of 168 h.p. one-hour rating, 122 continuous,

---

[1] Passenger door control, first tried in London on the Hammersmith and City line in 1936, made its début on the tubes on a standard stock train running on the Northern line in April 1938. It was used on 1938 stock trains as delivered, but there were many failures and it was taken out of use in January 1939. Further experiments were made on the Northern line in June 1939 and February 1940. It was also tried on the Central line up to the outbreak of war.

specially designed in a long and slim shape to fit beneath the car floor. There were ten such motors on a seven-car train, giving 1,680 h.p. compared with 1,440 h.p. on a similar train of standard stock.

The inner axles of the motor bogies were driven by steel single-helical pinions and gears, and carried spoked-steel running wheels. Trailing wheels were solid-forged discs. All wheels were of 2 feet 7 inches diameter when new, and the axle boxes had roller bearings.

Braking force was again provided by one cylinder per shoe, and was of the Westinghouse self-lapping interlocked electro-pneumatic type with retardation control. The latter feature was achieved through the medium of a mercury switch which kept the maximum braking pressure below a fixed limit and, as the train slowed down and braking friction rose, reduced the air pressure in the brake cylinders to avoid wheel locking.[1] Air at 90–100 lb. per square inch was supplied by one synchronized 24 c.f.m. rotary compressor on each non-driving car. Non-metallic brake blocks were used throughout.

The traction control gear, by BTH, was of the ingenious 'PCM' type, giving automatic acceleration and bridge transition, used in the USA since about 1928. Two pistons, one worked by oil pressure and the other by air, were connected by a rack. The latter meshed with a pinion mounted on a camshaft, and the cams controlled electro-pneumatic contactors which progressively cut out the amount of resistance in circuit. A star wheel and an electrically controlled roller arm restricted the camshaft revolutions to definite steps.

Under control of the accelerating relay, oil under pressure was admitted to one cylinder, and through the action of piston, rack, pinion and camshaft, caused resistance to be removed in nine steps, until the motors were operating in series with no resistance in circuit. After that, an electric-pneumatic series-parallel transfer switch operated and all resistance was again brought into circuit. Compressed air was then admitted to the second cylinder, which ejected the oil from the first and caused the camshaft to revolve in the reverse direction, again cutting out resistance in nine stages until the full-parallel position was reached. A pair of electro-pneumatic line switches made or broke the flow of current to the motors. If the driver had operated his weak-field 'flag-switch', two weak field notches were available after the full-parallel position, one controlled by the accelerating relay and the final notch by a field shunt relay.

There was one controller position for shunting, and two acceleration settings for wet or dry rails.

The PCM equipment was assembled and wired in two units at the BTH works. These units were attached to the underframe, and all space

[1] Retardation control was also fitted to the standard stock motor cars. Work began in 1938 but was curtailed during the war and not completed until after 1948.

above the floors was available for passenger (or guard's) accommodation or drivers' cabs.[1]

Following the supply of new cars, there was widespread reallocation of stock between the various lines. Deliveries were at first concentrated on the Northern line, but when that line had been fully equipped, many of its cars were sent to the Bakerloo and replaced by further new cars. Release of standard stock from the Northern enabled the 1920 Cammell-Laird trailers on the Bakerloo to be withdrawn. They were stored for some years, but most had been scrapped by 1947. Five survivors were converted to a mobile instruction school in 1949, with working examples of PCM equipment and e.p. brakes. One car was a lecture room. The external livery of this unusual train was orange with black lining-out.

Some of the standard stock released from the Northern and Bakerloo lines was used to replace the original stock of the Central and Northern City lines, whilst some went to the Piccadilly line, replacing the 1936 stock. Much of the displaced standard stock was stored at various depots and sidings pending the completion of the Central line extensions, and 48 driving motor cars and 35 non-driving motor cars of 1938 stock were also placed in store upon delivery. Fifty-eight 1927 Metropolitan-Cammell trailers were converted to run with 1938 stock by the addition of 50-volt lighting, rotary compressors, 1938-type couplers and certain auxiliary equipment, and marshalled in Bakerloo trains. The Bakerloo also kept some complete trains of standard stock. Until 1946, the Bakerloo 1938-stock trains were of six cars, of the formation M-T-M + M-T-M.

The increased capacity of the new stock and its improved acceleration and braking were of great value in reducing the overcrowding on the Northern line. The Edgware section was also relieved by the opening of the High Barnet and Stanmore extensions.

The 1938 stock probably represents the optimum combination of door space and seating capacity for London tube conditions, and the layout has been retained without substantial alteration on all subsequent deliveries of London Transport tube stock. For the three lines with seven-car trains (Bakerloo, Northern and Piccadilly), the same combination of driving and trailer cars has been perpetuated, as has the PCM control.

MODERNIZING THE DRAIN

Owned by a main-line railway company, the Waterloo & City tube railway, known to its patrons as 'The Drain' and held in scant affection,

---

[1] Longitudinal seats had to be fitted where the wheels came above the car floor. Furthermore, some ancillary items of equipment, particularly those forming part of the air-brake and air-door system, were housed above the floor but below the seats in both the standard and the 1938 stock.

had not fallen within the scope of the 1935–1940 New Works Programme. The Southern had inherited it from the LSWR in 1923, and it had been operating almost unchanged since its acquisition by the LSWR in 1907.

By comparison with the modernized London Transport tubes, the Waterloo & City was a museum of early electric traction, and travelling conditions in peak hours were distinctly unpleasant. Letters of complaint appeared in *The Times* and other papers, and by the middle 'thirties public opinion became more restive than ever. In 1931 the SR had obtained powers for an escalator subway at the City terminus but nothing had been done. Answering a Parliamentary Question in November 1937, the Minister of Transport indicated that modernization was being investigated.

A month later, the Southern issued a curious statement blaming the delay in modernization on to the formation of the LPTB, and the consequent possibility of new tube construction which would affect the Waterloo & City. Unctuously, the statement added 'we have every sympathy for travellers on this effete line'—as though the matter was nothing at all to do with the Southern. This utterance was promptly and indignantly contradicted by London Transport. In the same month a users' protest committee was formed, with representatives from each station from Surbiton to Guildford. On December 20, 1937, the SR managers received a deputation from the Protest Committee. After their grievances had been aired, they were told of a sweeping series of improvements—new rolling stock with larger doors and more standing space, new signalling, welded rails, anti-noise shields, and the transfer of ticket-issue and collection from the trains to the stations. It was hoped to complete the works within two years, but in the meantime some alterations would be made to the strap-hanging facilities on the existing stock, as requested by the Committee. The 1931 powers for escalators at City had not been used because after the formation of the LPTB it was necessary to look at the possibility of a low-level subway connection at the Bank with the Central London Railway and to consider the most advantageous siting of the escalators in relation to the other tube railways at that station—a final report on the matter was expected shortly. The Company realized 'that the public generally had been very patient in the matter'. Two months later, at the Southern's Annual General Meeting, the chairman referred to the line and confessed: 'We are ashamed of its present condition.'

The Protest Committee sent a further deputation to the Southern on October 31, 1938, and a statement was issued giving details of progress so far. Tenders had been received for the new rolling stock, and a contract sanctioned by the board of directors; some new signalling equipment had been delivered and progress had been made with the design of new conductor rail; experimental ticket issuing machines had been installed at each terminus. On the question of access to the City station, the deputation was informed that agreement had been reached with the LPTB for a mysterious-sounding 'three-way' escalator, and a deep-level subway connection to the Central line. Parliamentary powers would be sought in the Southern's 1939 Bill.

It had been hoped to complete the modernization in 1939, but the outbreak of war delayed the scheme. Progress continued through the anxious year of 1940, and on the evening of Friday October 25th the old rolling stock was removed for scrapping. Power was applied to the new conductor rail in the standard Ministry of Transport position, 1 foot 4 inches outside the running rails (replacing the old rail in the four-foot way) and a complete new fleet of 28 cars (12 motor cars, 16 trailers), built by the English Electric Co. to Southern specifications, was brought down the lift and into the yard, ready for start of traffic on Monday October 28th.

The new cars were of smart appearance, but in several features of their design they were less advanced than London Transport's 1938 stock, and in many respects resembled a development of the Underground's standard stock to Southern Railway ideas.[1] Each motor bogie had two 190-h.p. axle-hung motors, and beneath each motor car there was a power bogie at one end and a trailing bogie at the other. Where the floor was raised over the motor bogie, part of the space was used for control gear and part for longitudinal passenger seats, harking back to the Bakerloo motor cars of 1906 or to the original Waterloo & City stock. However, the main resistances and the air compressor were underframe-mounted.

The steel underframes were all-welded, but a combination of welding and riveting was used for the bogies, and also for the bodies, which were 47 feet long. The inside of the welded steel outer body panels and the outside of the wood panelling were sprayed with asbestos fibre. Trailers had two double sliding doors per side, motor cars one double and one single. These doors were air-operated and electrically controlled, and worked by the guard from the driving cab at the rear end of the train. The motor cars, which were designed for single-car operation in the off-peak, had a cab at each end, and 40 passenger seats. There were 52 seats in the trailers. Theoretical standing capacity of trailers was 80, motor cars 60, giving a total of 596 in each five-car train of M-T-T-T-M. In practice this figure is of course greatly exceeded and each rush-hour train probably carries about 700 passengers. In both types of car the seats over the bogies were longitudinal and those in the centre bay transverse. Lighting was at line voltage, half the lights on each car being fed from the motor car at either end. When the motor

---

[1] Operating conditions on the Waterloo & City were different from those on the L.T. lines, and standardization of the equipment with SR surface stock was achieved where possible.

Passing at Brent—the morning 'non-stop' from Edgware to Kennington passing another train at Brent in 1927. Both trains are composed of standard stock.—*London Transport*.

A streamlined tube train—a two-car s[et] of 1936 stock (nos 11000 and 10000) near South Ealing in November 1936.
London Transport.

Transition stage at Finchley Central— an LNER train from Barnet hauled by an 'N2' 0-6-2T meets a trial run tube train composed of 1938 stock. The Edgware line is on the left and the rail way substitution bus for Edgware is seen in the station yard. Platforms ha[ve] been lowered for tu[be] trains but tempora[ry] wooden staging to former height is in position. Photograp[h] taken in 1940.—
London Transport.

Waterloo & City modernised—a 194[0] stock train at the Bank in March 195[?]
—*Alan A. Jackson.*

cars were operating singly, all lights were connected to the same supply.

Automatic electro-pneumatic contactor control, with nine series and six parallel notches, and bridge transition, was fitted. The 70-volt supply for the control circuits was obtained from a potential divider energized from the 600-volt traction supply. The motor cars had a shoe beam and collector shoes on each side of the bogie.

The main suspension was by laminated springs of reversed camber under normal load; the spring hangers were cushioned by rubber springs. Braking was of the normal Southern Railway Westinghouse compressed-air type. Externally the cars were finished in light Southern green, with unpainted aluminium doors and cabs.

Full provision was made for communication in emergencies. Loudaphone equipment enabled the driver and guard to converse, and a telephone handset could be clipped to tunnel wires for the driver to speak to the substation attendant. The action of clipping, or of pressing the wires together, cut off the current in the section occupied by the train concerned. With the loss of traction current the tunnel lights (which had been modernized) came on automatically.

The remaining three reciprocating steam-engine and dynamo generating sets, which had been retained for emergency supply, were scrapped, as was a 600-volt battery installed in 1913 to clear the trains from the tunnels if the main supply failed. An additional feeder cable was laid from the main-line substation at Waterloo to the Waterloo & City switchhouse, where the new switchgear was installed, with automatic circuit breakers which could be re-set from the substation.

Removal of the conductor rail to the standard position enabled the four-foot way to be ballasted to form a walkway for emergency use, and for maintenance, and the corresponding outside position of the collector shoes permitted the new stock to be tested on the Southern's surface lines. The running rails, which were restricted to 45-foot lengths by the size of the lift shaft at Waterloo, were welded on site to 315 feet by the Boutet process.

The signalling was completely renewed, with modern colour lights, a.c. track circuits and electro-pneumatic train stops. The operation of the Bank terminus was made wholly automatic, with e.p. point operation and a master switch allowing the use of any one terminal road (for the morning peak or normal hours) or both (for the evening peak). At Waterloo, mechanical point operation was retained, but the lever frame was fitted with electric locks and circuit controllers.

Three new booking offices were built at Waterloo station and one at Bank; at Waterloo new barriers and ticket collectors' boxes were also installed. The construction of the escalators and subway to the Central line at the Bank (authorized by the Southern Railway's 1939 Act) was postponed by the outbreak of war.

On December 9, 1940, a bomb flooded the line and it was closed until the following March 3rd, when it reopened with peak-hour trains only (Saturday service 7 a.m.–2 p.m.). Full service was restored from April 15th, but further damage at Waterloo on the night of April 16–17th and May 10–11th closed the tube again, for a short time. For the benefit of servicemen, Sunday trains ran from 5 to 10 p.m. from March 7, 1943, but were withdrawn during the fuel crisis (February 23, 1947) and not restored.

After the war, peak-hour traffic increased to over 10,000 passengers passing through the Bank station between 9 and 10 a.m. (compared with 6,000 in 1939), and conditions on the ramp between the platforms and the LT circulating area became very bad. In 1951 an attempt was made to alleviate matters by installing crush barriers, but these caused such lengthy queues and delayed the departure of trains so much that they were removed. Soon afterwards about one-third of the width of the ramp was railed off for the use of passengers proceeding against the main stream. The BTC decided in the summer of 1955 to proceed with the 1939 escalator scheme as part of British Railways' Modernization Plan, but two years later announced that an American device known as a 'Travolator' would be used instead. Basically this was a stepless escalator on a much easier incline.

Construction began in June 1957. The main task was the driving of a new 282 foot, 16 feet 6 inches diameter cast-iron-lined tunnel north of and parallel to the famous ramp. The new tunnel was built from a vertical shaft sunk from the Poultry/Walbrook subway. At its lower end it widened to 19 feet 6 inches for a further 52 feet, to house the travolator return mechanism, then to 29 feet 6 inches for a length of 19 feet 9 inches at platform level. At the upper end, the work involved the construction of a new booking hall and new pipe subway, diversions of sewer and service mains, and the reconstruction of the Poultry/Walbrook subway, which was tiled in blue. The station tunnels were re-tiled in blue-grey glass mosaic, with the station name incorporated in white letters at 18-foot intervals; six additional 9-foot-wide cross-passages were provided between the platforms, and the timber platforms replaced by concrete. The old pedestrian subway was also re-tiled in blue, and apart from some tungsten lights in the booking hall, fluorescent lighting was used throughout.

The economic crisis of December 1957 slowed work almost to a standstill, but permission was given in August 1958 for full-scale work to be resumed, and the travolators were brought into use on September 27, 1960. They were supplied by the Otis Elevator Co., and the working part of each consists of 488 40 inch by 16 inch connected metal platforms, surfaced with hard-wearing, close-cleated aluminium and each supported on four reinforced-plastic wheels. They are driven by roller chains from motor-driven sprockets at the upper end. Service speeds of 90 to 180 foot per minute are available, and the gradient is 1 in 7, instead of the 1 in $1\frac{5}{7}$ normally used for a London Transport escalator.

Both travolators run in the direction of the flow in each peak (up in the morning, down in the evening), leaving the old subway for passengers going against the main stream. As the Bank platforms can now be cleared more rapidly, the peak-hour train-service interval has been reduced from three minutes to $2\frac{1}{2}$. To greet the new installation, the cars were refurbished during 1960 with plastic panelling and new upholstery.

'C' FOR 'COLD COMFORT'

When the LPTB was formed, its 'C' stock replaced the ordinary stocks and shares of the acquired companies and a standard rate of interest was laid down in the Act—5% for each of the first two financial years and $5\frac{1}{2}$% thereafter. If $5\frac{1}{2}$% were not paid for any three consecutive years, the holders of not less than 5% of the 'C' stock then outstanding were legally entitled to apply to the High Court for the appointment of a receiver.

In the event, the 'C' stock interest payments were:

Year ending June 30	%	As a % of gross Revenue
1934	$3\frac{1}{2}$	3·0
1935	4	3·3
1936	4	3·3
1937	$4\frac{1}{4}$	3·4
1938	4	3·2
1939	$1\frac{1}{4}$	1·2

During the Joint Select Committee's consideration of the London Passenger Transport Bill in 1931, a pro-forma profit and loss account was presented, showing on the information then available, that 5% or $5\frac{1}{2}$% could be paid on the 'C' stock. This account is compared with the actual results for 1938 and 1939 as Appendix 4B, but the main differences were:

	Compared with Pro-forma	
	1938 (£)	1939 (£)
Additional interest on prior-charge stocks	+ 169,731	+ 168,170
Shortfall of net revenue	− 396,957[1]	− 892,669
Combined effect of above	566,688	1,060,839
Shortfall of 'C' Stock interest paid	− 443,486	− 1,085,956

([1] After allowing a credit of £230,000 provided for income tax in previous years and not now required.)

When considering the 'C' stock interest it must be borne in mind that the amount needed to pay 5½% (£1,413,434) was only about a third of that needed to meet the interest charges on the prior charge stocks, which were invariably and necessarily paid in full. The 'C' stock gave the only element of flexibility in the accounts, and had to bear the full brunt of any shortfall. If gross receipts in the year to June 30, 1938, had been 1·25% higher (and expenses unchanged) the standard 'C' stock interest could have been paid in full; a 3·3% increase in the following year would have achieved the same result for 1939.

In a letter to *The Times* in July 1938, Ashfield reviewed the Board's first five years, and suggested that as London Transport stock could be considered as offering equivalent security to gilt-edged stocks, a more realistic rate of interest for 'C' stock would be 3%. The time was opportune to revise the financial regulations.

The *Transport World* agreed that a well-secured 5½%–6% had become an anachronism in view of changed circumstances since 1933, and suggested 4% as the new standard. The following year, when the 1½% interest was announced, the *Railway Gazette* contended that the whole financial structure needed revising—the 1933 terms were wholly inappropriate in present conditions.

The holders of 'C' stock were legally entitled to apply for the appointment of a receiver after the 1937/8 results were announced. Some of them suggested higher fares, and they met as a committee on December 22, 1938, to review the position, but made no firm decision. There was a further meeting on January 13, 1939, but by then the proposals to increase fares had been announced, and the stockholders decided to wait and see.

By the time that the 1½% interest for the full year 1938/9 was announced, Britain was at war and London Transport under Government control, so that any demand for a receivership was out of the question.

The dismal results of 1938/9 were due to a decline in the rate of growth of traffic and to the continued increase in working expenses. In these circumstances the Board was obliged to seek an increase in fares.

Outside London, the main-line railways had increased their fares by 5% from October 1, 1937, but within the London Transport area the position was complicated by the varying scales in operation, by the difficulty of trying to add 5% to fares that went up by 1d stages, and by the need to equate London Transport rail and road fares. Negotiations were initiated with the main lines through the Standing Joint Committee to secure an agreed fare policy and on January 10, 1939, the Board and the main lines applied to the Railway Rates Tribunal for increased fares within the London area. The hearing occupied sixteen days between March 20th and May 4th, and the judgment approving the application was given on May 24th. The estimated additional gross revenue per annum was £664,000 for London Transport and £515,000 for the originating receipts of the main lines, but of these sums, only £5,000 and £300,000 respectively depended on the authority of the Railway Rates Tribunal's judgment, the remainder being obtained under existing charging powers. After allowing for the operation of the pooling scheme, the net additional annual revenue for the Board was estimated at £732,000.

The new fares came into operation on June 11, 1939. The 1d-a-mile basis for London Transport ordinary fares was retained, but some substandard fares were increased.[1] On the other hand, the fares to Uxbridge and Stanmore were brought down from 1½d to 1d a mile, so that Wembley Park–Moorgate went down from 1s to 9d, but Morden–Moorgate rose from 6d to 7d. On the road services some exceptionally long fare stages were split into two, and some cheap fares introduced by the LCC tramways were withdrawn or increased.

The travelling public, accustomed to fares remaining unchanged for years at a time, did not accept the increases without protest. This was registered at a meeting held at Central Hall on June 8th. Mr W. J. Brown, then general secretary of the Civil Service Clerical Association, led the deputation which placed the protest before the Minister of Transport on June 15th. They asked the Minister to consider a public inquiry into London Transport's financial structure, and to look into amending legislation on the Railway Rates Tribunal's terms of reference in relation to London fares.

The Minister, Captain Euan Wallace, could not agree that any alteration to the *status quo* was justified; he added that it would be unfair to holders of 'C'-stock compulsorily to alter their rate of interest, and a Government subsidy was undesirable.

From the results of the revised fares in the short period to the outbreak of war, it appeared that the estimated increase in revenue would be earned with a margin to spare, and, but for Hitler, the Board's finances would probably have been restored to a state nearer equilibrium, with 'C' stock interest again at about 4%.

[1] Certain 'superstandard' fares had been reduced between 1933 and 1936, and some progress had been made with standardizing and revising cheap day and workmen's fares and season-ticket rates. Substandard (i.e. sharply-tapering) fares had been adopted on the Edgware, Morden and Cockfosters extensions as part of a deliberate policy to build up traffic, if need be at the expense of competitive services. Sometimes this policy was too successful, and in an address to the Royal Society of Arts in December 1935, Frank Pick admitted that passengers at Balham, Clapham and Stockwell were inconvenienced or entirely squeezed out by the immense volume of traffic coming through from Morden. He concluded with the observation: 'Here a well-meant mistake has had unhappy consequences for the development of London in an orderly and predetermined way.'

CHAPTER 12

The Tubes in Wartime (II)
1939-1945

PRECAUTIONS

WARNED by the experiences of 1917-1918 and the rapid development of aircraft, the postwar governments had devoted continuous and detailed consideration to air-raid precautions. As early as January 1924 an Air Raid Precautions Sub-Committee of the Committee of Imperial Defence was formed, under the chairmanship of the Permanent Under-Secretary of the Home Office. One of the decisions of this Committee was that the tube railways would be kept running during sustained air attacks as they would be needed to evacuate casualties and maintain other essential transport. Although it was recognized that most tube stations would offer sound protection, their use for sheltering was to be forbidden as this would interfere with the operation of trains. There was also a fear that the use of the tubes as shelters would foster a 'deep shelter mentality' which would sap civilian morale, the expectation being that people would stay down and refuse to come out, thus producing insoluble problems of food supply and hygiene and paralysing the daily life of the city. Under the direction of the Committee, certain anti-gas features were included in all new tube stations after 1931 and the the Post Office made plans for an emergency communications system which used the tube tunnels. In 1938, the London Passenger Transport Act authorized the Board to make agreements with the Central Electricity Board for the supply of electricity in emergencies. Part of the plan was to enlarge the old LCC tramway power station at Greenwich, bringing its capacity to 112 MW and changing its frequency to the industrial 50 c/s so that the Board could supply the Grid when required; in return, LT was to take power from the Grid if war damage affected its own power stations. Work on this scheme had begun before September 1939, and the extra capacity, when not required for the Grid, was to be used for the eastern extensions of the Central line.

In the late 'thirties, the Board advised the Committee that it intended to keep open in air raids all but twenty of the more vulnerable tube stations, and the Committee, recognizing the difficulty of distinguishing between a genuine passenger and the shelterer with a penny ticket in his hand, agreed to this, but insisted that at the outbreak of war an announcement be made that the tubes were not to be used as shelters. Some of the shallower tube tunnels, particularly those under the Thames, were vulnerable to the more powerful bombs available from about 1936, and the Board's attention was drawn to the fact that serious flooding of the system might follow a breach in these tunnels. In September 1938, at the time of the Munich crisis, makeshift measures were taken to prevent flooding at Charing Cross and Waterloo. Puzzled passengers found the Bakerloo and Northern lines under the river closed from 8 p.m. on September 27th for 'urgent structural works' and were unable to use Trafalgar Square, Charing Cross, Waterloo, Lambeth North, Elephant & Castle and Kennington (Charing Cross line). Down below, the tunnels at each side of the Thames were hurriedly stopped up with concrete plugs. When Neville Chamberlain returned from Munich with his uneasy peace, the concrete was drilled away and the stations were reopened on October 8th.

During the crisis a review of all stations requiring anti-flood precautions was completed, and early in 1939 work on these began in earnest. It was decided to build floodgates across the tracks each side of the Thames at Charing Cross and Waterloo, so that in between air raids trains could run under the river. The Bakerloo gates were ready when war broke out on September 3, 1939, and were 13 inches thick, sliding across the tunnel in a cast-iron framework. Weighing just under 6 tons, they could withstand water pressure of over 800 tons. They were electrically operated and could be closed in 30–60 seconds after orders were received from the control room at Leicester Square; in emergency, they could be moved by hand. Metal sealing blocks with rubber linings were placed in the gaps around the rails before closing. To prevent the trapping of a train between the two gates or the closing of a gate in the path of a train, the closing operation was interlocked with the signalling. Operating staff were always on duty and the gates were closed on the sounding of the air-raid warning. There were two gates at Waterloo, at the north end of each Bakerloo platform, and four at Charing Cross, one at each end of the Bakerloo platforms. The line was closed between Piccadilly and Elephant & Castle on August 27, 1939, for that day only, to allow completion of the work. On the Northern line, the gates had been completed at Waterloo but not at Strand, so on September 1, 1939, trains ceased to run between Kennington and Strand, and as an interim precaution, 35-ton concrete plugs were again placed in the tunnels on the north side. There was some hesitation as to what was to be done about the under-river tunnels at London Bridge—they were deeper than those at Charing Cross, which is probably the reason why they were not closed in 1938 nor until September 7, 1939, when concrete plugs were temporarily inserted. Northern line trains from northern termini then reversed at Moorgate and Strand; south of the river, trains turned back at London Bridge or Kennington. Apart from the tunnels under the Thames, many sub-surface booking halls and subways were liable to flooding if water mains and sewers were breached. At thirty-one of the stations thus affected it was possible to provide protection without interfering with traffic, and good progress had been made by the outbreak of war. At a further nineteen stations more complex works were necessary, involving severe disruption, and operations were deferred until war seemed imminent. From September 1, 1939, these nineteen stations were closed and the passageways temporarily sealed with concrete. The plugs were replaced as quickly as possible by watertight doors unless traffic could be diverted through other subways.

An impossible load was thrown upon the buses by this drastic closing of stations and the interruption of the Northern line service. Fortunately the evacuation of many offices and the closure of others prevented a complete breakdown. The work was tackled with vigour and the closed stations were reopened from November 15th onwards [1] The Moorgate–Kennington section was reopened with floodgates in position on May 19, 1940. From that date the full prewar tube system was again available to passengers. Before leaving the subject of the under-river sections, it should be added that from the end of 1943 diaphragms were added as a second line of defence should the gates be damaged by bombs. When it was realized that delayed-action bombs or acoustic mines might damage the tunnels after an alert had ended, a watch was kept along the Embankment and eventually hydrophones were installed in the river bed. Finally, to contend with dangers of attack without warning by 'sneak' raiders, flying bombs and rockets, devices were fitted to give automatic indication of the entry of river water into the tunnels.[2]

On September 1, 1939, the railways passed into Government control and the Railway Executive Committee assumed its powers. Frank Pick

[1] *November 15th*: Marble Arch, Tottenham Court Road. *November 17th*: King's Cross. *November 20th*: Oxford Circus. *November 24th*: Clapham Common, Oval. *December 1st*: Arsenal, Green Park, Knightsbridge. *December 6th*: Bond Street. *December 8th*: Hyde Park Corner, Old Street. *December 15th*: Chancery Lane, Balham. *December 22nd*: Trinity Road, Bank (CL). *January 9, 1940*: Maida Vale. (Charing Cross and Waterloo (Northern) with the restored Northern line service under the river, were reopened on December 17, 1939.)

[2] The only under-river tunnel breached by enemy action was the disused Hampstead loop at Charing Cross which received a direct hit in September 1940. The concrete seals at each end confined the water to a 200-yard section of the loop itself.

was a member of the Committee, but when he retired from the Board on May 17, 1940, Lord Ashfield took his place. The tubes thus came under central control, with the other railways, in contrast to 1914–1918 when only the District and Metropolitan were taken over. From January 1, 1940, the London Passenger Pooling Scheme was suspended and the controlled railways pooled their net revenue from September 1, 1939. The Government guaranteed a net annual revenue of £40 million, based on the main-line companies' annual net revenues of 1935, 1936, and 1937 plus the LPTB revenue for the year ended June 30, 1939, and an allowance for the controlled minor railways. Revenue could be retained up to £43 million, but one-half of any further excess up to £56 million had to go to the Treasury. There were special provisions to compensate London Transport for high air-raid precautions expenditure and the disruption of traffic that followed evacuation of the capital.[1] The main-line companies strongly resisted the inclusion of London Transport, pointing out that LT traffic had suffered a heavy decline which would have to be balanced from the increased freight receipts on the main lines. A new agreement between the Government and the railways came into force on January 1, 1941, by which the railways were guaranteed fixed annual amounts, the basic LT share being £4,835,705.[2]

The evacuation of schoolchildren and others began on Friday, September 1, 1939, and lasted for four days. Underground trains carried nearly 200,000 evacuees from the central area to outer suburban stations where they transferred to buses or main-line trains. Ealing Broadway was the busiest transfer point, and there was also much activity at Oakwood and Edgware whence shuttle buses took children to New Barnet and Mill Hill main-line stations. Frank Pick planned this movement expecting that bombs would be falling as it proceeded, but it went off smoothly beneath undisturbed skies. Soon the bored and restless evacuees were returning home and smaller evacuation schemes had to be organized after the invasion of the Low Countries and when bombing started in the autumn of 1940.

A complete blackout of artificial lighting was imposed when war began, and at first the tube trains ran on the open sections without lights, but after September 23rd passengers were just able to see each other through a cheerless dark-blue glimmer emitted by three low-power 'Osglim' lamps attached to battens along the ceilings of each car. Station lighting where visible from the air, was dimmed by special shades or 'Osglim' lamps. The sliding doors of the tube cars made it difficult to design a suitable reading lamp which could not be seen from the sky, but by the end of the year London Transport was installing

---

[1] The financial arrangements were set out in a White Paper of February 7, 1940 (Cmd. 6168).
[2] The Railways Agreement (Powers) Order 1941 (SR&O 1941/2074).

15-watt bulbs in three troughs attached to the battens holding the blue lamps. These troughs directed the light on to the seats, but cut it off from the windows and doors, producing a narrow flat beam at reading level which was completely blocked out if there were any standing passengers. During air-raid warnings, the guard switched out these reading lights, leaving only the blue glimmer. The first train with the new lights appeared on January 15, 1940.

THE BEST SHELTERS OF ALL

In accordance with plan, the Government announced on September 3, 1939, that the tube stations would not be available as air-raid shelters. Trains would run during air raids and most stations would remain open to genuine passengers. Posters to this effect were prominently exhibited at the stations, and in November further emphasis was given by the signs affixed at stations which gave particulars of 'the nearest air-raid shelter'.

This was all very well during the 'phoney war' of the winter and spring of 1939–1940 when the German air force was occupied elsewhere, but as soon as the air assault began in the autumn of 1940, Londoners flocked to the tube stations just as they had done in 1917 and, short of fixed bayonets at each station entrance, there was little hope of stopping them.

The reasons for the popularity of the tubes as shelters were obvious enough: the stations were easily accessible to many, and provided companionship and warmth with a very high level of security against the bombs then in use; the noise of the explosions above could not be heard, and it was possible to sleep, or at least rest, in preparation for the work of the following day. There was nothing comparable, as the public shelters provided by the Government were either trenches, which were often flooded, or reinforced basements with much less security. An insistent demand for the construction of deep shelters had arisen in 1938 and was supported by both the Labour and Liberal parties, but the Conservative Government had resisted it owing to the enormous expense of construction and the shortage of both time and labour. Fear of the 'deep-shelter mentality', which we have already mentioned, still persisted in official circles. The shelter policy had been based upon the expectation of short but intense raids with a high casualty rate.

The heavy raids (the *Blitz*) began on the night of September 7, 1940, and lasted with little relief until the next summer when the attacks tailed off following the diversion of the German air weapon towards the USSR. When it was seen that the sirens now really meant something, crowds of people bought low value tickets and refused to leave the platforms until the morning 'all clear', even though trains had stopped for the night and the staff were ready to book off duty. Stretched out on rugs, propped up on pillows, and surrounded by the more portable of their most precious possessions, 177,500 Londoners were estimated to be in the tube stations on the night of September 27, 1940, when the first rough count was made. Ten days earlier, the Government had appealed to the public not to use the tube stations as shelters 'except in urgent necessity', but by the end of the month it recognized the *fait accompli* and set about the task of organization. The reaction of London Transport suggested they were a little hurt about the whole business.[1]

During October 1940, the nightly average of shelterers was over 138,000. Of course, this was only a minute proportion of the total population of London, large sections of which lived miles from tube stations, but it was as many as the stations could comfortably accommodate.[2]

For the first few nights, chaos reigned and the stench was foul, but there was no panic. Spaces were reserved on a free-for-all basis by layabouts who queued for hours before the warning sounded. Every corner was occupied and ordinary passengers had to pick their way as best they could through the tangle of seated and reclining bodies. There were no lavatories at platform level and the ventilation fans were, as a precaution, all operating at pressure to exert a plenum on the tunnels and stations to minimize the entry of any poison gas at the entrance nearest the point where a gas bomb might fall. Other wartime measures, such as blast walls and buildings in subways, restricted the flow of air. The first arrangements were made by the police, who endeavoured to clear the more overcrowded stations by sending people to other points along the line. Late in September, chemical closets began to arrive and some of the pungency went out of the atmosphere.

In true journalistic tradition, many 'able-bodied men' were found in the tubes at night by reporters and, as in 1917, these were said to be mostly 'pacifists' and 'aliens'. It is likely that many were war workers who gained a night's rest they might not otherwise have had, and the national effort benefited accordingly. At the beginning of October, white lines were painted on the platforms at distances of 4 and 8 feet from the edge; no 'reservations' were allowed before 4 p.m. but after then, spaces could be taken up to the 8-foot line and, from 7.30 p.m., up to the 4-foot line. Grateful for the hospitality offered, the shelterers conformed with pathetic good behaviour.

Gradually more amenities were provided, refreshments, sleeping bunks and better sanitation receiving priority. Organized by the LPTB

---

[1] 'The Board were called upon to undertake a further unexpected task ... the new situation was promptly accepted.' (Annual Report, 1945.)

[2] A census taken at the end of November 1940 showed that 9% of the London population was spending the night in public shelters, 4% in tube stations, and 27% in household shelters.

at the request of the Ministry of Food, refreshment facilities were available at some stations from October 29th. The food was delivered to the stations by six four-car trains labelled TUBE REFRESHMENTS SPECIAL, the first of which ran on the Central line on November 6th. Each train carried seven tons of food in fifty containers and set out every day at 1 p.m., spending about thirty seconds at each shelter station to unload and then take on the previous night's empties. The food was warmed up at 'feeding points' equipped with electric ovens and boilers, and distributed to the shelterers, who brought their own utensils, in the mid-evening and early morning. By December 9th, the food service was in full operation and the erection of three-tier bunks was well advanced. All but three stations had the bunks by the end of February 1941, and eventually there was sleeping accommodation for 22,800 in the tube stations.

A rational ticket system for regular shelterers was arranged at the end of November 1940 and in the same month the first medical services were provided by the local authorities. Beginning in November, a much-improved sanitation system was introduced, involving ingenious use of the railway compressed air supply to force the sewage to ground level. Some stations had libraries and others were equipped with amplifiers which played recorded music; there were ENSA concerts and shelter newspapers. The medical staff found plenty of work; babies were born on the platforms and in the subways, and there were always minor injuries or ailments needing attention. Among the latter were mosquito bites, for the species *culex molestus*, which had bred in the tubes for years, had a fine time among hundreds of sitting blood donors. It was eventually subdued, but not conquered, by cresol-spraying of its breeding grounds. Fortunately, although its bite was vicious, it was not a disease carrier. The expected epidemics did not materialize, partly because the people all came from the same districts and were in good fettle after the fine summer of 1940, and partly because of anti-infection precautions taken on the advice of Lord Horder.

By Christmas 1940 the whole scheme was settling down and tube sheltering was no longer regarded with fear and suspicion by those in authority. The average nightly population was now smaller (102,000 in December) and conditions were more comfortable. Passengers no longer found any difficulty in moving about, and the air was cleaner and sweeter as a result of disinfectant spraying and better sanitation. Shelterers took a pride in keeping their 'spaces' clean, and some women even scrubbed the platforms around their bunks with the same alacrity as they did their own doorsteps.

The organization of the amenities with such speed and ingenuity was largely due to the efforts of Mr J. P. Thomas, the former Operating Manager whom the Board recalled from retirement on November 11th to co-ordinate its own work with that of the central and local governmen authorities.

Sheltering was eventually permitted at all the seventy-nine deep-tube stations, and in addition the Board were required to open up certain stations and sections not in railway use. These were: the tunnels and stations of the Aldwych branch (which had been closed to traffic on September 21, 1940); the constructed, but unopened, tunnels east of Liverpool Street as far as Bethnal Green (accommodating over 10,000); the abandoned CSLR spur between Borough Junction and King William Street, City;[1] the closed stations at South Kentish Town, British Museum and City Road; and the unfinished tube station below the LNER station at Highgate (to which the shelterers travelled in trains which were specially stopped).

Other stations and tunnels were converted for special purposes. On the Piccadilly Line, the disused station at Down Street was used as the headquarters of the Railway Executive Committee and there was also accommodation for Winston Churchill and the War Cabinet, which often met there. The closed station at Dover Street became offices for LPTB staff. The latter were also accommodated at Holborn, Hyde Park Corner and Knightsbridge. GWR traffic control staff were fitted into Paddington, Bakerloo and the abortive District Railway tube platforms at South Kensington housed LPTB engineering services emergency control headquarters. British Museum treasures, including the Elgin Marbles, were safely tucked away in the disused Aldwych tunnel. Most remarkable of all special uses was the conversion of over two route-miles of the unfinished Central line tunnels under Eastern Avenue at Wanstead and Ilford to an aircraft component factory. Here the Plessey Co. employed shifts of 2,000 workers day and night in 300,000 square feet of floor space stretched along 25,000 feet of single tunnel. Extra access shafts were built for entrances, machinery and ventilation, and the work on the conversion, which began late in 1940, was finished in March 1942. Each tunnel had an eighteen-inch gauge railway, operated with battery locomotives, to assist the transport of materials and finished parts.

As the raiding thinned out in July 1941, the number of shelterers fell to about 10,000 a night, often less, and throughout 1942 the figure was between 3,000 and 5,000. Renewed raiding in 1943 and again in January 1944 brought the shelterers back, and in the 'Little Blitz' of Janu-

[1] The junction at Borough had been sealed off with concrete in 1938, and in 1939 the under-river section of the abandoned spur was similarly sealed. This left two sections of twin tunnels available for shelterers. That on the north bank was let to the owners of the offices above King William Street station, and the southern section was let to Southwark Borough Council who constructed eight new entrances from street level (with stairs in precast concrete rings) and opened the shelter on June 24, 1940 (its capacity was 14,000—one of the largest in Britain).

ary–March 1944, the nightly population rose to 50,000. The V1 Flying Bomb and V2 Rocket attacks of 1944–1945 filled the tubes with shelterers for the last time, and will be mentioned again.

GOVERNMENT DEEP SHELTERS

The bombings of 1940 forced a reappraisal of deep-shelter policy, and at the end of October the Government decided to construct a system of deep shelters linked to existing tube stations. London Transport was consulted about the sites and required to build the tunnels at the public expense with the understanding that they were to have the option of taking them over for railway use after the war. With the latter point in mind, positions were chosen on routes of possible north–south and east–west express tube railways. It was decided that each shelter would comprise two parallel tubes 16 feet 6 inches internal diameter and 1,200 feet long, and would be placed below existing station tunnels at Clapham South, Clapham Common, Clapham North, Stockwell, Oval, Goodge Street, Camden Town, Belsize Park, Chancery Lane and St Paul's.[1] Each tube would have two decks, fully equipped with bunks, medical posts, kitchens and sanitation, and each installation would accommodate 9,600 at an approximate construction cost of £15 a head. In the event, the capacity was reduced to 8,000 as a result of improved accommodation standards and the actual cost varied between £35 and £42 a head. Work began on November 27, 1940, and it was hoped to have the first shelters ready by the following summer. There were great difficulties in obtaining labour and material, and when the Blitz abated, the Government had second thoughts. The old bogey of 'deep shelter mentality' was brought out of the cupboard by those who opposed the lavish expenditure of money and labour on this project, and in the middle of 1941 a Select Committee on National Expenditure recommended that no further deep shelters be built, but those started should be completed. Work at St Paul's was abandoned in August 1941 as it was feared that the foundations of the cathedral might be affected. Oval was also abandoned shortly after this as large quantities of water were encountered. The first complete shelter was ready in March 1942 and the other seven were finished in that year. The Board then urged the Government to open the shelters to relieve the strain on the stations, but the Cabinet were alive to the great cost of maintaining the deep shelters once they were opened and decided to keep them in reserve pending an intensification of the bombing. Towards the end of 1942, part of Goodge Street shelter was made available for General Eisenhower's headquarters and later two others were adapted for Government use. Another was converted to a hostel for American troops, and sections of the remaining four were used to billet British soldiers. These

[1] It may be assumed that at these points the deep-level express tubes would have no stations as the diameter was too small.

uses were maintained throughout 1943 despite agitation that the shelters be opened for their proper purpose.

At the beginning of 1944, the air attack warmed up again, and on June 13th the V1 assault began, to be followed on September 8th by the V2 rockets which then came over intermittently until March 27, 1945. The arrival of the flying bombs finally moved the Government to open the shelters to the public; Stockwell was available from July 9, 1944, Clapham North from July 13th, Camden Town from July 16th, Clapham South from July 19th and Belsize Park from July 23rd. The other three remained in Government use. Regular shelterers at nearby tube stations and homeless people were given admission tickets, but demand was not high and by September some of the space was made available to troops on leave. The highest recorded nightly population was 12,297 on July 24, 1944, about one-third of total capacity. On October 21st, two of the shelters were closed again and nightly use fell until by January 1945 only about 25,000 people were using the tube stations and deep shelters. The last air-raid warning of the war was sounded on March 28, 1945 (the European war ended on May 8th), but about 12,000 homeless and 'squatters' continued to sleep in the tubes until May, when the bunks on the platforms were removed and a start was made on tidying up the stations. The seventy-nine shelter stations were closed to 'shelterers' after the night of May 6th and the Board breathed a deep sigh of relief.

After the war, various uses were found for the Government deep shelters, including the storage of documents and the provision of overnight accommodation for students and troops. Goodge Street continued in use as an Army Transit Centre until it was damaged by fire on the night of May 21, 1956. The fire coincided with Parliamentary consideration of a Government Bill seeking power to take over the shelters (The Underground Works (London) Bill), and the Minister of Works assured the Commons they would not again be used for human occupation in peacetime (although no one was killed, the fire had caused some alarm and proved difficult to put out). During the progress of the Bill, it was revealed that the option for railway use had been retained only on the three Clapham shelters and the adjacent one at Stockwell.

TUBE TRAGEDIES

The average tube station, 60 feet or more below ground, offered secure enough shelter against the bombs in use between 1940 and 1945, and the combination of circumstances which could lead to devastation at platform level was improbable enough to make the possibility remote. All the incidents involving casualties in tube shelters took place at stations with less than 35-foot cover or as a result of direct hits on booking halls and subways just below the street level.

The worst bomb disaster was at Balham, where just before 8 o'clock in the evening of October 14, 1940, a bomb penetrated the northbound station tunnel at its northern end, rupturing water mains and sewers. The explosion was immediately followed by an alarming inrush of water and gravel which half-filled both station tunnels, by then plunged in darkness. Escaping coal gas heightened the horror of the situation and those still alive on the platforms were sealed off by watertight doors at the base of the escalator shaft. One by one, they were evacuated through emergency escape hatches from this grim station-tomb. The bodies of 64 shelterers and 4 railwaymen were found in the debris and it was three months before the line could be reopened. Nearly seven million gallons of water had to be pumped from the vast hole.

On the day before the Balham disaster, 19 shelterers were killed and 52 injured at Bounds Green when a bomb reached the platforms of the Piccadilly line, which is ascending at this point to reach the surface just beyond the station.

At the Bank, on the evening of January 11, 1941, 56 people were killed and 69 injured by a bomb which cut through the concourse just below the street and exploded in the escalator machine room immediately beneath it. The escalators to the Central line were completely wrecked and the blast effects damaged two trains at the platforms 62 feet below. Shelterers left without panic via the stairway and interchange passages to the Northern line lifts. The street fell into the booking hall, forming a crater 120 feet long by 100 feet wide. Royal Engineers and New Zealand Engineers assisted ARP workers to clear the débris and erect a bridge across the hole to carry road traffic. A temporary wooden staircase was built at one side of the escalator shaft and the station was reopened on March 17, 1941. To ease the burden of the ascending passengers, an MH-type escalator was fitted alongside the staircase and opened on December 13th. A second escalator, for down traffic, together with a steel staircase, was provided on April 2, 1944. This second escalator was taken from the centre of Chancery Lane's upper flight. The Central line platforms at the Bank never closed, as access was always available by using the Northern line lifts and booking hall.

Bomb incidents involving casualties occurred at six other shelter stations:

October 12, 1940: Trafalgar Square (7 killed).
October 14, 1940: Camden Town (1 killed).
December 24, 1940: Tottenham Court Road (1 killed).
January 11, 1941: St Paul's (7 injured).
January 11 1941: Green Park (2 injured).
January 16, 1941: Lambeth North (20 injured).

Altogether, 152 shelterers, employees and passengers were killed in tube stations, all but four in the incidents mentioned above.

In March and April 1943, stairwell entrances at Piccadilly Circus and other tube stations were hooded to enable more adequate lighting to be placed over the stairs. Some of the more intelligent members of the public were mystified by this sudden burst of activity nearly four years after the beginning of the war, and a few may have linked it with rumours of a terrible disaster in the East End, so bad that the Government had hushed it up. When the news was eventually released twenty-two months later, the nation learnt that what had happened at Bethnal Green just before 8.30 on the evening of March 3rd was the most dreadful accident ever to occur in a tube station, yet it was caused only indirectly by enemy action. At this unfinished station on the eastern extension of the Central line, used as a shelter since 1940, access to tube level was gained by a poorly lit wooden stairway in the incomplete escalator shaft. These stairs had no central handrail and there were no crush barriers at the top. A salvo of anti-aircraft rockets were fired off about a mile from the station at 8.25 p.m. causing a frightened crowd to surge towards the station entrance in the belief that air attack was imminent. Some people were already groping their way down the stairs, and the fierce pushing from behind knocked them forward until a woman carrying a baby tripped and fell on the third step from the bottom. A man fell with her, and others tumbled on top of them, the pile of fallen people growing larger as the insistent pressure from above continued. Those at the top, finding that they could make no progress, feared that the platform doors had been closed against them, and pushed all the harder in their blind desire to reach safety. Pushing continued for fifteen minutes and then the police at last gained some control of the situation. In the tangled human mass at the bottom of the stairs, 27 men, 84 women and 62 children were found crushed to death amid 62 injured.

With this sad exception, accidents in shelter stations were few. In the early days at least one restless sleeper rolled off the edge of the platform to be killed under an approaching train, but this danger disappeared when the white lines were painted. A serious fire broke out in the escalator shaft at Paddington on December 24, 1944, but the shelterers were evacuated through the railway tunnels without injury. It took some time to remove the water, and the station was not fully reopened until January 1st.

The open sections of the tube railways received their share of bombs and rockets, and special repair squads were organized at strategic points. Composed of picked men from the permanent-way staff, these squads were moved quickly to bomb incidents under the direction of a central headquarters housed in deep shelters. All open tracks were patrolled throughout air raids, and craters and unexploded bombs were quickly reported. Many odd and sometimes amusing stories are told of those days, when shuttle services had to be improvised either side of 'incidents', and an emergency fleet of 600 double-deck buses stood by to provide assistance when rail service was interrupted. Sometimes things happened so quickly that drivers had to make their own decisions. One Saturday afternoon in the autumn of 1940, the driver of a Northern line train speeding through the long stationless stretch between Hampstead and Golders Green gazed idly at the tunnel mouth growing larger and larger ahead. Suddenly he spotted a bomb crater right across the tracks just outside the tunnel; the train ground to a halt, its head poking from the tunnel, into which it soon returned, like a frightened rabbit. Unexploded bombs were a considerable nuisance on open sections, and all the more so because the official attitude towards them tended to be over-cautious. One of these fell on the line between Hendon and Brent in September 1940, causing the train service between these stations to be suspended for fourteen days. When it eventually exploded, it did not damage the railway at all and the trains were running again within ten minutes. The most serious damage on an open section was the destruction of Colindale booking hall in the same month.

WARTIME CHANGES

The conditions of wartime brought many changes to the tube railways, and apart from the events just described, there will be a strange jumble of recollections for those who can remember these years—Billy Brown, dim lights, the missing rubber grips, escalators that only worked upwards—and window netting. Yes, above all, that depressing gluey yellow netting over the car windows that was 'there for our protection' . . . it blocked the view completely until they placed diamonds of cleared glass in the centre of each pane; it engendered the very basest of feelings in the passengers, who longed to rip it off. Alas, some succumbed to this temptation, and 'Billy Brown', a London Transport publicity hero,[1] who epitomized the type of passenger LT hoped to have and never got, was brought to bear on this grave social problem:

> In the train a fellow sits,
> And pulls the window net to bits
> Because the view is somewhat dim,
> A fact that seems to worry him.
> As Billy cannot bear the sight
> He says, 'My man, that is not right,
> I trust you'll pardon my correction,
> That stuff is there for your protection.'

To which 'Bee' of the *Daily Mail* replied:

[1] Billy was drawn by David Langdon and spoke through the verse of Richard Usborne and Joan Chapman (a bus conductress). He had an equally righteous female companion named Susan Sensible.

> ... But you chronic putter-righter,
> Interfering little blighter,
> Some day very soon by heck,
> Billy Brown—I'll wring your neck.

—which neatly summed up the average Londoner's feelings about this irritating character.

And then there was the strange matter of the Rubber Grips. These were a feature of the 1936 and 1938 stock, and did not wear well in service although they were more comfortable to hold than the old type of strap. During the war many were found to have been wrenched off their springs, and although some of this was accidental, a certain type of tube traveller had realized the grips could be put to use as coshes to aid nocturnal activities. London Transport discovered that one had been found in a street in Algiers and proudly announced that they had been used as Commando weapons. After 1943, the rubber grips were replaced by another type, similarly shaped with a plastic ball suspended from a stronger, spring-loaded fastening. These were much more difficult to detach and have been seen to bear the full weight of a man.

As the war years passed, the pattern of 1914–1918 appeared again: reduced maintenance; increasing economies in the use of fuel and materials; and the ever-growing traffic across and within London which nearly all went by the underground railways.

In 1942, lighting in trains and stations was reduced to a third of prewar strength in an endeavour to save some 5,000 tons of coal a year —the counterbalancing wear and tear on the eyes and tempers of passengers and staff was not calculated. A year later, down escalators were halted at slack hours (and on Sundays) at twenty-four stations to save coal.

LT railways handled about 333 million passengers in 1941, a figure which increased to 491·5 in 1944 and 543 in 1945 (traffic originating at LT stations). The congestion was especially noticeable at main-line stations where servicemen and women transferred to and from the tubes. At Waterloo in 1944, over 100,000 were changing between the tubes and the Southern Railway in one day, whilst at King's Cross 54,000 made a change. There was heavy transfer traffic between tube lines at Leicester Square and Piccadilly Circus, the 1944 figures for one day being 61,000 and 40,500. As most of these passengers were not Londoners, they were unfamiliar with escalator etiquette; loaded with the paraphernalia of war, and unsure of their feet as they moved up and down, they caused much congestion at the interchange points. LT began a 'PLEASE STAND ON THE RIGHT' campaign, erecting 15-foot banners bearing this legend across the escalator shafts. With a pathetic faith in British obedience to passive commands, the Board installed thermal flashers opposite the platform ends of some busy stations, blinking PASS ALONG THE PLATFORM. These still survive, but regular travellers have long since ceased to be aware of them, let alone obey them. Another measure to counter wartime congestion and assist strangers through the tube warrens was the fitting of loudspeakers at twelve stations in 1944. The weird acoustics of the station tunnels somewhat damaged the clarity of the announcements. One of the more practical steps to counter the increased traffic was the addition to the Bakerloo service of 17 three-car sets (M-T-M) of 1938 stock, brought from store in 1944. From the summer of 1945, Bakerloo trains were lengthened to seven cars by bringing non-driving motor cars from store, and there were 16 seven-car trains in service by the end of August. From that date, one train was strengthened to seven cars each week, and the winter peak-hour service to Stanmore was increased from 14 to 16 trains an hour. By June 1946, all Bakerloo trains were seven-car, 42 of 1938 stock and 5 of earlier stock.

Women porters and ticket collectors appeared on the tube stations from September 17, 1940. At first they wore white dust coats and grey *képi* hats, but these were eventually replaced by blue tunics and slacks (or divided skirts) and the ugly *képis* by berets. After the war many women remained in service and they are now a permanent feature.

In 1940, the Board made an application for a fare increase to cover rising costs, and from July 3rd, 1d fares, comprising almost 40% of total receipts, became 1½d; 1½d fares were raised to 2d (fares of 5d and over, and seasons, went up by 10% on 1 May). The 2d fare remained, but the famous 1d fare, which had been restored with some publicity in 1925, had gone for ever. The original financial agreement between the Government and the railways had allowed fare adjustments to meet variations in working expenses, but the new arrangements operative from January 1, 1941, rescinded this and no more fare increases could be obtained as long as Government control lasted, a factor which materially disturbed the fare structure. There was no bar on the withdrawal of facilities, and road-rail bookings other than seasons disappeared from August 3, 1941, cheap day tickets suffering the same fate from October 5, 1942.

WARTIME WORKS

Works already in hand on September 3rd were tailed off as described in the last chapter, and apart from the construction of deep shelters and other defence measures, no new capital works were begun. Escalators were opened at Marylebone on the Bakerloo line on February 1, 1943, but this was a 1937 project stimulated by the bombing of the old surface station in Harewood Avenue. Two escalators gave direct connection between the tube platforms and the main-line station circulating area where a temporary booking office was provided to permit the closing of the old station and lifts.

One important and apparently unexpected work had to be urgently undertaken in the midst of war. After routine inspections of the Bakerloo tunnels under the Thames in 1943 and 1944 had revealed a disquieting state of affairs, the Board was faced with the tremendous task of removing the defective concrete track foundation. This foundation rested on an armour-plate lining placed inside the original tunnel in 1919–1920 as a belated protective measure against bomb damage, 540 feet of it in the southbound tunnel and 480 feet in the northbound. It was decided to cut away the lower section of the lining and support the top part with 1,000 corbel plates whilst constructing a fresh concrete foundation. To complete the work 500 men were employed in three eight-hour shifts and the line closed between Piccadilly Circus and Elephant & Castle from August 14–24, 1944 (inclusive). The floodgates at either end were left open whilst the men worked, but the remainder of the railway was sealed against flood damage by closing the gates at the north end of Charing Cross station and erecting a temporary waterproof diaphragm at the south end of Waterloo. Although a fan was provided at Waterloo, the heat, cement dust and noise of the pneumatic picks in the narrow confines of the tube made working conditions extremely trying. Track was relayed with transverse Jarrah wood sleepers carrying 95 lb. rail and 85 lb. conductor rail. German propaganda claimed it was all due to a flying bomb.

PLANNING FOR PEACE

There was the usual wartime spate of high-minded planning for the better world that was to come with the peace, and this time the work was undertaken with great fervour, although the wiser and older must have known that many of the plans would not mature for many years, if ever.

The County of London Plan by Mr J. H. Forshaw, Architect to the LCC, and Patrick Abercrombie, Professor of Town Planning at the University College of London, was published in 1943. It suggested a specialist investigation of railway planning and, as far as the tubes were concerned, did little more than envisage duplication of central-area tunnels to provide relief. It was also suggested that the tubes should be more closely knit to the main-line suburban railways with low-level interchange stations between the two types of line. Only tentative suggestions were made regarding possible new routes in the form of loops embracing all the main-line termini and large enough to accommodate full-size trains. In Professor Abercrombie's Greater London Plan of 1944, the need for a specialist study of railway planning was again stressed and there were no special recommendations for tube railways.

The recommended specialist investigation was conducted by The

Railway (London Plan) Committee, which was appointed by the Minister of War Transport in February 1944. Reports were presented on January 21, 1946, and March 3, 1948, and the proposals reviewed and co-ordinated by the London Plan Working Party, which reported on October 29, 1948. By this time, the British Transport Commission had been formed, and it approved the proposals for new small-diameter tube railways as follows:

First Priority:

ROUTE C Edmonton (Angel Road) to Streatham and Norbury (and possibly East Croydon) via Tottenham Hale, Finsbury Park, King's Cross, Euston, Oxford Circus, Green Park, Victoria, Vauxhall, Stockwell and Brixton. Possible branches to Walthamstow from South Tottenham and also along the alignment of the Great Cambridge Road.

This was visualized as an *urban* railway. To exchange traffic with existing tubes at Oxford Circus and other points where space was restricted, the use of ordinary tube stock was obligatory. Relief would be given to the overcrowded Piccadilly line and the railway would provide the long-desired connection between the West End and the heavily populated areas of Tottenham and Edmonton. Cross-platform interchange would be provided at Edmonton with British Railways, Eastern Region, which would be electrified.

ROUTE D From a connection with the Chingford branch railway in the Clapton area to Victoria via Liverpool Street, Bank, Ludgate Circus, Aldwych and Trafalgar Square. A low priority western outlet via Knightsbridge, Olympia and Uxbridge Road to Yeading Lane, Hayes.

The tube trains would run over the Chingford branch and also over the Enfield Town branch (from Hackney Downs). Another *urban* line, 'D', would distribute traffic concentrated at Liverpool Street station and relieve the Fleet Street–Strand axis. The route bears a remarkable resemblance to the frustrated Hammersmith and North East London schemes of 1902–1905 mentioned in Chapter 5.

ROUTE H Extension of the Bakerloo line to Camberwell, with possible low priority extension thence in the direction of Herne Hill.

Completion of 1935–1940 New Works Programme. Electrification of the Alexandra Palace branch and its connection to the Northern City tube line. Completion of the electrification and doubling of the Finchley–Edgware branch and the extension to Bushey Heath.

Other recommended small-diameter tube railways as follows were accorded lower priority:

ROUTE E A new tube paralleling the Northern line between Kennington and South Wimbledon, with one branch south-west to Raynes Park and Motspur Park whence trains would run over the Southern Electric to Chessington South. Another branch south to Morden and North Cheam.

Trains were to run non-stop between Kennington and Tooting and there would be interchange provision at Raynes Park and Motspur Park. This line would give relief to road traffic feeding Morden station. It would apparently make use of the deep-shelter tunnels at Stockwell and Clapham.

ROUTE J ⎱ Alternative schemes for small-diameter tubes extending the
ROUTE K ⎰ main-line size Northern City line southwards from Moorgate. Route J would take it via New Cross and Woolwich to Plumstead with possible further extension back across the river to Dagenham, whilst Route K would run via Bricklayers Arms and Peckham to Crystal Palace.

ROUTE L Extension of the Piccadilly line Aldwych branch to Waterloo.

In addition to the above, the final recommendations included some proposals for tubes large enough to take main-line stock, as follows:

First Priority:

ROUTE A From a junction with the Southern Electric at Loughborough Junction to Euston via Elephant & Castle, Ludgate Circus, Chancery Lane, Russell Square. Branches from Euston to Finsbury Park (junction with suburban lines) and West Hampstead (junction with suburban line out of St Pancras).

In connection with Route B, this tube would allow the removal of Blackfriars Bridge and would take suburban trains from Hertford, Welwyn, St Albans and Luton, etc., in the north and those from the Chatham main line, Catford Loop, and Tooting/Wimbledon/Sutton services in the south. There would be interchange facilities at points of crossing small-diameter tubes (e.g. Russell Square).

ROUTE B A new freight-carrying tube between West Ham and Lewisham via Greenwich.

Lower Priority:

ROUTE F A tube to take the Southern Electric services from Hither Green and Lewisham to Kilburn via Fenchurch Street, Bank, Ludgate Circus, Trafalgar Square, Marble Arch and Marylebone. There would be junctions at Neasden and Kenton with the suburban lines to High Wycombe, Aylesbury and Tring and interchange with tube railways at points of crossing (e.g. Trafalgar Square).

ROUTE G A tube from Fenchurch Street to Waterloo via Bank and Southwark Street carrying the London, Tilbury and Southend line trains into south-west London via a surface connection near Battersea. Convenient cross-platform interchanges within London.

These ambitious plans amounted to a grand total of 102¾ route-miles of new tube railways estimated to cost some £238 million (110% above prewar prices) including depots and rolling stock. Of this total, 39½ miles were to be in 17-foot tunnels to take main-line stock and 63¼ miles in the conventional 12-feet diameter tubes. The main-line tubes were to have 26-feet diameter station tunnels with 650-foot platforms to take ten-car trains. Small tube platforms would generally be 450 feet to take eight-car trains. Some of the tunnels would be as much as 120 feet below the surface in the central area and two banks of 50-foot rise escalators were envisaged at such points. It was recognized that much of the new tunnelling would need to be driven under compressed air, especially in south London.

An interesting distinction was made between *urban* and *suburban* types of service. It was stressed that the routes with urban characteristics should not extend too far from the central area, as the low average speed due to frequent stops made the journey time excessive and increased congestion. It was suggested that a reasonable radius for an urban service was 12–14 miles from the centre, with a journey time of 35–40 minutes, and such services should be self-contained and not subject to interference from other types of railway. The small-diameter tube lines, existing and proposed, fell into the urban category and it was observed that the Bakerloo service to Watford was overlong, and should be cut back to Harrow & Wealdstone after the opening of Route F.

The maps accompanying the Report gave a startling demonstration of the complexity of the proposals and recalled maps published during the London Railway Manias of 1864 and 1901–1902.

CHAPTER 13
London Transport
Since 1945

RETURN TO NORMAL

ON October 2, 1945, the dim filament-bulb lighting of the westbound Piccadilly line platforms at Piccadilly Circus was replaced by fluorescent tubes providing three times as much light. This gesture, in the station that serves the traditional centre of the British Commonwealth, might be taken as a symbolic beginning of the postwar era for London's tube railways. It was followed by a complete conversion of the whole station to the new lighting. From 1946 to 1948 the war-begrimed stations were restored to something of their former selves, as all the clutter and rubbish of shelterers and ARP were removed. Late-night trains, withdrawn in October 1942, ran again from February 18, 1946. Down escalators which had been stationary in slack hours were restored to all-day movement in January of that year. On July 1st, the Aldwych branch was reopened to traffic Mondays to Saturdays (no Sunday trains had run since April 15, 1917). Twelve years later, on June 7, 1958, all-day service on this tiny branch ceased; after then, the shuttle train ran only between 7 and 10.30 a.m. and between 3.30 and 7 p.m. (7 a.m. to 2 p.m. Saturdays). From June 18, 1962, trains ran only Mondays to Fridays, starting 15 minutes later and ending 15 minutes earlier.

After a temporary setback caused by the serious fuel crisis of February 1947, with its 10% cut in train service, 50% reduction of lighting and almost complete standstill of down escalators, the train services were gradually built up again. The running of frequent, but short trains in the slack hours was started again on the Bakerloo line on June 5, 1950, followed by the Central on February 19, 1951, the Northern on November 12, 1951, and the Piccadilly on May 12, 1952. Short trains (usually four cars) ran from 10 a.m. to 4 p.m. and 7.30 p.m. to 11 p.m., and after they started, the slack-hour service interval generally became $2\frac{1}{2}$ minutes instead of 3 or $3\frac{3}{4}$, and the evening headway 3 minutes instead of 4 or 5. On Saturdays, the service between 6 p.m.

and 10 p.m. was usually 2½ minutes and the short trains also ran on Sunday mornings. Increased off-peak traffic in central London, and staff problems associated with the coupling and uncoupling of trains, finally led to the cancellation of short-train working in 1960–1961.

The increase in off-peak traffic was purely local to the busiest parts of the central area; generally over the system, off-peak travel declined sharply after the war. Servicemen moving across London on leave and duty became a mere trickle, private car ownership gradually and then steeply increased, and television became a substitute for cinema and theatre visits. Rush-hour loading maintained its level, even increased; worse still from the traffic point of view, it became more concentrated. Noble, but not very successful efforts were made to persuade employers to stagger arrival and departure times of staff. By 1960, over half a million people were travelling home by LT railways, almost three times as many as by bus. As the five-day week became widespread, Saturday morning and midday peaks diminished and eventually disappeared entirely.

CHANGES AT THE TOP

The Transport Act of 1947 set up a British Transport Commission with a number of Executives to act as its agents in the day-to-day operation and administration of the various types of public transport nationalized by the Act. Established on January 1, 1948, the London Transport Executive took over the functions of the London Passenger Transport Board, and the wartime Government control ended on that day. The Transport Commission retained the formulation of fare policy and charges schemes and also directed that their prior approval must be obtained for all capital expenditure exceeding £50,000, all appointments and salaries of senior officers, important agreements on wages and working conditions, and certain other matters. In practice there was little change in policy or atmosphere and the LPTB officials remained in post to operate and administer the railways and road transport.

Lord Latham, a member of the LPTB, and a former leader of the LCC, was the first chairman of the Executive. The seventy-three-year-old Lord Ashfield, chairman of the Board, left 55 Broadway to become a full-time member of the Transport Commission from November 1, 1947, an office which he held until his death on November 4, 1948. He thus completed an almost unbroken service of forty-one years guiding and controlling London's underground, and ended his career in harness, giving his wisdom and experience to the newly born national transport undertaking. The other members of the new London Transport Executive were former officers of the LPTB, supported by a trio of part-time members—the managing director of Harrod's store, a director of the Co-operative Wholesale Society and the chairman of the Kent County Council.

From its inception, the Executive took over the Ealing & Shepherd's Bush Railway (except Ealing Broadway station), the GWR-owned tube railway extension to Greenford (except Greenford station) and the LNER-owned lines from East Finchley to High Barnet and Mill Hill East, and from Leyton Junction to Newbury Park via Grange Hill. The SR-owned tracks between Studland Road Junction, Hammersmith, and Turnham Green, used by Piccadilly line trains since 1932, were formally transferred to the Executive on January 23, 1950.

Lord Latham was succeeded on October 1, 1953, by Mr John Blumenfeld Elliot, who had been chairman of the Railway Executive since 1951 and before that, general manager of the Southern Railway. Created a Knight in the 1954 New Year Honours, he served until 1959, when he was succeeded by Mr Alexander Balmain Bruce Valentine, a former officer of the LPTB, member of the Railway Executive Committee and the BTC.

The 1953 Transport Act which made substantial changes in the functions of the BTC did not materially affect the LTE, which was left as the sole survivor of the Executives established by the 1947 Act.

In 1953, the Government appointed a Committee of Inquiry under the chairmanship of S. P. Chambers, CBE, CIE, to investigate the conduct of London Transport and recommend measures that might be taken to secure greater efficiency and economy. The Report, made in January 1955, included a recommendation that London Transport should either be constituted a separate body, responsible directly to the Minister of Transport, or be abolished as a statutory body and vested in the British Transport Commission, who would have full power over the Board of Management appointed in place of the existing Executive.

This proposal was not followed up at the time, but in 1960 the Minister of Transport, Ernest Marples, announced a reorganization plan for the whole of nationalized transport which included the establishment of a London Transport Board to replace the Executive. This Board, established by the Transport Act, 1962, is responsible direct to the Minister who appoints its members; as formerly, it works in close concert with the London lines of British Railways (by the same Act, BR were organized as a similar Board, and the British Transport Commission was abolished).

Other interesting points in the Chambers Committee Report included a comment that the indirect advantages to London transport and London's economy arising from the construction of the Route 'C' tube railway (Victoria–Walthamstow) were so important that the project should not be abandoned or postponed purely on grounds of potential unprofitability on the basis of direct receipts and expenditure. It was noted that checking of tickets at inward barriers could be made more efficient and that co-ordination between London Transport and British Railways was poor, especially in the fields of comprehensive publicity and interchange facilities.

Nothing very much has been done to implement these recommendations. In 1956, the first official map to show all the railways in the London area, together with bus roads, was published by the LTE, and this useful map has since been in continuous issue in poster and pocket form. An interesting aspect of it is that the LTE stations are printed more prominently than those of British Railways. London Transport posters still advise the public to use bus or Green Line coach services to reach towns such as Sevenoaks or St Albans, when the most expeditious route is provided by British Railways.

One recommendation of the Chambers Committee was pursued vigorously. Renewed attempts were made to persuade employers to stagger working hours and special committees appointed for the task have made some progress; by the end of 1961, about 56,000 had changed their hours of a total of 630,000 whose hours were known to the committees.

PROMISES FULFILLED : THE CENTRAL LINE EXTENSIONS

Such was the importance attached to the eastern and western extensions of the Central line that work was resumed late in 1945, almost as soon as the war ended, when circumstances were hardly propitious. On the eastern extension, the main task was the preparation of the tunnels for railway use and completion of the stations. More complicated work was involved on the western side, including the construction of a new station and layout at Wood Lane, the replacement of tracks taken up for war use and the erection of new surface stations.

The first section to be completed was the $4\frac{1}{4}$ miles between Liverpool Street and Stratford, which were formally opened by the Minister of Transport, Alfred Barnes, on December 3, 1946, and in public service on the following day. There were intermediate stations at Bethnal Green and Mile End; the latter was 1·77 miles from Stratford, the longest tunnel distance between any two stations on the tube system (the previous record length was the 1·44 miles between Manor House and Turnpike Lane, Piccadilly line).

At Stratford, the tunnels debouched into a new surface station arranged to provide cross-platform interchange with the LNER electric trains from Liverpool Street to Shenfield (these were to have started within nine months of the tube extension, but postwar difficulties delayed completion until September 26, 1949). At the eastern end of the Stratford platforms, the Central line tracks descended steeply into tubes, to emerge into daylight once more $\frac{1}{4}$ mile south of Leyton LNER station. The tube trains terminating at Stratford continued through these tunnels to reverse on to the westbound track over an automatically operated crossover at Leyton.

Liverpool Street's platforms had been extended to take eight-car trains, and two shunting necks were provided east of the station. Bethnal Green, scene of the wartime tragedy, had an under-street booking hall, three escalators and fluorescent lighting throughout. A crossover was sited just west of the two 420-foot platforms. Just before Mile End, the tubes rose steeply to the level of the Whitechapel & Bow Railway, immediately below the busy Mile End Road, and the tube trains used the outer edges of the existing platforms, which thus became islands. The enlargement of the old station to allow for this cross-platform interchange had involved some skilful surgery to avoid disturbance to the street and surface buildings. Beyond Mile End, the tunnels descended, and swerved north-east to cross the Lee marshes directly under the old Great Eastern main line. All tunnels were eventually sound-proofed with the type of horizontal concrete fin which had been used on the Northern and Bakerloo extensions in 1938–1939.

A four-minute service was provided in the rush hours, lengthening to five minutes off-peak. Liverpool Street was reached in nine minutes from Stratford, Oxford Circus in twenty.

Public service was extended to Leytonstone LNER station, 2·31 miles from Stratford, on May 5, 1947. On that day the fares on the Loughton–Epping–Ongar line and over the Hainault loop were reduced to LT scale. From a point just west of the LNER station at Leyton, the tube trains ran over the tracks of the LNER Loughton and Ongar line. LNER steam trains worked a shuttle service from Leytonstone to Epping and Ongar instead of running to and from Liverpool Street and Fenchurch Street. Tube trains ran every five minutes from Leytonstone, every three minutes in rush hours. Leytonstone's level crossing was abolished and the tracks walled-off, the road being diverted through a new underpass. The old station was substantially reconstructed, with a ticket hall and pedestrian subway below the tracks and the up platform converted to an island to give a platform road for trains of each of the two future branch services. In the meantime, this new platform road was used for the steam shuttle service. The goods yard became a car park. At Leyton, the old station was retained, somewhat tidied up.

The new service was immediately popular and the happy residents of Leyton, Leytonstone, Woodford, Loughton and North Ilford swarmed into the trains which gave them a direct run to the West End of London for the first time. Less happy were those unfortunates still obliged to change from LNER suburban trains to the tube at Liverpool Street, for when they sought the Central line trains, they found them so tightly packed they could not gain a foothold.

The first section westwards was opened on June 30, 1947, when trains began to run on two new tracks alongside the GWR Birmingham main line for 3·9 miles between North Acton and Greenford, with new intermediate stations at Perivale and Hanger Lane. Greenford had a five- to six-minute service at peak hours, $7\frac{1}{2}$ to ten minutes at other times. Great Western steam railmotors ceased to run between Westbourne Park and Greenford and their intermediate halts at Old Oak Lane, North Acton (GWR), Park Royal West, Brentham and Perivale were closed. Fares on the GWR suburban line as far out as West Ruislip were brought down to the London Transport scale.

This extension had involved the quadrupling of the Ealing & Shepherd's Bush line between North Acton and the Wood Lane flyover as a first step. The Cleveland Bridge & Engineering Company received the contract, and from June 19, 1938, separate pairs of tracks were used by the LT and the GW trains. A burrowing junction was then built west of North Acton station to carry the new westbound tube track under the Ealing line. Westwards from this junction, the GWR laid a double track alongside its main line, but had to lift it during the war. The whole line from White City to West Ruislip was built to District line loading gauge.

All three stations were still incomplete on the opening day and temporary staircases and offices had to serve meanwhile. GWR architects and engineers supervised the design and construction of these stations, and they were arranged for the utmost economy in maintenance. Perivale and Hanger Lane had simple 440-foot island platforms with temporary asbestos sheet roofing. The temporary booking hall at Hanger Lane was replaced from January 2, 1949, by a pleasant circular hall with a reinforced-concrete slab roof. At Perivale, the line was on an embankment, and the permanent booking hall, at street level, was a graceful curved structure of brick and glass. Greenford was also on an embankment, with platforms 33 feet above the street, reached by a single escalator of 30-foot rise—the first on LT railways to take passengers *up* to the trains. The island platform, with temporary roofing, had a single-track bay at the London end for the future accommodation of GWR shuttle trains running from Ealing Broadway via South Greenford. Meanwhile, GWR railmotors and some other trains on the main line continued to use the old Greenford station, which was connected to the tube booking hall by a subway. All three stations were lit by fluorescent tubes sited parallel to the track under the canopies and supported in reinforced-concrete **T** standards at right angles to the track elsewhere. The main contractors for the Greenford extension were George Wimpey & Co. Ltd.

Wood Lane was the key point in the whole scheme for modernizing and extending the Central line. An awkward triangular station, which could not handle the eight-car trains planned, it was not possible to extend it on the existing site, and it was decided to build a new station

350 yards further north, almost opposite the White City Stadium. Work began on May 6, 1946, and involved the widening of the existing cutting to take a new four-platform station (three roads) and improvement of the approaches to the south by realigning the westbound track in a new 220-yard covered way to the east of the old lines. This work was delicate as the track had to be brought beneath the depot, the exhibition arcade and the Hammersmith & City Railway viaduct.

The old Wood Lane station was something of a railway curiosity; apart from single platforms on the eastbound and westbound through lines, there were the platforms either side of the single-line loop joining the through lines which had been built when the station was first opened in 1908 to serve the Franco-British Exhibition. These latter platforms were used by trains reversing at Wood Lane by running round the loop. As constructed, the southernmost platform (No. 1) was too short to take all the doors of a rebuilt six-car train, and the only possible space for extension was blocked by a depot turnout. In March 1928, as part of a scheme for improving the station to handle traffic from the new White City Greyhound and Cycle Racing Stadium, the problem of extension was solved by constructing a swivelling wooden platform, 35 feet long, which could be pushed out to a maximum of 3 feet over the nearside rail of the turnout, providing a platform face for the rear car of a six-car train. Electro-pneumatically worked from the signal box, the movable platform was interlocked with the signals and worked well for almost twenty years.

The new station was called White City and had two island platforms, 444 and 465 feet long, with three tracks, the eastern one for westbound trains, the western for eastbound, and a centre road, normally used for trains terminating, which ended in a 447-foot reversing stub. Wood Lane was closed on November 22, 1947, and the new station opened with a temporary booking hall on the following day. All four platform faces were not in use until July 4, 1948; the new westbound approach road was ready on July 18th and the complete new layout, with the old westbound line converted to a depot access road, came into use on August 7, 1949. A Festival of Britain Architectural Award was given for the design of the station, with its stark uncluttered brick buildings and linking concrete footbridge. A new signal cabin, with a 47-lever power locking frame, sited on the east side of the new station, was opened on July 4, 1948. This box controlled route setting through the junction at North Acton as well as movements within the White City station and depot area.

Central line trains made further penetrations eastwards into LNER territory on Sunday, December 14, 1947, when tube train service was extended from Leytonstone to Woodford (2·89 miles) and through the new tunnels under Eastern Avenue from Leytonstone to Newbury Park (4·11 miles). North of Newbury Park, as far as Grange Hill, conductor rails were energized to allow empty stock to reach the new depot at Hainault. The intermediate stations to Woodford were substantially the former GER stations of Snaresbrook and South Woodford (George Lane) and, as everywhere on the eastern extensions, platforms were lowered to a compromise height of 2 feet 9 inches. South Woodford (which lost the appellation 'George Lane' in 1950) and Snaresbrook were both provided with an additional booking hall on the westbound side in 1948. Between Leytonstone and Newbury Park were the entirely new tunnel stations of Wanstead, Redbridge and Gants Hill. The new tunnels began at the end of 1 in 45 ramps either side of the existing LNER tracks north of Leytonstone station and then swung east through Wanstead and North Ilford. The removal of the wartime factory in these tunnels and preparation for railway use was in itself a lengthy task. At Newbury Park, the tubes turned south and then curved north to debouch either side of the LNER Ilford–Woodford loop line just south of Newbury Park station.

Built solidly, with handsome and closely spaced stations, the Ilford–Hainault–Woodford loop had been opened on May 1, 1903, and for many years failed to attract the building development that had been envisaged. At Ilford, the loop made junction with the GER main line in both London and Shenfield directions (the western spur was cut by an extension of the Ilford carriage depot in 1948 and the eastern one, which had continued in use for freight, was finally cut in 1956; all tracks south of Newbury Park have since been removed). Steam passenger trains between Ilford and Woodford via Newbury Park ran for the last time on November 29, 1947, and were replaced by special buses from Ilford, calling at all stations to Roding Valley. With the opening of the tube service to Newbury Park, the bus service worked between that station and Roding Valley via Woodford. The LNER steam trains to Epping and Ongar reversed at Woodford, platform 2, and tube trains normally terminated at platform 1 to provide easy interchange.

Wanstead, the first of the new stations on the Eastern Avenue tube, was of conventional pattern below ground. Two escalators of 58 foot 6 inch rise led to a surface ticket hall of 40 square feet which was an adaptation of one of the wartime factory entrances. An ugly building with a square tower and concrete-framed windows, it was strategically sited at the junction of Wanstead High Street and Eastern Avenue. Redbridge station, just east of the river, had a temporary surface booking hall with stairs leading to an island platform about 15 feet below in a reinforced-concrete cut-and-cover tunnel. Gants Hill was nicely placed at the important intersection of Eastern Avenue, Cranbrook Road and Woodford Avenue and had a booking hall with ten entrances below the road roundabout. From this, three 32 foot 3 inch rise escalators led to an underground concourse 150 feet long and 20 feet high.

Tube trains on the Metropolitan—1938 and standard stock cars at Stanmore terminus in July 1958.—*Alan A. Jackson.*

Steam and electric at Epping—standard stock Central line train for West Ruislip and ex-GER 2-4-2T (BR 67203) on two-car Epping–Ongar push-pull train, June 12, 1956.—*Alan A. Jackson.*

Central Line in the west—a standard stock Central line train for West Ruislip burrowing under the Ealing & Shepherd's Bush line at North Acton. The Birmingham main line of the former GWR is on the right, and the eastbound tube line from West Ruislip is seen curving round by the signal box in the centre. Beyond, the two tube tracks continue north-westwards alongside the main line. August 24, 1957.—
Alan A. Jackson.

Three types of stock at Ruislip depot, June 1961. *Left to right:* Craven 1960 prototype; 1927 standard; 1959 stock. All are motor cars.

Situated between the platforms, this underground hall, not fully opened until April 4, 1948, had a domed roof supported by two rows of eight tiled columns. The effect was spacious if a little extravagant for a suburban station and was a rather pale imitation of the plan used for many Moscow Metro stations. None of these three stations was fully completed for the opening day.

At Newbury Park, a temporary booking hall on the east side of the station replaced the original overbridge building, and there was a new concrete footbridge to both platforms. A pleasing bus station with a copper-covered barrel-vault roof of reinforced-concrete, 30 feet high and 150 feet long was opened on July 6, 1949, at the same time as the permanent booking hall. Situated on the east of the line, alongside the booking hall, the bus station is used by services 66 (Leytonstone–Hornchurch) and 139 (Gants Hill–Dagenham). A new dual-carriageway bridge was built in 1957 to carry Eastern Avenue over the line, and the old GER booking hall was demolished. North of Newbury Park, a fan of car-storage sidings was provided and a similar one was built east of the bay road at Woodford.

Hainault depot now came into partial use, but 54 cars were stabled at Woodford and 18 at Newbury Park. The peak-hour service interval on each of the new sections was four minutes, increasing to $7\frac{1}{2}$ minutes in slack hours. Oxford Circus was placed within 34 minutes of Woodford and 39 of Newbury Park. Following Inner Circle practice, the lines of the Hainault loop became *inner* and *outer rail*, Leytonstone–Newbury Park–Woodford constituting the *inner rail*.

At the official opening ceremony on December 12, 1947, at Wanstead, Sir Ronald Matthews, chairman of the LNER, hailed 'with acclamation this further peaceful penetration of the trains of the Central line into what was once our jealously guarded suburban territory'. Modestly in the background stood members of the Ilford and District Railway Users' Association watching the fulfilment of their fifteen-year campaign.

But the quiet satisfaction of fulfilled hopes was to be rudely shattered. The new extensions attracted traffic on to a line that was already under great strain. Everyone who could now reasonably do so deserted the steam-operated Chingford branch and made for the Central line trains, which were already crammed with existing passengers from the Loughton and Hainault lines and former bus passengers hoping for a quicker run to the West End. Complaints about overcrowding were made in the House of Commons early in 1948 when Epping's MP described the conditions as 'absolutely disgraceful, a danger to life, limb and the nervous system'. There were some interesting comparisons to be made with the old steam service. In 1939, 11 steam trains left Liverpool Street and Fenchurch Street for Woodford and beyond between 5 and 6 p.m., Mondays to Fridays. Assuming each train was normally composed of two quintuple articulated sets, this would represent 9,592 seats. The 15 tube trains leaving Liverpool Street in the same hour after December 14, 1947 (assuming 13 seven-car trains with 278 seats and 2 eight-car with 308) offered 4,230 seats. By February 1961, the scheduled Woodford service in the same period was 18 eight-car trains, and if all were of the 1959 stock then coming into use, this would offer 5,904 seats. Figures of the number of 'places' available for standing passengers must inevitably be imprecise, but there is little doubt that in equivalent conditions of crowding, the old steam service also provided greater *total* capacity.

The congestion was not made easier by the fact that the tube trains were composed of standard stock dating from 1923–1927 which had been stored in the open air throughout the war. Breakdowns were frequent and maintenance could not be properly organized until the new depots at Ruislip and Hainault were fully operational; it was not possible to run the planned number of eight-car trains. The situation had been foreseen to some extent. The original plan had been to open the line as far as Loughton in April 1940 and the Loughton–Ongar and Hainault loop sections were to be opened concurrently with the Shenfield line electrification in January 1941. This sequence had to be dropped after the war in deference to the urgent transport needs of the eastern suburbs (particularly North Ilford) and the effects of inadequate terminal accommodation and maintenance facilities had been accepted as the price for earlier opening dates.

Some relief was available after the installation of speed-control signalling at Liverpool Street in April 1948. After the opening of the eastern extensions, station stops at this station had increased to fifty or sixty seconds, and a queue of waiting trains formed up behind the station at peak hours. Speed-control signalling allowed another five trains through the station during the busy period. In the autumn of 1948, work was resumed on the prewar scheme to modernize the top station and interchange facilities at Liverpool Street. This involved a new booking hall below Liverpool Street, with two new escalators to the Central line lower landing, stairways to the street and to a redesigned Metropolitan line booking hall. There was also a new subway to the old CLR booking hall below the Broad Street forecourt to allow closure of the old escalators. These new facilities came into use on November 6, 1950, but the new ticket offices were not ready until February 19, 1951.

Passenger service was extended from Newbury Park to Hainault (1·91 miles) on May 31, 1948, through the intermediate stations of Barkingside and Fairlop, which were hardly altered apart from the retiling of the booking halls. At Hainault a new island platform replaced the old down platform and a new booking hall was constructed in the embankment at street level. Peak-hour service between White City and

Leytonstone was now increased by four trains an hour, two of which terminated at Woodford and two at Hainault, giving each branch seventeen trains an hour at peak periods. Basic service to Hainault was 7½–10 minutes.

Hainault depot, for 344 cars, had been completed in 1939 and used during the war by the U.S. Army Transportation Corps for railway rolling-stock assembly. It was now fully operational and comprised a nine-road car shed, a cleaning shed, washing plant and sixteen open storage roads.

Final completion of the westward extension was achieved on November 21, 1948, when the Central line trains reached West Ruislip. On the same day, in the east, tube trains ran for the first time between Woodford and Loughton and between Hainault and Woodford. Altogether, more tube railway mileage had not been opened on one day since the Watford extension of April 1917.

The 4·38-mile western section had received a formal inauguration two days earlier at the hand of Alfred Barnes. It was mainly built by Caffin & Co. Ltd and had new stations, adjacent to the old, at Northolt, South Ruislip, Ruislip Gardens and West Ruislip (formerly Ruislip & Ickenham). The original intention of terminating at Denham, with an intermediate station at South Harefield, was dropped as that section was now in the London Green Belt. With the arrival of the tube trains, the GWR (now Western Region of British Railways) railmotors ceased to run between Ealing Broadway and West Ruislip and were replaced by an augmented railmotor service shuttling between Ealing Broadway and Greenford, which terminated in the new bay road. There were no conflicting movements as the steam line left the station on a ramp, descending between the eastbound and westbound electric lines, to pass beneath the latter. Since August 25, 1958, this WR shuttle service has been worked with diesel railcars.

One of the largest works on the western extension was the widening of the Northolt cutting, requiring the removal of some 37,000 cubic yards of clay over half a mile. German prisoners of war, and later Polish labour, were used to complete the two new lines alongside the main tracks.

All four new stations were of the 440-foot island platform type with double-curved ('butterfly-shell') reinforced-concrete canopies 3¼ inches thick, supported on centre columns 24 feet apart. Platform offices were of simple brick construction. Fluorescent tubes were used as on the other western stations. At Northolt an austere temporary booking hall was built on the road overbridge, but at South Ruislip, where the platforms were on an embankment, an impressive circular booking hall which suggested a classic temple was partly built at street level. At Ruislip Gardens, a temporary booking hall was set in the embankment arches. The island platform at West Ruislip was adjacent to the Western

Region station and the two lines share a new combined booking hall on the road overbridge. Construction of this new building was left in a half-finished state. These stations remained as described until 1961 when work began on their completion. A permanent rectangular booking hall with a frontage of light-coloured bricks was built at Northolt and the circular structure at South Ruislip completed with elevations faced with dark blue-grey brick to contrast with the white translucent glazing of the upper section. A frieze of precast concrete panels to represent 'Passenger Flow' was included. Ruislip Gardens received a new station building with elevations in warm brown bricks and vertical panels of polished grey-green Broughton Moor slate. The West Ruislip booking hall was finished off and a canopy added over the pavement outside.

A reversing road for short workings was provided just west of Greenford station, but this was moved to Northolt on May 1, 1960, to allow extra workings to operate from there. The basic service to West Ruislip was ten minutes and the journey time to Oxford Circus thirty-nine minutes. At first, all the WR stations beyond Greenford except Northolt remained open, with subway or footbridge connection to the new platforms, but the ramshackle halt at Ruislip Gardens was finally closed in 1958. All the new stations except Northolt were staffed by British Railways.

Ruislip depot, with room for 472 cars (184 under cover on sixteen roads) had been completed in 1939 and used by the Admiralty during the war. Situated on the south side of the line between Ruislip Gardens and West Ruislip, it was designed, with Hainault, to handle the routine examinations and maintenance of the 627 Central line cars. The day before the Ruislip extension opened, 200 tons of machinery and stores were transferred from White City (Wood Lane) depot and part of the old sheds there became a stabling and storage point.

The new service between Woodford and Loughton (2·59 miles) was $3\frac{3}{4}$ minutes in peak hours, 16 trains per hour, giving 32 trains per hour on the in-town section (the 17 trains per hour announced at the opening of the Hainault extension had evidently proved too much for the capacity of the line). Basic service from Loughton was ten minutes and Oxford Circus was reached in forty minutes. Steam trains ran from No. 2 platform at Loughton to and from Ongar.

The only intermediate station on the Loughton section was Buckhurst Hill, which remained almost unaltered apart from the suppression of a level crossing. Site difficulties made alteration and enlargement of the old Loughton station impossible, and a new building had been opened by the LNER on April 28, 1940. The square booking hall, with its two arched windows of glass bricks revealing the line of the barrel vault ceiling, stood on the western side of the embankment. A subway and stairs from the hall led the passenger to two island platforms serving the three roads. Each platform was protected by concrete canopies of smooth and pleasing sweep. South of the station, on the east side, a fan of eight storage sidings was laid out.

Between Hainault and Woodford, at the top of the loop, tube trains worked a ten-minute shuttle service (7½ minutes in the rush hours) calling at the original GER stations of Grange Hill and Chigwell and at the LNER-built Roding Valley, which lost its halt status and underwent some rebuilding. On this section there were 3·8 route-miles of new electrified track.

Epping (4·9 miles from Loughton) was reached by the Central line trains on September 25, 1949. The original stations at Chigwell Lane (renamed Debden by London Transport), Theydon Bois and Epping were little altered apart from some rebuilding of booking halls, the installation of electric light and the provision of standard LT signs. Two reversing and storage roads were inserted between the running lines north of Debden. At Epping, the electric trains normally terminated at the eastbound (or down) platform; on the other side, steam trains, driven back to the last six miles of this long suburban branch, departed for Ongar. Since August 14, 1949, the single line between Epping and North Weald had been operated under non-token working, with track circuit control; the single line beyond North Weald continued under train staff and ticket regulations until September 25th. An additional loop and platform were opened at North Weald on August 14th to permit the running of a faster and more frequent steam service.

Epping had a 40-minute tube service to London, increased to 12–15 minutes in the peak periods, but there was a 20-minute service to Loughton and additional trains from Debden gave a 10–12 minute basic service inwards from there (6 minutes in rush hours). The 52 minutes running time from Epping to Oxford Circus was very little better than the combined steam and tube journey via Liverpool Street before the war—in 1939 Liverpool Street could be reached in 41 minutes from Epping by the best trains and the journey thence to Oxford Circus would take about 16 minutes.

The Central line extensions were signalled to London Transport open-air practice; two-aspect long-range colour-light stop signals (red and green), repeaters in the rear of stop signals (yellow and green) and fog repeaters (with 'F' in black on the yellow aspect) 400 feet in the rear of all running stop signals and switched in when required. Stop signals at junctions had indicators displaying a line of white lights to show the route for which the points were set. Steam- or diesel-hauled trains continued to provide freight service on the Leytonstone–Ongar line and also on the Hainault loop (via Woodford), working from Temple Mills, mostly in the small hours. Occasional steam- or diesel-hauled excursion trains are also worked. To give the necessary stopping distance for these trains, distant signals of the externally floodlit disc type were provided on the eastern extensions. These signals had a black swallow-tailed bar on a yellow ground.

New signal cabins with power frames and miniature levers were opened with electrification, or at the dates stated, at Leytonstone (59 levers), Newbury Park (59), Hainault (83, opened May 12, 1948), South Woodford (35), Woodford (59), Loughton (59), Debden (35), Epping (47, opened August 14, 1949), Greenford (11) and West Ruislip (59). Subsidiary cabins at Ruislip Gardens, Bethnal Green, Leyton and Grange Hill (opened October 29, 1948) were remotely controlled from West Ruislip, Liverpool Street, Leytonstone and Hainault respectively. On the eastern extension, access to goods yards was controlled by ground frames. Tube tunnel mouths were guarded by electric-train detectors, short lengths of current rail energized by the shoes of a tube train and so operating the appropriate signal-control circuit.

The western extension was fed by current supplied by the nationalized electricity authority through a switchboard at Old Oak Common substation to new substations at Brentham, Greenford, Northolt and Ruislip. These substations, which produced traction current at 630 volts d.c. through BTH water-cooled rectifiers, were remotely controlled from Old Oak Common. Cables were connected in rings so that a fault in either loop would still permit a reasonable service over the whole line. The eastern extensions took their supply from Greenwich power station, which provided 22kV a.c. at 50 cycles, converted to 630 volts d.c. by new substations at Bethnal Green, Bow, Leytonstone, Leyton, Redbridge, Newbury Park, Hainault, Roding Valley, South Woodford, Loughton and Epping. The latter was sixteen miles from the power station and beyond the economic limit for 11-kV transmission, hence the 22-kV supply to all these substations. All were remotely controlled from South Woodford and the rectifiers were of the GEC pumpless air-cooled type.

Flat-bottomed conductor rail of 150 lb. per yard was used on the open sections and 130-lb. rectangular type in the tunnels. This rail was brought to site in 300-foot welded lengths and then welded into lengths of up to $\frac{1}{2}$ mile. Running rails throughout were 95 lb. bullhead, welded into 300-foot lengths before delivery, then joined on site to approximately 2,700 feet with special fishplates. Machined rail ends and jigged holes ensured a tight joint and plastic-covered fishplates were employed to separate track circuits. In the open, there were adjustment switches between each 2,700-foot length..

For eight years, the odd survival of steam push-pull trains isolated at the end of this very long electric line provided a much-visited and photographed curiosity in the London railway scene, but finally the tip of the branch was electrified and tube trains ran through the Essex countryside to the remote Ongar terminus from Monday, November 18, 1957. For reasons of economy no substation was built and the line was worked by a shuttle service of specially modified two- and three-car trains. To avoid excessive voltage drop, it was necessary to ensure that trains passing on the loop at North Weald did not start simultaneously. Basic service was about forty minutes, reduced to twenty at peak hours. The track circuits were changed from d.c. to a.c. and the existing boxes at North Weald and Ongar were re-equipped and provided with illuminated diagrams. Signalling conformed with LT practice except that the mechanically operated points at North Weald, Blake Hall and Ongar remained in commission, together with the semaphore home signals at North Weald and Ongar. These old country stations were not altered, apart from the installation of LT signs and electric lighting.

Central line trains now worked over a continuous stretch of 40·2 miles straddling the whole of Greater London from east to west, a tube railway grandiose enough to deserve a place in the wildest dreams of Yerkes and Pierpont Morgan.

PROMISES UNFULFILLED: ANTI-CLIMAX IN NORTH LONDON

Except for the short section between West Ruislip and Denham, the 1935 Plans for the Central line reached fruition; not so those for the Northern line, whose sad story must now be told. At the end of the war, London Transport were ready to resume work on the Northern line extensions, and pronouncements made at that time showed that they had every intention of doing so at the earliest opportunity. The London Plan Working Party 1949 Report recommended that 'these works should be finished substantially as proposed' and powers were taken in the London Transport Act of 1947 for re-siting the proposed Bushey Heath station on the north side of Elstree Road. In March 1949, at a Rotary Lunch, Lord Latham referred to 'the uncompleted portions of the New Works Programme of 1935–1940, on which much money has already been spent'. In fact, between 1937 and 1940 about a million pounds' worth of work had been done on the Drayton Park–Alexandra Palace and Finchley–Edgware–Bushey Heath sections and details have been given in Chapter 11.

The general expectation was that with so much already finished, the completion of the Northern line works would not be long delayed. Parliamentary questions were asked from time to time, particularly about the Alexandra Palace line where the sight of conductor rail *in esse* irritated the patience of commuters waiting at Muswell Hill and Crouch End for their grubby, steam-hauled, gaslit trains.

Then, early in October 1950, London Transport announced that the Bushey Heath–Brockley Hill section would not be finished; there was 'no justification' for it 'in view of the introduction of the Green Belt proposals since the original extension plan was made, and the consequent restriction upon development'. It was added that it might be possible to consider Edgware–Brockley Hill and Mill Hill East–Edgware at a later date when the land to be retained for the Green Belt had been more precisely defined and traffic assessed. Any hopes engendered for this were dashed by another pronouncement in February 1954, which stated that the addition of these sections to the Northern line had been abandoned as it was 'now clear that no further major housing development is intended in the area that would be served by the extension and electrification originally proposed'. Costs of construction and operation had increased so much that the traffic arising from existing development would not justify the expenditure. It is not clear what 'major housing development' could be expected in the districts east of Edgwarebury Park or south of the Watford By-pass, or between Edgware and Mill Hill (The Hale) as all these areas were already fully built up. The extension of tube trains over the completely rural Epping–Ongar line in 1957 must have caused some wry smiles in Edgware and Mill Hill. It is interesting to note the remark in the earlier statement about the Green Belt, as the Bushey Heath extension was originally justified, in 1937, in the face of Green Belt opposition, as a means of reaching the depot at Aldenham, which was then said to be essential to the whole New Works Scheme for the Northern line. The Parliamentary Secretary to the Minister of Transport had said that the extra passengers on the extension would be mainly those then using Edgware station, and the primary aim of the scheme was to relieve the over-burdened Edgware–Morden line by diverting some traffic to the City via Finsbury Park, avoiding the very busy Camden Town junctions. In other words, the 1937 story was that the plans had been drawn up principally to afford relief to the existing line rather than develop new traffic in the fields north of Edgware.

The Alexandra Palace–Drayton Park section was treated as a separate issue, perhaps on the principle of *divide et impera*. In December 1953, the Alexandra Palace Trustees and the local authorities were told by the British Transport Commission that 'reduced traffic' had resulted in a decision to close the Alexandra Palace branch, subject to the approval of the Transport Users' Consultative Committee. A reduced traffic flow towards the City would not support the proposed through services to Moorgate, and electrification, costing upwards of £2 million, would not be justified as the line would still run at a loss. It seems possible that some of the traffic reduction alleged might not be unconnected with the rather moth-eaten steam service which had been provided on the branch since 1942, and the statement that the line would still run at a loss after electrification should be considered in the light of the threefold traffic increase which occurred after tube trains took over the High Barnet line. Before the war, both the LNER and the LMSR had served the Alexandra Palace branch intensively, operating about forty-three trains each way on weekdays. From September 7, 1942, the service was cut to three trains an hour in peak hours and no service at all after 7 p.m. (after 5 p.m. Saturdays). This meagre ration was suspended entirely from October 29, 1951, to January 6, 1952, inclusive and by 1953, when the announcement was made, there were only nine up and six down trains each day.

Closure of the branch was approved by the Consultative Committee and the last trains ran on July 3, 1954. For a time, goods trains still ran to the coal yards at Muswell Hill and Cranley Gardens, but track was removed early in 1958. The section between Finsbury Park and Wellington Sidings, Highgate, from which the unused conductor rail was removed in January 1954, remains open for freight traffic and transfers of tube stock from Drayton Park to the Northern line at Wellington Sidings. It now seems that the Northern City line between Finsbury Park and Moorgate will be transferred to British Railways, to become part of the electrification of the GN suburban lines, the purpose of its construction sixty years ago.

About 500 tunnel segments which had remained at Elstree Hill since 1937 were removed in May 1953, and in August the Elstree tunnel mouths were bricked up. Material dumping from the Aldenham bus depot site began to hide the earthworks and the partly built substation. Late in 1954, London Transport offered the railway right-of-way between Glendale Avenue and Hillside Gardens, Edgware, to nearby residents for garden extensions; early in 1958, a block of flats was built right across the line of the extension just north of Edgware station, and in 1959–1960, maisonettes were erected on the alignment just south of the proposed Brockley Hill station. That somehow seemed to finalize it.

BLOW HOT, BLOW COLD FOR CAMBERWELL

It will be recalled that powers were obtained for a Bakerloo extension to Camberwell in 1931, but the financial situation had prevented a start. The local authorities of central south London, led by Camberwell, pressed hard for construction after the war and in 1947 sent deputations to the Minister of Transport. Political pressure was strong, and the 1949 British Transport Commission Act once again renewed the 1931 powers and authorized a station below Camberwell Green, a site barred by the original Act. At a Press conference in March 1949, Lord Latham made the heartening announcement that the extension would be undertaken, describing it as 'a scheme of medium size, urgently needed, which it is possible to envisage in the present economic situation'. The cost would be £3½ million with another million pounds for the fourteen extra trains which would be required. Bakerloo line capacity would be increased as much as 25% by the improved reversing facilities which would be available at Camberwell Green. As part of the scheme, a new

£600,000 car depot would be built at Stanmore. In July, it was stated by London Transport that the sinking of five working shafts would begin in January 1950 and completion could be expected in 1953; the station at Camberwell Green was to have three platforms and a booking hall below the Green.

Alas, on September 29, 1950, all these brave hopes were dashed away. 'With great regret', London Transport announced, 'it has been found necessary to defer the proposed extension.' It was explained that the cost had been reassessed at £6¼ million; such a large expenditure could not be justified until more favourable conditions prevailed.

With admirable persistence, the Camberwell Council continued its campaign, and when in 1955 the BTC proposed to extend the powers until December 31, 1961, the Council unsuccessfully opposed the Bill in order to gain a hearing in Parliament. Since then, the campaigners have not relaxed, but in July 1960, the Central Transport Consultative Committee offered no encouragement, approving the London Committee's conclusion that the extension would be an 'uneconomic proposition.'

Neither the Underground group nor London Transport had ever displayed much enthusiasm for the extension *per se*; it was regarded merely as a convenient way of meeting other needs. In 1931, the extension was suggested partly because an LCC development scheme would require the complete rebuilding of the stations at Elephant & Castle and partly because the 1930 programme of Treasury-assisted works had been found to cost less than estimated and the Underground felt morally bound to find further new works in order to keep faith with the Government. In 1949, the revived interest came when the extension was seen as a means of increasing Bakerloo line capacity. Camberwell, like the rich spinster, was courted for what she could provide for other purposes rather than for her intrinsic attractions.

THE VICTORIA LINE

Perhaps the most obvious deficiency in London's electric railway network is the lack of a direct line between the populous north-eastern districts of Tottenham, Walthamstow, Chingford, Enfield and Edmonton and the West End. Increasing office development in the west central districts has underlined the need for better communication across the West End from Victoria to King's Cross. Route 'C' of the 1948 London Railway Plan met both these needs and was therefore regarded as the most important of all the postwar schemes. It was made the subject of detailed planning and included in the British Transport Commission Act of 1955. This Act sanctioned a tube railway from Victoria to Walthamstow, 11½ miles, via Finsbury Park and Tottenham, but the Minister of Transport would not at that time commit the Government to any contribution in respect of the estimated construction cost of £50 million (later revised to £56 million). The BTC Act of 1956 authorized the purchase of land for working sites, and small deviations were included in the 1957 Act. Detail planning was completed whilst MPs continually pressed the Minister for a decision. In July 1959, the London Travel Committee, appointed by the Minister, considered the line 'essential to meet present and expected future demands for travel on the Underground system. It would benefit thousands of Londoners and would enhance the value of the Underground by filling a gap in the present network.' The Committee stressed that construction 'should be started as quickly as possible'. Just over six months elapsed without any comment from the Minister and then the news leaked out in a London evening newspaper that construction of a mile of twin-tube tunnels along the route between Finsbury Park and Tottenham (Netherton Road) had begun in January 1960. This work is now completed and the purpose of it was to try out new types of shield and cheaper methods of tunnel lining, using concrete blocks expanded by keystones and, on another section, groutless cast-iron segments. One tunnel was connected to the Northern City southbound track at Finsbury Park, and rails laid in it for tests. Government approval to the necessary loan was finally given in August 1962 and construction of the whole line was then begun.

Because the new tube will draw about 70% of its traffic from existing rail and bus services, it is expected to run at a 'loss' when taken with the whole London Transport system, but balanced in the economy of London as a whole, it will indeed be a valuable asset, saving much expensive road reconstruction and easing surface congestion. If the value of its benefits were assessed in money terms as were those of the M1 motorway, they would surely blot out the 'accountants' loss' estimated at two and a half to three million pounds a year, nearly all of which would be interest charges on capital.

The main route of this tube, which London Transport christened the 'Victoria line', has already been described in Chapter 12, but as now planned, the eastern end will run below Seven Sisters Road to Page Green, Tottenham, thence to Tottenham Hale and beneath Forest Road to a point near Blackhorse Road station. From here, the final section will cut across Walthamstow to a terminal station at Hoe Street, beneath the British Railways station. At Tottenham Hale, a spur will climb to the surface and run alongside the Liverpool Street–Cambridge main line as far as Northumberland Park, where the running depot will be built. The route south of Victoria, not yet authorized, remains nebulous. There will be 12 stations: Victoria, Green Park, Oxford Circus, Warren Street, Euston, King's Cross, Highbury, Finsbury Park, Seven Sisters, Tottenham Hale, Blackhorse Road and Walthamstow Hoe Street. All but Blackhorse Road will have interchange facilities with other tubes and suburban railways, including the electrified Chingford and Enfield branches. Automatically driven trains with cab signalling, manned only by a driver, are a possibility.

Other schemes remain on the shelves gathering dust whilst street congestion in London grows worse each week. The London Passenger Transport Act, 1947, authorized station improvements at Holborn, Bank and King's Cross, and the BTC Act of 1949 sanctioned a new Piccadilly line depot at Ickenham and widening of the line between Acton Town North Junction and North Ealing, to segregate Piccadilly and District trains. The Piccadilly line scheme also included a flyover junction on land already owned at Rayners Lane and envisaged increased frequencies north of Acton Town. Apart from the modernization of Bank Central line booking hall, none of these works has yet been carried out, but at the end of 1948 the section between Uxbridge and Rayners Lane was resignalled to increase its train capacity to thirty an hour. This work involved replacing the existing three-aspect colour-light signalling by standard LT two-aspect signals; Rayners Lane–Ruislip Manor was completed on October 17, 1948, the remainder on December 12th.

RETRENCHMENT—AND RUCTIONS

It was not only new works desirable on social and traffic grounds that had to be sacrificed to the nationalization Act's requirement that London Transport should be self-supporting. The Executive now began to seek permission to reduce unprofitable railway activities. The restriction of the Aldwych branch service to Monday to Friday peak hours has already been mentioned. It was followed on October 5, 1958, by six more Sunday station closings including the tube stations at Chancery Lane, Essex Road and Fairlop. In that summer, proposals had been submitted to the London Area Transport Users' Consultative Committee to close completely the Aldwych and South Acton branches and Mornington Crescent tube station. Permission was also sought to shorten the traffic day so that first trains would arrive in central London about 6.15 a.m. instead of 5.45 a.m. (on certain lines) and leave about midnight instead of about 12.30 a.m. on all lines.

This move stimulated strong opposition, particularly to the reduced traffic day, and the Central Committee, after considering the London Committee's recommendations, advised that the Aldwych proposal be deferred until higher fares on the branch had been considered, and that the Mornington Crescent closure be deferred for the time being. As for South Acton, the London Committee were overruled and the abandonment was agreed. The objectors to the shorter hours were persuaded to entrust their case to the LCC, the City Corporation and the London Trades Council, and representatives of these three met and discussed details with London Transport. The Executive were then persuaded to withdraw the proposals but they warned that the matter would be kept under review. Some reductions were made in early morning and late-night services without affecting first and last trains, and Sunday services were reduced in March 1959. In an endeavour to increase patronage of last trains, an owl-adorned poster was produced drawing attention to the fact that the Underground runs until 12.30 a.m.

Whilst these reductions in service were not designed to increase public goodwill, they affected only a minority and passed without much fuss; but a symptom that all was not well in the goodwill department now appeared. Stay-in gestures by passengers refusing to leave trains which were reversed or terminated short of their advertised destinations owing to equipment defects or late running had occurred, as already recorded, in 1909–1914 and in 1937. A positive rash of them broke out in 1959. The demonstration is foolish as it merely increases the delay, except in the few cases where the passengers are allowed to win the game, but it captures newspaper headlines and elicits sympathy from commuters who are not directly affected. More seriously, it is an outward sign of a muddled feeling of irritation and dissatisfaction with the whole service. Outcomes of the 1959 straphangers' revolt were the appointment of 'Information Controllers' to ensure that passengers quickly learned the reasons for delays, the installation of platform loudspeakers at fifty-two more stations and a poster explaining why passengers should be prompt in leaving defective trains. As a result of these measures and the introduction of new rolling stock, the public dissatisfaction has died down; another contributory factor may well be the refreshing courtesy of the large number of West Indian uniformed staff now employed.

STATION AND LAYOUT IMPROVEMENTS

Since the war, there has been much activity in the field of station reconstruction.

For the Festival of Britain South Bank Exhibition a new booking hall was constructed for the Waterloo tube stations on the west side of York Road, and was joined to the lower landing by a bank of three new escalators opened on May 4, 1951. This entrance remained open after the Exhibition closed, was shut temporarily from October 5, 1957, during the erection of the Shell Building and reopened on May 13, 1962, with a new booking hall finished externally in marble and mosaic and internally in dove-grey tiles, plastic panelling and aluminium fittings. The waterlogged soil which had given trouble in 1927 was encountered during the construction of the escalator shaft, but it was consolidated by the Joosten chemical process. New staircases to the Northern line platforms were built from the extended lower concourse. At Charing Cross, two new escalators, for descending passengers, from the street level to the lower concourse, were opened on May 3, 1951, in time for the Festival. Construction of a new booking hall and the installation of two escalators at Euston, part of the total rebuilding of the main line and tube stations, was started in the autumn of 1961.

On the Central line, an improved station at Notting Hill Gate was mooted in 1929 and a combined Metropolitan and Central line station, with escalators, was authorized in the LPTB Act, 1937. Work began in the summer of 1938, but was stopped by the war; it was to have been resumed in 1947, but restrictions on capital investment caused postponement. An LCC road improvement scheme prompted the long-delayed completion and the major part of the new station was opened on March 1, 1959. The remainder was finished in 1960. A combined booking hall below the street had direct stair access to the Circle line, and two escalators, flanking fixed stairs, led from it to an intermediate landing. From this, a 200-foot-long passage for interchange traffic ran to the Circle line platforms and a further pair of escalators and fixed stairs linked the landing with the westbound Central line platform. A third pair of escalators, opened on July 31, 1960, connected the lower landing with the eastbound platform. The vertical rise of the top flight was 31 feet, the middle flight 37 feet $7\frac{1}{2}$ inches, the lower flight 16 feet 9 inches. All escalators had a fixed speed of 120 feet per minute except the up machine of the intermediate flight and the lowest pair, which had speedray control decelerating the escalator from 120 feet to 60 feet per minute when no passengers were being carried. Escalator panelling was satin-finished aluminium and the ceilings of the shafts and landings were lined with grey, pink and yellow pastel-coloured plastic. The tiling marked a notable break from the semi-matt biscuit tradition of the LPTB era; high-gloss light-grey tiles were used in the station tunnels, and interchange subways and landings were tiled either in rich wine-red with black relief, or sky-blue. Fluorescent lighting was used throughout, the whole effect being bright, spacious and cool. Civil engineering was undertaken by Balfour, Beatty & Co. Ltd.

At Ealing Broadway, station reconstruction to provide a combined booking hall for all lines was undertaken by BR in 1961. Rehabilitation of the war-torn Central line booking hall at Bank was also started in that year.

On the Piccadilly line, Green Park was given a third escalator on September 5, 1955, to cope with greatly increased peak-hour traffic following the wartime and postwar migration of business establishments from the City to the West End. This station now handles almost 10 million passengers a year, a 50% increase on the 1938 figure. The escalator came from the bank of three installed at Waterloo station for the Festival of Britain. Alperton also benefited from the Festival when the Dome of Discovery escalator was installed at that station on November 27, 1955, to carry passengers up the 21-foot rise from the booking hall to the eastbound platform. A 100-yard passenger subway at Leicester Square, connecting the lower landing of the Piccadilly escalators with the Northern line platforms, was opened on July 5, 1948, to afford relief to the original interchange passages around the old lower lift landing. The three top flight escalators at King's Cross have proved inadequate to handle the flow of passengers at peak periods to both Northern and Piccadilly lines, and an improvement scheme has been mentioned. Meanwhile, two of the old Piccadilly lifts were brought back into use in rush hours from September 15, 1947, for upwards traffic.

On the Bakerloo, Stonebridge Park station, partly burnt out for the second time in its short life in September 1945, was rebuilt in modern style during 1948. Carpenders Park, opened by the LNWR as a wooden halt on April 1, 1914, to serve Oxhey golf course, was rebuilt as a concrete island platform station, 450 feet long, and formally opened on September 27, 1954 (the new platform came into use on November 17, 1952). In both cases, the work and design were carried out by the London Midland Region of British Railways.

The lifts at Highbury, on the Northern City line, were modernized in 1952 with new cars and rehabilitated gear from the redundant 1906 installation at Baker Street. The old lifts at this station were the last hydraulically worked set on the Underground. One of the very deep shafts at Hampstead, Northern line, was embellished with two new high-speed lifts performing the 181-foot journey in eighteen seconds (maximum speed 800 feet per minute, or almost four times as fast as the 1907 equipment). In use on April 11, 1954, the new cars provided a more frequent service and were fully automatic. Their delicate electrical equipment was upset by the entry of dust carried on the strong winds that blow down below and the many breakdowns culminated in the imprisonment of a prominent Labour MP in a car halfway down the shaft. A subsequent debate on the Adjournment (January 31, 1957) was followed by a visit to the station by the Parliamentary Secretary to the Minister of Transport. Pairs of 500-feet-per-minute high-speed automatic lifts were installed at Oxford Circus in 1942 and at Queensway in 1956. The former installation is for ascending traffic only and takes passengers direct from the Central line to the pavement of Argyll Street. The reconditioning of the lifts at Holland Park was completed in July 1959 and the remaining Sprague electric lifts, at Lancaster Gate, were similarly treated in 1960–1961. New motors and other equipment, new doors and door-operating gear were fitted and the car interiors re-panelled.

An important track rearrangement was carried out at Wembley Park in 1953–1954 to eliminate track sharing by Metropolitan and Bakerloo trains between the station and the Stanmore branch junction. Complete segregation of the Metropolitan fast and local and the Bakerloo tracks was provided, but to make room for a new passing loop the seven roads in the car shed had to be reduced to five. Before this,

Wembley Park station had been altered to handle the Olympics traffic in 1948. An additional footbridge was built at the west end, with stairs from all six platforms, and led to an open-sided gallery parallel to the tracks. This gallery opened into a new ticket hall over the steam lines, joined to the old hall alongside. A 280-foot subway was built from the new ticket hall to Olympic Way and the Stadium. All was ready for the start of the games on July 29, 1948.

SIGNALLING IMPROVEMENTS

Speed-control signalling, introduced before the war at certain points on the Northern and Bakerloo lines, was much developed and improved. The equipment brought into service at Liverpool Street on April 11, 1948, was followed by an extensive installation on the Piccadilly line between Green Park and King's Cross (eastbound) and between Finsbury Park and Piccadilly Circus (westbound) in 1949–1950. By this method, additional home signals are installed and controlled by time-element relays. If the approaching train is travelling below the speed indicated on an illuminated sign placed before the timing section, the signals ahead clear, allowing it to slowly roll forward until it is very close to the train in front. Trains are thus enabled to draw into stations upon the heels of their predecessors, often without halting and restarting, and the cumulative effect of long station stops on the column of trains behind the busy stations is minimized to a remarkable degree.[1]

The few remaining semaphore signals controlling tube-train movements on the open sections disappeared after the war. Those west of Wood Lane went in 1946–1948 (the whole line from Wood Lane to Ealing was resignalled to LT standards from December 19, 1948.) Those at Golders Green were replaced on October 14, 1950, the section between Barons Court and Acton Town was converted in 1951–1952, the Queen's Park installation was modernized from September 22, 1953, and the last semaphores on LT running lines, those at Hanger Lane Junction on the District and Piccadilly lines, were removed on November 21, 1953. The old type of tunnel signals, with moving spectacles controlled by compressed air, had all given way to colour lights by 1946, and the original d.c. track circuits were finally ousted by a.c. in 1951. Route-setting frames with remote-control interlocking were installed at Ealing Broadway box in 1952, at Wembley Park in 1954 and at Golders Green in 1961. The latter will eventually control the signals at Edgware, Colindale, Barnet and Finchley Central.

In recent years much progress has been made towards completely automatic signalling of regular movements through junctions. The first

---

[1] Full descriptions of LT speed-control signalling and Mr R. Dell's instantaneous inductive speed detector introduced on the District line in 1955 will be found in the *Railway Gazette* of June 4, 1948, and August 12, 1955.

equipment was brought into use at the Camden Town junctions on September 18, 1955, when the forty-six-lever frame of 1924 was replaced by interlocking machines remotely controlled through the train-description apparatus, using descriptions sent out by the signalmen at Kennington, Edgware and East Finchley. Two of the new machines were located at Camden Town station and one at Mornington Crescent. As the descriptions were received in the apparatus they were stored, and later the relays caused the interlocking machines to set up and signal the appropriate route through the Camden Town maze. At converging junctions, the first train to arrive was given precedence and no unsafe or conflicting movements could be set up. This was the first step in rendering the normal operation of the 1,200 daily trains on the whole Northern line complex entirely automatic as far as signalling was concerned. The second stage involved provision of ingenious 'programme machines' based on the same principle as punched-card accounting systems. A roll of Melinex plastic passes through each of these machines once each day and on this roll are punched entries for each train to run that day. As soon as the entry for a train is in position, feelers pass through the holes, closing all the contacts necessary to set up the appropriate routeing through the interlocking machines.[1] Four of these programme machines were brought into use at Kennington on January 26, 1958, and four more at Camden Town on June 15th. Further installations were made at Euston (November 16, 1958), East Finchley (June 25, 1961), Morden and Tooting (1962). A central Regulating Room was established at Leicester Square station where supervisors watch repeaters showing the operation of the machines. In an emergency, they can take over by switching out the machines and using push-button route controls, and it is also possible, in the event of dislocation, to operate the junctions automatically on a 'first come, first served' basis, when the signals at splitting junctions will operate in accordance with destination, and signals at converging junctions in accordance with time of arrival. There are also buttons in the control room which will make a programme machine 'rub out' a cancelled train and the machines themselves can 'memorize' delayed trains whilst they pass later trains through the junctions. Interlocking machines at Archway, Clapham Common, London Bridge, King's Cross, Moorgate and Strand are remotely controlled from Leicester Square and used for reversing trains when required.

DRICO

The telephone wires in the tube tunnels were originally designed for emergency use by the driver, who could clip a telephone handset to them and speak to the substation attendant, requesting him to switch off the traction current. The Yerkes tubes had this system from the be-

[1] For a full description see *Modern Transport*, February 22, 1958.

ginning, as did the Waterloo & City, and it was installed on the CSLR about 1900. In some cases at first the telephone connected only to adjacent stations, but later all were to substations. It was found in practice that conversation on these telephones was often made difficult by extraneous noises and it was realized that it would be more flexible to permit emergency action by gatemen and guards as well as drivers. This was achieved about 1921 by arranging that mere pinching together of the telephone wires or the attachment of the handset would be the signal for the traction current to be cut off. About three years later, the speed of the operation was increased by making the contact of the two wires operate the substation cut-out and simultaneously switch on the tunnel lighting.

When train service is disrupted by an emergency, the handling of the situation (and detraining of passengers if necessary) is the responsibility of the Control Staff. Formerly, all telephone messages from Control to drivers had to be passed through the substation attendant, a process which led to delay and, more seriously, to the distortion of instructions. Detraining of passengers often took an unnecessarily long time. This undesirable state of affairs led to the development of a direct system of telephone communication between drivers and the Traffic Controller known as *Drico* (*Driver-Controller*). Worked out in practice in 1946–1947, it was introduced on the Northern line on March 9, 1952. The driver's hands are left free, as he speaks to the Controller through the train telephone microphone after clipping connections to the tunnel telephone wires. The driver's message is heard through a loudspeaker in the Control Room and the Traffic Controller's replies are passed to the driver through an additional loudspeaker in his cab. Existing procedures for cutting off current by nipping the tunnel wires were not changed [1] and the driver could still use his handset to speak to the substation attendant if required.[2] In the Ministry of Transport Inspecting Officer's Report on the Stratford accident of April 8, 1953, Colonel McMullen criticized the old system and recommended that Drico should be extended to all lines as soon as possible. The Central line installation was completed on September 1, 1954, and the system is now operating on all tube lines.

WINTER PRECAUTIONS

Many miles of open-air track were added to the tube railway system by the extensions under the 1935 New Works Programme and it became necessary to consider improvements to existing cold-weather procedures. Since 1940 sleet locomotives had been used. Built from pairs of 1903 Central London motor cars, fitted with two extra trucks carry-

---

[1] Drico connection does not cut traction current.
[2] Where the substation is remote-controlled, the connection is to the electrical control room.

ing cogwheels, wire brushes and jets to spray anti-freeze solution, these machines patrolled the open-air sections during the worst nights, in an endeavour to keep the tracks free for traffic next morning. The snag about this system was that traction current had to be kept on all night and essential maintenance was restricted. In any case, the large mileage of open track could not always be covered in time to prevent ice formation. These disadvantages were overcome by the development of a satisfactory de-icing track bath in 1941, followed by a full-scale provision of this equipment on the Stanmore branch in 1944–1945. All lines were fitted with baths for the winter of 1948–1949.

The 18-inch-long metal baths, containing ¾ gallon of anti-freeze solution, were spaced in gaps along the conductor rail at quarter- to half-mile intervals. Set in the bath was a rubber roller which painted the solution on the collector shoes of the trains, so spreading it in a film over the conductor rail ahead and preventing ice formation. At first it was thought that the baths could be switched on continuously in frosty weather and the sleet locomotives scrapped, but it was found that continuous use led to rolling-stock troubles caused by splashing of the fluid, and the cost of the solution rose steeply in the postwar years. The baths were then only switched in when weather forecasts dictated precautionary measures, and the sleet locomotives were retained for those nights when frost descended suddenly and the baths had not been switched in long enough to give protection. Other measures such as the installation of alcohol anti-freeze devices in car brake systems, drainage receptacles for the water condensing in brake pipes, and metal covers for brakes also played their part in reducing cold-weather delays. Thermostatically controlled electric and oil-circulation point heaters were also adopted on a large scale and put into use from 1942 onwards.

In 1957, two experimental sleet tenders were constructed to be propelled or hauled by empty trains. These were simply a four-wheeled truck equipped with the same brushes, rollers and sprays as the sleet locomotives.

POWER STATIONS

At the request of the Government, oil firing was adopted for four boilers at Neasden power station in 1947–1948 and two more boilers were converted later. The scheme for enlarging and modernizing Greenwich had been interrupted by the war. Two 50-c/s 20-MW turbo generators were installed in 1941, one in 1943 and the fourth in 1951. This left three 25-c/s 20-MW generators dating from 1932 and 1937; the latter was converted to 50 c/s in 1958 and one of the two 1932 sets is being converted at the time of writing. The planned connection with the national grid was made at Aldgate.

A policy of replacing substation rotary converters with modern rectifiers has been pursued steadily since the war, and the scheme included the complete rebuilding of the important Charing Cross substation in 1957.

ROLLING STOCK SINCE 1945

The immediate rolling-stock problem after the war was the provision of extra trains for the Central line extensions. Put into store pending the opening of these extensions were 340 cars of standard stock displaced from the Northern line by the 1938 stock. A few of these stored cars had been used for the 'Tube Refreshment Specials' described in the last chapter, but the bulk stood in the open throughout the war in sidings at Hainault, Edgware, Morden, Golders Green, Neasden and Ruislip, suffering severe damage from weather, hooliganism and, in some cases, shrapnel. A small maintenance staff did their best to patch up the worst wounds, but the battle was almost hopeless.

As soon as space became available at Acton Works after the cessation of war work, renovation of these cars was started. The first ones entered the works in April 1946. The cars were found to be in an appalling condition and the cost of the rehabilitation, which included complete re-wiring, was almost as much as that of repairing the 1,050 LT railway cars damaged in air raids. Continuous shrinkage of window mouldings had let in rainwater to rot the car floors and force wood off the floor plates; door runners had rusted and jammed the doors, whose rubber edges had perished to powder. Gutter pipes were choked with dirt, and rust and grime rested more than half an inch thick on all flat surfaces. Electrical equipment and door engines were rusted up and severely corroded. In the midst of all this, the trimming and upholstery remained in a fairly reasonable condition. Many of the cars had to be stripped right down to the steel panels to allow the removal of deep rust.

After the cars entered service they gave much trouble, adding to the difficulties of working the heavily loaded eastern extensions. At the opening to Woodford and Newbury Park on December 14, 1947, an effort was made to work the Central with 55 seven-car and 12 eight-car trains, but in the following month this had to be altered to a more realistic figure of 65 seven-car and 2 eight-car. In September 1949, when all the extensions were completed (except Epping–Ongar), the Central was worked with 67 seven-car, 12 eight-car, 4 four-car and 1 two-car trains (583 cars). The number of eight-car trains was later increased as mentioned below. No other tube line has trains longer than seven cars.

Fourteen additional trains were needed for the Bakerloo extension to Camberwell and service augmentation, whilst the rolling-stock requirements for the Central line exceeded the estimate, and more cars were also required for the Piccadilly line which had been speeded up in 1949–1950 by speed-control signalling and the reintroduction of weak field running on parts of the open sections. In March 1949, 91 cars were ordered from Birmingham Railway Carriage & Wagon, the plan being to equip the Bakerloo line entirely with 1938 and 1949 stock, moving the displaced standard cars to the Piccadilly and Central. The Northern City line was also to have 1949 stock and give up its standard cars to those lines. Delivered from November 16, 1951, the 1949 stock consisted of 70 'uncoupling, non-driving' motor cars and 21 trailers. Designed to replace a normal driving motor car at the coupling-up end of a 1938 three-car set, the new motor cars had simplified driving equipment built into a cabinet in the panelling at one end, for shunting purposes. Other new features were the lack of tip-up seats in the end vestibules, grooveless door tracks and mercury door interlocks (switches proving that the doors have closed properly), but, generally, these cars were very similar to the 1938 stock.

Twenty-two of the 1938-stock non-driving motor cars which had been used in the Northern line nine-car train experiment were rebuilt as uncoupling, non-driving motor cars, similar to the 1949 stock. Eighteen cars of the 1936 experimental streamlined trains were rebuilt in 1950–51 much as 1938-type trailers, and the six motor cars of the fourth (non-streamlined) 1936 train were fitted with standard PCM control gear, end brake pipes and Ward couplers, and formed into two-car shuttle trains for the Aldwych branch and the outer sections of the Central line eastern extensions. Subsequently a 1927-stock trailer was marshalled between the two motor cars, and all three sets were concentrated on the Epping–Ongar section.

With the 17 surplus trailers already in hand, this rolling-stock programme provided 18 additional seven-car trains. New cars had been ordered in appropriate quantities to rectify the lack of balance in the types of 1938 stock due to the all-motor car trains that remained from the nine-car experiment. It also released driving motor cars from existing stock by substituting the less expensive uncoupling non-driving motor cars in the three-car sets. In the nine-car trains, the guards' compartments were in the third and seventh cars of a train (in the non-driving motor cars instead of the driving motor cars). Both types of car were converted to standard in the 1949 programme.

From early in 1952 onwards, the 1949 cars and the others just mentioned were put into service on the Northern and Bakerloo lines. A transfer of 15 seven-car 1938 trains to the Piccadilly line from the Northern in 1951–1953 released standard stock to make up more eight-car trains for the Central and also allowed a beginning to be made on the scrapping of the oldest cars on that line. By February 1959 there were 52 eight-car trains on the Central, and further increases took place from February 1960 onwards as 1959 stock entered service on the Piccadilly line, releasing more standard stock. All Central trains were of eight-car length by September 1960 except the shuttle trains.[1] Another measure to alleviate Central line overcrowding was the conversion to passenger use in 1955–1956 of the driving compartments of 70 disused control trailers marshalled in standard stock trains as ordinary trailers. This provided four more seats in each of these cars.

Several interesting improvements and experiments were made to rolling stock in this period, some in preparation for new fleets of cars. 'Passenger-open' doors were introduced on the Central line on October 25, 1948, on the Bakerloo on December 19, 1949, and on the Northern line from April 18, 1950. Although it was popular with passengers as it reduced the winter heat loss inside the trains, this feature proved troublesome by complicating operation and maintenance. Abolition had been mooted for some time when its fate was dramatically settled as the by-product of a train fire near Holland Park station on July 28, 1958. The Ministry of Transport Inspecting Officer found that if, in an emergency, the guard isolated the passenger door control circuits by removing his auxiliary position switch key, this prevented the use of the train telephone, cut off the Drico and extinguished the emergency lights. Yet this was correct procedure if the guard had to leave his post. Although this wiring peculiarity applied only to standard stock that had been modified for passenger-open control, it was decided to abolish the feature from all stock after March 1959.

During hot weather, tube cars can become unpleasantly stuffy, especially in the rush hours when standing passengers block the large windows in the communicating doors at the ends. Air conditioning has always been considered too expensive, indeed a luxury in the English climate. In 1947, four fans were fitted to a 1938-stock car on the Bakerloo line. These fans were inverted, with rotor blades working in the cavity between the ceiling and the car roof, sucking air into the roof space, whence it was swept back into the car through vane control apertures. This car later ran on the Northern line, and in November 1949 a seven-car train on that line was fitted with four fans to each car, all under the control of the guard. This train, together with another introduced on the Piccadilly line in the same month, also had a further innovation. Each car was fitted with door fault-detection lights of an orange colour, operating in connection with the door-signalling circuits and enabling station staff to locate improperly closed doors without delay. The door lights and fans were removed at the end of 1952, and whilst the latter were not considered a success, door lights were judged worthy of general adoption.

In 1949, a 1938-stock driving motor car was rebuilt with windows which curved up into the roof space and gave standing passengers a

[1] In January 1962 a batch of standard-stock motor cars had to be withdrawn prematurely for scrapping, and as a result 15 trains reverted to seven-car formation.

clear view of station name signs. Door windows were also extended upwards, and to give more rigidity to the door structure, the rectangular windows each side of the doors were replaced by circular ones set in a steel panel. This additional glazing produced a pleasing car interior with a more spacious impression, but the design was not perpetuated.

Anyone who has travelled on the tube in busy periods will have noticed at once how the arrangement of the draught screens at right angles to the door openings encourages standing passengers to lean against them. This standing position is a favourite one for the short-journey passenger who thus ensures himself a quick getaway at his destination stop, but it is irksome, to put it mildly, to boarding passengers, for these 'door sentries' effectively reduce the door opening by well over one-third. Loading is slowed down and station times prolonged. To meet this problem a 1938-stock trailer was rebuilt at the end of 1953 with draught screens set back and four 'L'-shaped padded ledge seats in the space around each of the double doors. Whilst eight normal seats were lost, room was made for at least another sixteen standing passengers, none of whom could now maintain a tenable foothold in the door opening. Regrettably, this sensible rearrangement was not adopted for the large new fleet of cars ordered in 1959 and 1960, but the few 'prototype' 1960-stock motor cars for the Central line had draught screens set back at the door openings with good effect.

After successful experiments on the District and Metropolitan lines, a 1938-stock driving motor car was fitted with fluorescent lighting tubes along the centre of the ceilings and this lighting is now standard on all new cars. Experiments with rubber suspension took place on the Piccadilly line between 1953 and 1956, when various designs were fitted to bogie bolsters and axle boxes of 1938 stock.

Another modification had its origins in the Holland Park fire already mentioned and a similar fire near Redbridge in 1960. Both fires occurred in tunnel whilst the trains were in passenger service and both began in the power-cable receptacle boxes of standard-stock motor cars (these boxes take the cable which trails from overhead lines in the depots to provide current for moving the cars). After Holland Park, one of the two boxes in each pre-1938-stock motor car was taken out and in August 1960 the boxes were completely removed from all middle motor cars of standard-stock trains. Shortly afterwards, the batteries, Drico and emergency lighting equipment were taken from these cars and the switches blanked off. In 1961 train formation of standard stock on the Central line was rearranged M-T-M-T-T-M-T-M.

Between January and June 1951 a trial was given on the Piccadilly line to the idea of placing the guard in the centre of the train instead of the rear, as it was thought supervision of door operation might be easier from this position. The disadvantages were greater than the advantages and the scheme has not been pursued.

Before the main orders were placed for the new fleets of cars it was decided to gain running experience of the proposed design by ordering three prototype seven-car trains. The first was delivered in June 1957 and entered service on the Piccadilly line on September 9th. The three trains, known as 1956 stock, were built by Metropolitan-Cammell, Birmingham Railway Carriage & Wagon and Gloucester Railway Carriage & Wagon. The 1938-type front end was restyled to incorporate roller-blind destination indicators above the train door in the cab (a reversion to 1923–1925 practice) and the aluminium alloy bodies were unpainted. Internally, new features were: fluorescent lighting; stainless steel, aluminium and matt chrome fittings; a colour scheme of dove grey with maroon and grey moquette seating; and transverse seats between the main doors arranged in two bays of four seats on each side. Train formation was: driving motor-trailer-non driving motor-uncoupling driving motor-uncoupling driving motor-trailer-driving motor. The successful features of door-fault detection lights and rubber suspension were incorporated. Other features were similar to 1949 stock. All three trains were in service by April 26, 1958.

The first seven-car train of what was intended to be the new fleet for the Piccadilly line began work on that line on December 14, 1959. Delivery of the other 75 trains of this 1959 stock was completed in 1962. Bodies and trucks were by Metropolitan-Cammell, traction control equipment by AEI and traction motors by GEC. The original plan was to use all 76 trains on the Piccadilly, together with the existing 15 1938-stock and 3 1956-stock trains. The Central line was to benefit by the transfer of the later standard stock from the Piccadilly, some of which would be used to complete the eight-car train programme. Later, the Central would have its own new stock, 350 motor cars, to run with an equal number of reconditioned standard-stock trailers.

Unfortunately this plan was overtaken by events. The delivery of the three prototype trains of the new Central line stock was delayed, the reconditioning of the trailers proved very expensive, there were increasing breakdowns and troubles with the old standard stock, and the Eastern Region suburban electrifications of 1960 threatened new strains on capacity west of Liverpool Street. In April 1960 it was announced that after 19 of the 1959-stock trains had been delivered to the Piccadilly line (in August), the whole of the flow from the manufacturers would be diverted to the Central line together with 57 more Metropolitan-Cammell non-driving motor cars[1] to make up eight-car trains. The first seven-car train of 1959 stock was put into service on the Central line on April 19, 1960; the first of the additional 57 non-driving motor cars was delivered early in 1961. By asking the manufacturers to concentrate on four-car sets it was possible to run the first new eight-car train on the Central line on July 25, 1960. In the autumn of 1960 619 more cars[1] of the 1959 type were ordered from the Birmingham Railway

---

[1] Although virtually identical to 1959 stock, both these batches are designated '1962 stock'.

Carriage & Wagon Co.[1] (338 motors, 112 non-driving motors) and the Derby works of British Railways (169 trailers). When these were delivered, the first batch of 1959 stock was returned to the Piccadilly line, but the 57 additional cars remained on the Central.

The 1959 stock was similar in all main respects to the three prototype trains of 1956 stock, but detail changes included the provision of automatic couplers at both ends of each unit instead of at the inner ends only. The 1956 cars were altered to be operationally interchangeable. Each 1959 motor car had two self-ventilating 80-h.p. (continuous rating) series-wound traction motors and the well-tried PCM control equipment with two weak field notches. Instead of both weak field notches being under the overriding control of the driver's flag-switch as in the 1938 stock, one step of field shunting came in automatically, and only the second step was under optional manual control. As the first step was reached at speeds corresponding to those of the earlier stock running at full field, 1959 stock reached full field at lower speed and the maximum demand for current at starting was lower. Elimination of passenger door control permitted simplification—there was only one set of door valves on each side of the car, that much less to go wrong.

The first of three prototype trains of what was to have been the new Central line stock entered service on November 9, 1960. The bodies and bogies of the new motor cars were by Cravens Ltd, traction control equipment by AEI and traction motors by GEC. The motor cars measured 52 feet $0\frac{3}{8}$ inches over end panels and the two-motor bogies had a 6 foot 3 inch wheelbase. There were two double sliding doors each side with a 4 foot 6 inch opening and single doors at one end with an opening of 2 feet 3 inches. Like the 1959 stock, these cars had unpainted aluminium alloy bodies and roof panelling, destination blinds, rearranged transverse seats, fluorescent lighting, PCM control, electric speedometers, automatic operation of one stage of field weakening, steel underframes and bogies, rubber suspension, electro-pneumatic brakes and resilient mounting of motor noses.

Each traction motor was of 300-volt, 60-h.p. continuous rating and the two motors on each bogie were permanently connected in series. Many new features derived from the use of four motors on each car. Contactors and cables had to be of greater capacity to handle the increased maximum current, and a new type of rubber-neoprene power cabling was used experimentally. Instead of the asymmetric arrangement of single motor bogies, the bogie pivots were central and protection from single-axle spin was given by a differential voltage relay. Hand brakes acted on discs on each axle at the driving end.

In appearance, both inside and out, the cars made a refreshing break

---

[1] The Birmingham Co. was released from its contract in April 1961 and these cars were built by Metropolitan-Cammell.

from the 1938 tradition. The driver's windscreen sloped backwards and there were six large windows on each car side. Interior hinged-frame casement lighting, forming pockets for the doors, gave the effect of double glazing. Line diagrams were carried on the quadrant-type ventilator covers. At the sacrifice of two seats a car, draught screens were set back from the doorways to frustrate the 'door sentries' mentioned earlier. Each car had forty seats. A peg-board pattern ceiling helped to reduce the noise level, but introduced the atmosphere of a 'do-it-yourself' kitchen modernization.

Emergency lighting was fluorescent and fed from the battery through a transistor-inverter; the 115-volt main lighting received its current from an inductor-type motor-driven alternator (220 volts) and transformer. The 50-volt d.c. supply came from a separate transformer winding and a germanium rectifier. An improved non-handed automatic coupler was fitted, with sixty-four studs, allowing an 'A' end to couple to a 'D' end, or another 'A', unlike all other tube stock on which an 'A' end can only couple to a 'D'. If required, this feature could, for example, permit a train to uncouple into two four-car sets at Leytonstone, and the front unit could run round the Hainault loop and couple its front end to the northern end of the other set at Leytonstone or Woodford.

The prototype motor cars ran with twelve rehabilitated standard-stock trailers fitted with fluorescent lighting, rubber auxiliary side springs (instead of coiled metal) and painted in aluminium paint to match the new cars. Four trailers were taken from the 1927 MC&W batch, three from the 1931 Gloucester and five from the 1931 Birmingham.

THE INCREASING IMPORTANCE OF TUBE RAILWAYS

In 1951, for the first time in the history of co-ordinated passenger transport in London, the Underground railways showed better working results than the road services, whose loadings had been affected by the mounting street congestion caused by the increasing ownership of private cars and the marked growth of 'C' licence vehicles. More people were finding out that the railways were the fastest means of crossing or getting about the metropolis. By 1953, the ownership and use of private cars had so increased that all LT traffic was falling, but the smallest decline was on the railways. The trend towards fuller use of the Underground lines was accelerated by a bus strike in October 1954, which introduced many newcomers to the convenience of Underground travel. Early in 1955, cuts in the bus services were announced and at the same time it was added that as more people were using the Underground, slack-hour services would be increased. In fact, the new time-tables virtually restored the 1952 frequencies given on page 313. The new services began on the Central line on February 14, 1955, on the

Northern on April 25th and on the Piccadilly and Bakerloo on May 2nd.[1]

Following the new trend, London Transport became a little more rail-conscious and the old trade tag : 'UndergrounD', which had been suppressed for many years, came back in October 1955, at Sir John Elliot's instigation. A good deal of money had been spent removing the 'UndergrounD' signs at stations and replacing them with 'London Transport', but now the process was reversed. New signs appeared bearing such phrases as 'It's quicker by Tube'. Increased accommodation for parked cars was provided at suburban stations to encourage motorists to travel into town by train.

Since 1955, peak-hour traffic figures have shown a continued decrease in bus patronage and an increase in rail travel. Further stimulation was given by the Suez Crisis early in 1957 when the oil shortage caused cuts in bus services, and also by the long bus strike of May and June 1958. The strike showed the Underground lines were quite able to carry the additional load of bus passengers, and many thousands of people were borne without incident, even if in some discomfort (much of which was due to the continued running of short trains in the off-peak because the railwaymen would not alter their normal routines during the strike).

The future of London's tube railways seems bright; the first completely new tube railway to be built since 1907 is now under way; stations and rolling stock are undergoing modernization, and signalling techniques show promising development towards automatic operation.

---

[1] In 1962, Monday to Friday central area midday services were at 3 minute intervals and evening services 3½ to 6 minutes.

CHAPTER 14

The Tubes and London

THE SHAPE OF THE SYSTEM

THE shape of the London tube railway system has been influenced by geological, political and economic factors. London's geological structure, with its extensive areas of stiff blue clay subsoil upraised from the floor of an ancient sea, made deep shield-driven tunnels a cheaper and more practical proposition than the more conventional 'cut-and-cover' subway lines built in other large cities: only a few miles of tube tunnel have been constructed in those districts where the clay is thin or non-existent.

All the early promotions were by private companies, each promoter choosing what he hoped would be a lucrative route across the city, paying little or no attention to rival lines and not concerning himself in the least with the general transport needs of London. In October 1902 Lord Ribblesdale told the LCC that: 'As a rule, the questions before the Committees centre on competition and finance, and no attempt is made to consider the necessities of London as a whole. The tube railways are all parts of a puzzle and it is left to destiny to see how they can be pieced together.' Three years later, the Report of the Royal Commission on London Traffic recommended the establishment of a London Traffic Board to co-ordinate all forms of transport, but the rivalries and jealousies between Parliament and the LCC and the other local authorities effectively stifled such a logical development. For many years the public transport development of the metropolis was allowed to proceed in the familiar environment of drift and *ad hoc* action that is sometimes kindly described as the English genius for compromise. The lack of any overall plan or of even lukewarm official encouragement increased the final cost of providing the tube system, and resulted in a haphazard and inadequate group of lines. There was also much waste. When two different companies served one centre such as Oxford Circus or Bank, two distinctly separate stations were built and in later years much money had to be spent in combining the duplicated facilities. It is interesting to compare the map of the London tube railways with that of the Paris Métro, which was planned and constructed by the city authorities as an integrated whole. Fortunately for London, there were men like Yerkes, Speyer, Gibb and Stanley, men big enough to master many of the difficulties and make the best of a bad job.

But these men were capitalists whose primary interest in tube railways was to make money (not that they were very lucky in this—they soon found that the heavy initial capital investment made tube railways a well-nigh hopeless commercial proposition). In Paris and New York the tunnels were built with public money for the use of trains owned and operated by private companies. Had the same system applied in London there is little doubt that tube railways would have been constructed in the east, north-east and south-east districts, but private companies, already scorching their fingers with their existing liabilities, would not waste their substance tunnelling expensively through treacherous subsoil merely to carry workmen and clerks at cheap fares for an hour at each end of the day. Transport charity of this kind was left to the Great Eastern and the South Eastern & Chatham. It is true that some optimists tried to promote tube railways into these unfashionable areas, particularly the north-east, but financial support was not forthcoming and the schemes had bad luck in Parliament. After the First World War, another factor tended to discourage tube construction south of the Thames. Southern Railway suburban electrification proceeded rapidly in the 'twenties and early 'thirties until the whole of the elaborate network was served by multiple-unit trains. To duplicate or supplement these extensive facilities appeared extravagant, yet there existed a transport need. These Southern trains provide an excellent and fast service to the outer suburbs, but the service afforded to the inner districts compares unfavourably with that given by a tube railway. The stations are neither as attractive nor as accessible, and the routeings are complicated and often indirect; the trains are not so frequent, and the inner termini, with the single notable exception of Charing Cross, are inconveniently situated for the off-peak traveller. Small wonder that tube-spoilt North Londoners tend to undertake journeys through South London with trepidation.

The result of this interplay of geology, politics and economics can be observed on a scale map of the tube railways (page 24). No even pattern emerges: the City and West End and the western and north-western suburbs are well favoured; the north-east and south-west have but a single route each; the south-east is void.

The alignment of the tunnels themselves was directly influenced by the highly developed British regard for the sanctity of private property. The pioneer City & South London was almost entirely under the line of streets, and the 1892 Joint Select Committee effectively ensured that this principle should be followed in future construction by recommending that a free wayleave should be granted beneath public streets provided the railway company accepted an obligation to run an adequate number of cheap and convenient trains. If the line were to pass under private property the Committee recommended that the companies should be required to purchase a wayleave only and not the freehold, subject to compensation for damage. But private landowners were apprehensive of vibration and other troubles, and negotiations for wayleaves were fraught with difficulties and obstruction. So it came about that up to the end of the tube construction boom in 1907 the lines followed the streets almost everywhere, and where the highway above was narrow, the two tunnels were built one above the other to avoid the necessity for seeking a wayleave. This led to some fierce curves, notably the double reverse curves just east of South Kensington where each curve is of 5 chains radius. Nearly 30% of the Bakerloo was on curves of 15 chains or less and there was a 4·87-chain curve north of Trafalgar Square. As time went by and tube railways were seen to be innocuous to property above, easements became less difficult to obtain, and for many years the policy has been to design layouts with a minimum radius of 20 chains. The legacy of the original limitations remains in permanent speed restrictions, heavy rail wear, increased maintenance costs and higher tractive resistance.

CHANGING LONDON

In the first decade of this century, the tube railway, conventional electric railway, electric tramway and motor omnibus were established in London almost side by side. Together they stimulated important changes in the social pattern and allowed the city to expand into the sprawling mass we know today. It is not easy to apportion particular changes to the influence of tube railways alone, but the transformation of the West End during the last fifty years from a superior residential area to a prosperous shopping, business and entertainment district was probably aided more by the tubes than by any other form of transport. Before the tubes were built, the daily tidal moves of traffic had one main objective—the City of London—and most surface railways catered directly for this. The tubes put the West End on the railway so that it has grown to rival the City as a centre of employment. Holborn, the intermediate area, has been similarly transformed. The Oxford Street and Regent Street quarter of the West End has become one of the leading shopping centres of Europe, a development greatly assisted by the excellent tube services provided to those streets since 1906. Even before 1914, the damage to some of the older retail centres of the inner suburbs, such as Islington, was evident. When Gordon Selfridge opened his department store in Oxford Street in 1909 he tried to construct a subway from the basement into Bond Street station, which he modestly suggested might be renamed *Selfridges*, but the CLR, a very British concern, were quite unmoved by the fearsome drive of the American store magnate and, it seems, unimpressed by the traffic potentialities of his vast shop. Later they relented a little under the nagging pressure, establishing a booking office inside the store and issuing five-shilling all-line season tickets for use by lady shoppers during the January sales. In 1923, ticket machines in the store issued 2d or 3d tickets which could be used either on tubes or buses.

The Underground Company and London Transport made great efforts by advertising and other means to encourage cinema and theatre traffic to the West End, as it was a useful source of off-peak income. When the central area stations were rebuilt, the various street exits were labelled with the names of theatres and cinemas they served, and at the new Leicester Square station a door from one of the street exits led directly into the foyer of the London Hippodrome. Television and the greatly increased use of private cars have killed off much of this traffic, which reached its maximum about 1937.

Most of the intensive suburban growth which occurred in central Middlesex and that part of Surrey between Epsom and Merton during the years 1928–1939 was directly assisted by the extension of tube railway services into these areas. This is a subject worthy of special consideration.

LUCRATIVE FIELDS, OR PERPETUAL EXPANSION

At first tube railways were thought of as a special kind of street providing rapid transit along traditional traffic routes, linking urban catchment areas. This concept soon changed into something more complex.

When Yerkes decided to alter the terminus of the projected Hampstead tube to Golders Green, his first thought was probably that the depot land would be cheaper and easier to obtain, but it is certain that he also appreciated the traffic potential of those city-skirting fields, then lying ripe and ready for the attentions of the builder. Two estate agents were quickly alive to the opportunity offered and set up shop at Golders Green about the same time as railway construction began, one in an isolated house near the pig farm on the station site, the other in a small wooden hut opposite. Land changed hands quickly and profitably, and the first of the new houses at Golders Green was erected at the corner of Finchley Road and Hoop Lane in October 1905.[1] Tube trains arrived on June 22, 1907, and thereafter development was rapid: 73 houses in that year, 340 in 1908 and 744 in 1911.[1] Passengers using Golders Green station rose from 1·18 million in 1907 to 10·3 million in 1914 and 11·18 million in 1921, but the Underground enjoyed no share in the six and sevenfold increase in land values that its trains had directly encouraged. This story was to be repeated many times in the

[1] *The Story of Golders Green and its Remarkable Development*, by F. Howkins. Ernest Owers Ltd, 1923.

next thirty years, as the tube extensions pushed into the rural environs.

In *The Times* of March 17, 1923, there appeared the following paragraph under the heading 'The Estate Market':

> For obvious reasons, the hitherto quiet and purely rural Edgware is seeing a new and steadily rising level of values . . . the district presents plenty of opportunities of acquiring . . . freeholds which are certain to improve in value by leaps and bounds as soon as the railway extension to Edgware is in operation . . . the fact that large profits are still being made at Golders Green proves that Edgware freeholds are very attractive at their present prices. So far comparatively little retailing of land has taken place at Edgware. However, as soon as the 'tube' takes tens of thousands out every day there will be a rush for sites. . . .

Some very important transactions had in fact already taken place. The Rectory, and four acres of garden and paddock, with a frontage of 700 feet to Hale Lane facing the site of the proposed tube terminus, had come into the market in 1919. At that time it seemed difficult to believe that the tube would ever come to sleepy Edgware, though many knew that the powers had been kept alive by the Company. An old lady bought the property for about £4,000, intending to spend the rest of her life in rural peace. She was not left alone for long. The significance of Lord Ashfield's attempt to obtain Government Guarantees was appreciated, and late in 1921 she was approached with suggestions that she might sell her new acquisition at a considerable profit. She was not in need of money and did not want to leave her house, which was everything she required. Unremitting pressure continued and finally she gave in. Mr George Cross, in his autobiography, *Suffolk Punch*,[1] concludes an account of this incident as follows: 'What profit she unwillingly accepted, I do not know, but I do know that she refused to consider £20,000 from me.' By this time, Mr Cross had purchased the Edgware Manor Estate of 70 acres, with almost a mile frontage to Hale Lane, from the last house in the old village to the Deans Brook. He paid £175 an acre for this neglected farmland on October 1, 1919, and about two years later received from the LER a notice to treat for a few acres of it for lines, sidings and station. He was overjoyed—'the day I received that notice I knew that I had made a huge fortune'.[2] Although he could have sold all his land at once with an immediate profit of £35,000, he held on to it and carefully developed it with houses and shops. In his book he relates in detail how this was done and shows how the land he bought in 1919 for £12,250 brought him a net profit of nearly £57,000.

Nor was this an isolated instance. At Southgate, land selling at £160 an acre in 1923 realized ten times that amount in 1937; a small bunga-

[1] Faber & Faber, April 1939.   [2] *Ibid.*

low built at North Ilford about 1923 cost £900 with land and changed hands in 1937 for £10,500 when the site was required for shops next to the proposed Central line station. Other land in this vicinity increased in value from £20 to £180 per foot frontage. Sometimes miscalculations were made—those who purchased land at Carshalton and North Cheam in anticipation of further extensions beyond Morden were disappointed.

These were the pioneer days before town planning legislation had any teeth and before the Green Belt concept had hardened into reality. Once land was bought, the speculator was free to do anything he wanted with it. Mr Cross, in his book, writes of the pleasures of 'moulding that important slice of the suburbs of London in any way I pleased, planning the roads as I wished, naming them as I fancied'.

Economists and others have sometimes expressed the view that the cost of suburban railway extensions could be met at least in part by passing on to the railway authority some portion of the profits made from the enhanced land and property values, and this has been tried in New York and other cities. Such measures have not always been either successful or easy to administer.

The early 1920s saw the beginning of a policy of expansion, with no limits adumbrated, so enthusiastically adopted that it might well be called a policy of perpetual expansion. Lord Ashfield set the mood year after year at the Underground Group annual meetings:

'An undertaking such as ours can never stand still.' (1923)

'No one will be more happy than we shall be to embark once more upon a progressive and expansive policy, for no one can realize more acutely the danger and difficulty of a policy of stagnation and the risk of eventual decadence which must flow from it.' (1925)

'We are compelled to expand—it is impossible for us to stand still. By some means or another, London will have to supply the resources for our expansion.' (1930)

And there was much more in similar vein.

After the Board was formed, the same policy continued, with a more scientific approach. It was now stated how large a population was needed to support the extensions and the undertaking as a whole. Thus, in 1934, we find Ashfield telling the students of the London School of Economics that the economic structure of the LPTB depended on a continual process of expansion, and that a density of twelve houses to the acre was not high enough to support a tube railway. The 1936 Report of the Board called for another 600,000 people to live in the areas served by the lines proposed in the New Works Scheme, and in 1938 Frank Pick said that half to three-quarters of a million people would be needed to support a second programme of New Works.[1] Pick thought it likely that a population of roundly 12 million would ultimately be required to support the whole undertaking.

There were several reasons for this constant urge to expand, but the main one was financial. On the tube lines that were confined to the inner area the volume of traffic was far below the maximum capacity and tailed off towards each terminus; the construction of cheaply-made surface extensions provided such lines with a heavier traffic load throughout their length and secured the full benefits of the original capital expenditure at the relatively low additional cost of improved stations, signalling and rolling stock. This argument was used by Ashfield in his evidence for the Morden extension Bill of 1923 and gave Counsel for the Southern Railway a few happy moments—'It is a very curious policy . . . it is promoted in order that the tube companies may rehabilitate their finances by an excursion into the country. . . . I should have thought that a policy founded upon such a proposition was little better than a gambler's throw.' As late as 1936 the Board was still expounding the same argument, stating in its Annual Report that until the existing railways were extended into the suburbs and had secured the maximum traffic they could carry, it would be financially impossible to build additional lines.

It will be noted that London was to be shaped to meet the economic needs of the Underground and the Board; that the planning was for transport and not for London—although it was piously hoped that the needs of both might somehow coincide. This topsy-turvy thinking was a direct result of the lack of any central authority linking transport administration with the patchwork of London government.

The expansion policy could not have worked but for the population increase that Greater London experienced between the wars and, of course, Ashfield and Pick knew all about that. People were moving out from the crowded inner districts to the new suburbs, and during the industrial depression thousands came to London to work in the new factories producing light consumer goods. Between 1919 and 1939 the population of Greater London increased by over a million and the largest increase was in Middlesex, very much a 'tube county', where the inhabitants increased by one-third. By 1951, the Middlesex population was nearly double that of 1921.

The practical expression of the expansion policy was seen in the extensions to Hendon (1923), Edgware (1924), Morden (1926), Arnos Grove (1932) and Cockfosters (1933), also in the projection of tube trains over existing railways to South Harrow (1932), Hounslow (1933), Uxbridge (1933), Stanmore (1939), Barnet (1940) and Mill Hill (1941). Builders and estate agents dutifully and enthusiastically followed in

[1] Evidence to the Royal Commission on the Distribution of the Industrial Population, 1938.

support and provided many thousands of small three-bedroomed houses adorned with mock timbering and stained or leaded glass to distinguish them from Council 'dwellings'. Local authorities were also active, particularly the LCC, who erected 4,033 houses on their Watling Estate, near Burnt Oak station, between 1926 and 1931, and 9,076 at St Helier, near Morden station, between 1930 and 1936. These LCC developments brought much traffic to the Edgware and Morden extensions and enabled Ashfield to speak of the Underground's 'notable contribution to the slum problem' (at the London School of Economics, March 5, 1934).

Census returns indicate the growth of typical 'tube suburbs' such as Merton & Morden Urban District which grew from 17,532 in 1921 to 41,227 in 1931 and 74,602 in 1951, and Hendon Borough whose figures for the same years are 57,566, 115,640 and 155,835. Between 1921 and 1951, Southgate expanded from 39,525 to 73,376.

In some areas housing development took place before any railway extension had been planned. Just because transport facilities were poor, land was cheap and readily available and builders took advantage of this. When railways were eventually provided, their construction was much more expensive than it might have been. The classic example of this process is seen in the development of North Ilford between 1925 and 1935. When an extension was built, it had to be placed in tunnel to avoid property demolition.

Developers who obtained land in areas beyond the outer tube termini encouraged prospective purchasers to believe that further extensions would be made to serve their houses, even when there were no very firm grounds for such hopes. At The Woodstock, North Cheam, about 1½ miles south of Morden station on the Epsom Road, a builder's hoarding of the late 'thirties depicted a tube train, adorned with the slogan COMING THIS WAY!

CHANGING THE TRAFFIC FLOWS

Traffic on the new extensions was encouraged by elaborate bus feeder services draining the surrounding districts, and by judicious provision of road-rail season tickets and substandard fares. The resulting heavy commuter load from the outer stations even became an embarrassment in later years as the trains were filled to capacity before they reached the older suburbs such as Balham, Clapham, Finsbury Park and Hampstead.

Not all the traffic was new. Much of it was abstracted from the suburban railway services of the main-line companies, a process greatly facilitated by the road feeder services the Underground was able to provide from its own resources without interference. Even so, many people who had formerly used the older lines walked long distances to tube stations, preferring the Underground service. Perhaps the best example of this traffic abstraction was the eight-station Cockfosters extension of the Piccadilly line which was skilfully arranged by clever station siting and road transport feeding to tap some twenty-five LNER and LMSR suburban stations in addition to its 'legitimate' local traffic.

Where the extensions served the same district as the steam railways the choice was obvious. The new stations were sited on main roads, usually at traffic centres, and their wide, welcoming entrances led the passenger quickly from the street, past automatic ticket machines to an escalator. In little more than a minute, with the minimum of effort, he was on the platform, where trains were arriving so frequently that he needed no timetable, and once in the clean and attractive train, he was carried speedily to the very centre of London. In contrast, the old steam railway stations were frequently in back streets away from trams and buses. Dirty and dismal in appearance, the approaches to their platforms were often tiring, and the long wait for a train was not lightened by the gloomy surroundings. The grey and gritty trains were mostly slow and the 'London' they took one to was a bus or tube journey from the burgeoning West End. Sad indeed was the case of Bowes Park, once one of the busiest stations on the GNR suburban lines. Tucked away in a side road, it lost almost all its traffic when Bounds Green tube station was opened on the main road a few hundred yards away.[1]

BARLOW AND AFTER

About the middle of the 1930s it became apparent that 'perpetual expansion' was doomed. After many years of unofficial activity, the Green Belt idea was accepted by the authorities. In 1935 the LCC initiated a scheme and three years later the Green Belt Act was passed; the County Councils concerned then began to take concerted action to preserve 'green' land around London's fringes. The rapid and uncontrolled growth which the Underground had helped to encourage now began to cause concern. The Report of the 1937 Royal Commission on the Distribution of the Industrial Population (the 'Barlow Commission') condemned the recent growth of London as unhealthy and recommended severe limitations on new industry. Professor Abercrombie's Greater London Plan of 1944 proposed firm restrictions to curb any further growth of suburbia; the existing 'Suburban Ring' was to be

[1] This competition between the tubes and the main lines marked a hostile phase in relationships which had at times been fruitful. The tubes had at first been welcomed as a way of distributing traffic from the main-line termini, and whilst the North London Railway was somewhat put out by the Hampstead line, it is noteworthy that both the GWR and the LNWR assisted financially with Bakerloo extensions and the GER welcomed the CLR to Liverpool Street. In 1919 the Underground Group was still thinking of more 'inter-running' on the Watford pattern, but the policy changed to straightforward extensions with the promise of Government Guarantees, and competition then prevailed until the Board was formed in 1933.

surrounded by a 'Green Belt Ring' as a barrier to further expansion and the 'Outer Country Ring' beyond would receive population and industry from inner London, absorbing them into existing communities and New Towns. This was adopted as Government policy. Professor Abercrombie summed up the prewar period in the following words:

> 'Modern transport attracts people to live away from their work, where houses cost less, and thus suburban spread is encouraged; but only too often, the housing spread, arriving first, creates a demand for further transport; and in each case the wheel turns a full and vicious circle. The London Passenger Transport Board, now pioneer, now camp follower, plays a vigorous, if sometimes uncertain, role. It creates new suburbs and then finds itself unable to cope with the traffic: extensions in other directions aim at a further spread of the population. On routes over-crowded beyond cure, it asks the straphanger to exercise patience beyond limit' (*paragraph 8*).

(This last phrase is interesting. It is a twist of one already coined by Pick in 1938 when he had told the Barlow Commission that the limit of expansion for London was 'the limit of the patience of the straphanger'—a 15-mile-ride from Charing Cross.)

Other influences were at work. After the formation of the Board, there was a community of interest with the main lines and the new works announced in 1935 were mainly for projection of tube services over or alongside existing railways (two exceptions were the sensible link across North Ilford between two LNER lines which were to be used by tubes, and the more controversial extension over the Watford & Edgware route to Bushey Heath). After the war, the 1935 programme was re-examined in the light of the Greater London Plan and, in deference to the Green Belt, the Bushey Heath and Denham sections were abandoned together with the Mill Hill East–Edgware electrification; the Central line projection over the existing line to Ongar, which penetrated the Green Belt, was completed and tube trains on this line run through undeveloped countryside.

CONGESTION IN THE CENTRE

Although much thought has been given to the planning of London's growth, for a long time no effective control was exercised over the construction of new office buildings in central London. From about 1954 a large number of new blocks was erected in the City, Holborn and West End, elbowing out more and more residential and entertainment accommodation and adding more pressure to peak-hour traffic at already overburdened tube stations. Some of this building replaced war damage or old-fashioned offices, but much was new provision, and the white-collared population of London has substantially increased in recent years.[1] Both the Government and the LCC grew concerned about the new monsters in their midst, and firms were enjoined to move out to the suburbs or the Outer Country Ring. Some office building near suburban tube and other railway stations began in the early 1960s and large buildings at places such as Morden and Rayners Lane will play a useful part in relieving rush-hour congestion.

Most of the office workers are tube users and such a large concentration of new buildings in the central area is only made possible by the existence of the tube railways; were they not there the streets would quickly become choked at rush hours. The relief that tube railways afford to street congestion is a hidden benefit—only a complete stoppage of all tube services can provide a visual indication of what is normally achieved every day by these submerged traffic lanes. Private cars are a very inefficient means of mass movement; a single motorway lane used by cars passes only about 2,000 people an hour, but in the same lane buses could move up to about 19,000 if unrestricted by other traffic. High-capacity tramcars on private track can manage about 25,000 an hour, but in contrast a single-track tube railway can carry at least 36,500, the equivalent of private cars on a one-way motorway *eighteen times as wide*. The speed at which these passengers can be moved across the urban area is about 3 to 8 m.p.h. on the existing streets, but an average of over 20 m.p.h. on the tube railway.

DESIGN AND ORDER

The Metropolis has received other benefits from the Underground. London Transport's railways are noted for quiet and orderly design. Functional but pleasing, the railway, its buildings and appurtenances form a harmonious pattern that integrates the urban landscape. This result was not achieved by chance or by haphazard applications of genius, but was due entirely to the impetus of one man whose influence lives on. Inspired and guided by him, artists, architects, industrial designers and typographers gave of their best. His efforts came to a climax in the new works of the 'thirties and by 1940 the familiar environment was completed. No detail was forgotten and even such accessories as door handles and clocks received attention.

Frank Pick (1878–1941), who was responsible for this great work for London, entered the service of the North Eastern Railway in 1902, after qualifying as a solicitor, and four years later came to the Under-

[1] The central London office population is now about 650,000. New office space approved by the LCC between 1948 and 1958 was estimated to provide employment for nearly a quarter of a million people in the central area alone, an increase of 25,000 a year. Some 21½ million square feet of new office building was completed in central London between 1945 and the middle of 1959. Reasons for this expansion are considered in paragraphs 335–341 of the Report of the Royal Commission on Local Government in Greater London, 1957–1960.

ground Company with Sir George Gibb, his general manager. At first working on traffic development and publicity, he became commercial manager in 1912 and during these early years arranged for Underground posters to be designed by leading artists such as Frank Taylor and Gregory Brown, whilst not neglecting younger men like McKnight Kauffer and Austin Cooper. In his concern for the appearance of Underground publicity, he became dissatisfied with the Grotesque type face then in use for posters and notices and sought a completely new design evolved specially for the Underground Company. This need was admirably filled by Edward Johnston, who in the summer of 1916 completed a wonderfully legible and simple display type of symmetrical and balanced form, based on the square and circle and the classical Roman capital.[1] This Johnston type was quickly adopted for all signs, notices and posters, and remains in use today as modern and fresh as ever. Delicate and distinctive, clear and immediate in impact, it is now the very hallmark of London Transport. Johnston also redesigned the familiar bull's-eye device used for station names and the original solid red disc crossed by a horizontal label became a more balanced ring and label.

Stations offered a great challenge. The original ones designed by Green fulfilled their purpose in that they stood out from surrounding property, easily recognizable, and were simple to erect and maintain. But their *art nouveau* and sombre dark rhubarb elevations soon palled. Next in sequence were S. A. Heaps' Edgware extension stations, in pleasant, if unoriginal neo-Palladian style, which settled down comfortably enough among the false-timbered villas and tree-fringed streets of the new suburbs. For the Morden extension, Pick, who had been appointed assistant managing director in 1921, determined to find an entirely new style. He engaged Dr Charles H. Holden of Adams, Holden & Pearson, an architect he had met on the committee of the Design and Industries Association in 1915. Holden's first tube stations were boldly modernist, and the strident angular forms with their bare white Portland stone were ideal for floodlighting, if a little garish and uncompromising in natural light.

In the summer of 1930, two years after he had become managing director, Pick visited north-west Europe with Holden and they were impressed by the simple but effective brick buildings of the new architects. Soon after their return, Holden produced a design for the new Sudbury Town station. This structure was completed in July 1931 and was a triumph. Of clean, undemonstrative and functional style, it set the fashion for a classic series of stations, some by Dr Holden, his pupils and associates, others by R. H. Uren, S. A. Heaps, L. A. Bucknell and James & Bywaters & Pearce. The patterns and forms were similar

[1] See *Edward Johnston*, by Priscilla Johnston. Faber & Faber, 1959, p. 198, et seq.

throughout, but the completed buildings were each distinctive, with variations suggested by site conditions or the architect's inspiration. Detail items such as lamps, clocks, doors, seats, etc., were to designs suggested by Holden himself. Perhaps the most successful of this great series were Arnos Grove (Holden, 1932), Oakwood (Holden, with James & Bywaters & Pearce, 1933) and Rayners Lane (Holden and Uren, 1938). Lapses from the near perfection of this trio were Boston Manor and Osterley, where ugly towers shout against the inherent quietness of the general style, and Leicester Square and Warren Street, where the office blocks and flats overhead are somewhat dreary. Some stations of lesser importance such as Sudbury Hill, South Harrow and Bounds Green seem a little uninspired. The sweeping majesty of Wood Green's curved front is spoilt by unsuccessful treatment of ventilation outlets. Almost everywhere, the concrete work has become stained and dirty with age, marring the original beauty of the design.

The general pattern of orderly grace was extended to staff quarters, signal cabin interiors and other places beyond the public eye. Passenger flows were studied, wall corners rounded and lighting skilfully arranged to blend with the smooth, good-natured interiors. Ease of movement, both real and apparent, from street to train, was the subject of special attention. After it was all done, Dr Holden described the designing of the London Transport railway structures as one of the greatest pleasures he had known. All those who use these buildings owe much to Frank Pick and his architects, for whether they perceive it or not, these fine stations make their daily journeys easier and more pleasant than they might otherwise be. Dr Holden, who designed many other large public buildings, including the London University Senate House (1937), was awarded the Royal Gold Medal of the Royal Institute of British Architects in 1936. He died in 1960, aged 85.

Tinges of the Holden influence can be seen in the stations of the postwar Central line extensions, though they were the work of other architects. Without doubt the best of this group is Loughton, designed for the LNER by J. Murray Easton. The barrel vault theme of the booking hall and the sweeping concrete canopies on the platforms are notable features—it is curious that the LNER instructed Easton to make it clearly distinguishable from contemporary LT stations. Greenford and Perivale (Brian Lewis) and the other GWR western extension stations (Dr F. F. Curtis) are pleasing, but show little originality. Much reinforced concrete was used in an endeavour to reduce maintenance costs, but again this material has aged badly.

THE LONDONER'S TUBE

Londoners look upon the tube as a commonplace. For most of the time it works so efficiently that it is taken for granted. The word itself is synonymous with quick, reliable transit and to live 'on the tube' is regarded as the acme of convenience. In the prewar speculative building boom an estate agent described his area as *Southgate-on-the-Tube* and to add 'near tube' to a small ad. is a quick way of disposing of a flat.

Unlike the trams and buses, the tubes were not constantly in the public eye, and in the early years they were not well patronized. For cheapness, many stations had been erected on second-best sites away from the main traffic streams (Down Street, Russell Square) and we have seen how these disabilities were counteracted with astute publicity after the arrival of Stanley. But it was not until the late 'twenties and early 'thirties, after the construction of extensions and modern stations, that the tube habit became widespread. As long as access to platforms was by the original lifts, street to train times were so prolonged that the tubes were not competitive for short distances; at unrebuilt stations of average depth, an active person can often reach the platform most quickly by running down the emergency staircase. As the stations were reconstructed and as street congestion increased, the knowledgeable Londoner found that for a journey involving two or more stations, the tube was usually faster than surface transport—even taxis can be beaten at most times of the day, although it is fair to say that, for those who can use one, a bicycle is probably the quickest means of covering short distances at rush hours. Quite early the democratic nature of tube travel became evident.[1] A station like Covent Garden is used alike by expensively adorned theatre-goers and hard-swearing market porters, and a tube car usually contains a good cross-section of London types. In the summer, the tubes are full of provincial and foreign visitors who, with map in hand, and one eye cocked for the excellent direction signs, are confident in a strange city. Visitors' comments are usually very favourable, for whilst the London tube railways have not the twenty-four-hour service of the New York Subway, the romance and ubiquity of the Paris Métro or the opulence and splendour of the Moscow system, they are more comfortable and cosy than the clinically austere American trains, much faster than the trundling old ladies of Paris and less forbidding, if less tidy, than Russia's monumental Metros.

The word 'tube' comes easily to the tongue, and for years the London Press, which, like the public, takes the system for granted until there is a few minutes' delay, has misused it; careless sub-editors apply the word to both the deep-level *and* the Metropolitan, District and Circle lines, which are not tubes at all. TUBE CHAOS makes a neat headline, and is to be found above lurid descriptions of minor incidents whether they be at Holborn or Whitechapel, Colindale or Croxley. It

---

[1] 'One thing that struck me from the start was the way in which the new (Central) line was patronized by both businessmen and the ordinary working folk. It was something new for them to share the same means of transport.'— John Burrows, retired stationmaster, Holborn, in *London Transport Magazine*, August 1960.

is truly a great compliment to the general efficiency of the London tubes that the Press makes so much fuss when something quite unsensational occurs.

Today these deep-level railways with their surface extensions form a vital part of the life of London: without them the central area would be immobilized, and the orderly daily comings and goings of the millions who maintain London's greatness would be utterly disrupted. In the future, new tubes in the centre could help defeat the ever-increasing surface congestion, whilst fast and safe access to suburbs and environs at present poorly served could be obtained by using the central reservations of new motorways as cheaply built extensions of tube lines.

We have seen how the tube traffic has grown as more cars clog the streets above. In North America, where great cities have almost died of car thrombosis (or smothered themselves with concrete in an effort to allay it), it is increasingly realized that urban electric railways which run through the heart of a city are essential if the city is to remain integrated and street congestion kept to a manageable level. Public transport in the streets is now so hampered by other traffic that it has lost much of its former importance, and this was recognized by Sir John Elliot in a statement made to the press on June 1, 1958:

'With the congestion in the streets, the heyday of the bus has gone. We must get people underground. It is our hope that some of the traffic which has gone to the Underground will stay there. Our policy is to get more and more people underground....'

This was during the bus strike, when the railways were able to handle all the diverted traffic, and many more people found the usefulness of the tubes and the subsurface lines for the first time. In a quieter and less strained atmosphere some nine months earlier, Sir John had spoken to the American Transit Association in Montreal on the same theme:

'During the last few years, when street congestion has been growing rapidly, much passenger traffic has been transferred from the road services to the Underground. This experience bears out the conviction of London Transport officers that for the modern city, as much traffic as possible must be taken underground. It seems that the underground railways will be able to hold their own against road competition in the years to come, and this is confirmed by experience in America and Europe—including Russia'.

A finer tribute to the wisdom and foresight of Pearson, Barlow, Mott, Greathead, Yerkes, Ashfield and the many others in this history could not be devised.

APPENDIX 1

Dates of Opening to Public Traffic

1. *Northern Line*
   City & South London Railway
Stockwell to King William Street	18.12.1890
Borough to Moorgate	25.2.1900
Stockwell to Clapham Common	3.6.1900
Moorgate to Angel	17.11.1901
Angel to Euston	12.5.1907
Euston to Camden Town (legally part of the LER)	20.4.1924
Clapham Common to Morden	13.9.1926

   Hampstead Line
Strand to Golders Green with branch from Camden Town to Archway	22.6.1907
Strand to Charing Cross	6.4.1914
Golders Green to Hendon	19.11.1923
Hendon to Edgware	18.8.1924
Charing Cross to Kennington	13.9.1926
Archway to East Finchley	3.7.1939
East Finchley to High Barnet	14.4.1940
Finchley Central to Mill Hill East	18.5.1941

   Northern City Line
Moorgate to Finsbury Park	14.2.1904

2. *Bakerloo Line*
Lambeth North to Baker Street	10.3.1906
Lambeth North to Elephant & Castle	5.8.1906
Baker Street to Marylebone	27.3.1907
Marylebone to Edgware Road	15.6.1907
Edgware Road to Paddington	1.12.1913
Paddington to Kilburn Park	31.1.1915
Kilburn Park to Queen's Park	11.2.1915
Queen's Park to Willesden Junction	10.5.1915
Willesden Junction to Watford Junction	16.4.1917
Baker Street to Stanmore	20.11.1939

3. *Piccadilly Line*
Hammersmith to Finsbury Park	15.12.1906
Holborn to Aldwych	30.11.1907
Hammersmith to South Harrow	4.7.1932
Finsbury Park to Arnos Grove	19.9.1932
Acton Town to Northfields	9.1.1933

Northfields to Hounslow West	13.3.1933
Arnos Grove to Oakwood	13.3.1933
Oakwood to Cockfosters	31.7.1933
South Harrow to Uxbridge	23.10.1933

4. *Central Line*

Shepherd's Bush to Bank	30.7.1900
Shepherd's Bush to Wood Lane	14.5.1908
Bank to Liverpool Street	28.7.1912
Wood Lane to Ealing Broadway	3.8.1920
Liverpool Street to Stratford	4.12.1946
Stratford to Leytonstone	5.5.1947
North Acton to Greenford	30.6.1947
Leytonstone to Woodford (direct)	14.12.1947
Leytonstone to Newbury Park	14.12.1947
Newbury Park to Hainault	31.5.1948
Greenford to West Ruislip	21.11.1948
Hainault to Woodford (via Grange Hill)	21.11.1948
Woodford to Loughton	21.11.1948
Loughton to Epping	25.9.1949
Epping to Ongar	18.11.1957

5. *Waterloo & City Line*

Waterloo to Bank	8.8.1898

[N.B. The station names are those in current use except for Wood Lane (now closed).]

APPENDIX 2

Accidents and Interruptions of Service

THIS list is not comprehensive; to restrict it to reasonable length, breakdowns of signals, track and rolling stock have been omitted (none of these has resulted in loss of life or serious injury). All important accidents are noted, as are many of the more interesting minor incidents. Enemy action and other wartime interruptions have been dealt with in the main text, together with the more important disruptions caused by engineering works.

The notes on the accidents are derived from the conclusions of the Government's Inspecting Officers where these were published.

* denotes an incident fatal to staff or passengers.

SECTION I: COLLISIONS

26.9.1900
End-on collision between London Bridge and Bank. The rear train passed the Bank up outer home signal at danger.

4.9.1912
End-on collision at Caledonian Road, 7.27 p.m. A train detained in the station by an electrical fault was run into at rear by the following train, which was due to pass Caledonian Road without stopping. The driver of the rear train passed a signal at danger and the train stop failed to operate properly owing to defective maintenance. Twenty-three passengers and six employees injured.

30.9.1913
End-on collision at Shepherd's Bush, 7.5 a.m. The signalman used the releasing 'spring' to counter the effect of the treadle, alleging that seven treadle failures had occurred in the previous week. Both inner and outer home signals were showing green behind the stationary train which was hit. Twelve passengers and two employees slightly injured.

25.2.1918
End-on collision between Maida Vale and Warwick Avenue due to a failure to observe the 'stop and proceed' rule when signals fail to clear. Accident occurred at 10.20 a.m. Two trains had stopped owing to fusing. There was some doubt as to whether the tail lamp on the first train was lit. Six passengers and three employees injured.

11.7.1922
Buffer stop collision at Ealing Broadway, owing to faulty manipulation of brakes. Train was being driven by a guard under supervision.

19.1.1925
End-on collision in the terminal loop between Strand and Charing Cross, 8.35 a.m. The rear train had passed a signal at danger under authority and was told to proceed normally on the assumption that the section ahead had been cleared. One passenger and four employees injured.

10.6.1932
Sideways-on collision at Hammersmith, Piccadilly line. 9.50 a.m. Driver misread signal and left a siding for the westbound main line; his train was halted by a train stop, but he descended, reset the tripcock and proceeded, colliding with a westbound service train at the points. No casualties.

10.3.1938
End-on collision on northbound track between Waterloo and Charing Cross, Northern line, 8.32 a.m. Trains were moving with caution whilst a signal fault was being repaired. The lineman rewired the signals incorrectly, shortening the clearing distance of a signal. Twelve passengers received minor injuries or shock. Long delays in detraining (the last passengers were not out until 11.57 a.m.) were criticized by the Inspecting Officer.

1.1.1940
* End-on collision between a Piccadilly line and a Metropolitan line train near Hillingdon early in the morning. Driver of the tube train was killed.

27.7.1946
Buffer stop collision at Edgware, 9.52 p.m. Driver collapsed and died from natural causes as his train was entering the terminus. The deadman's handle failed to operate as it was cut out by the driver's use of the reversing key (probably involuntary). Eight passengers suffered minor injuries or shock.

5.12.1946
* End-on collision of empty trains east of Stratford, 10.38 a.m. Trains were moving under the 'stop and proceed' rule after a signal failure. One employee killed and three seriously injured.

1.12.48
End-on collision in dense fog. Metropolitan train passed signal at danger between Rayners Lane and Eastcote, after being tripped, and collided with rear of Piccadilly train. No casualties.

19.8.1949
* End-on collision between empty trains in Queen's Park depot. Two drivers killed, one seriously injured.

2.11.1949
End-on collision in fog, a.m., at South Ealing.

23.7.1951
End-on collision, late evening, between Mornington Crescent and Euston. Service not normal until following day.

8.4.1953
* *First major tube railway accident*: End-on collision in tunnel east of Stratford, 6.56 p.m. Trains were moving under the 'stop and proceed' rule after a signal failure. The collision took place at about 20 m.p.h. on a descending gradient of 1 in 30 and a curve of 20 chains radius. Damage to both trains was extensive and far exceeded that of any previous tube accident. The Inspecting Officer concluded that the driver of the rear train 'failed to exercise the caution which is explicitly required when applying the "stop and proceed" rule'. Twelve passengers killed, forty-five passengers and one employee injured. Detraining was not completed

APPENDIX 2 367

until 9.50 p.m., and normal working not resumed until 10 a.m. on April 10, 1953.

3.8.1955
Buffer stop collision at Aldwych, 10.54 a.m. Five passengers and employees injured.

6.10.1960
Buffer stop collision in siding tunnel at Tooting Broadway. Driver trapped in cab and seriously injured.

16.10.1962
A southbound Bakerloo train run into at rear by an LMR up electric train between Watford Junction and Watford High Street. Three passengers admitted to hospital.

SECTION 2: ACCIDENTS DUE TO EXTERNAL CAUSES

27.11.1923
Northbound train struck poling boards projecting from the tunnel roof between Elephant & Castle and Borough stations, 5.12 p.m. (see page 175).

15.12.1930
Burst water main flooded Tottenham Court Road stations and affected Oxford Circus and Bond Street. No Central London service all morning, and trains did not run through Tottenham Court Road until 8 p.m. Hampstead line service interrupted for 1½ hours.

22.11.1934
Rayners Lane signal box demolished by runaway ballast train.

4.8.1935
Auxiliary Air Force 'plane crashed across the line near Colindale and caused a short-circuit fire in the signal box, burning it down. Service north of Golders Green suspended for several hours.

6.1.1939
Train struck a piece of timber near Chancery Lane, 9.30 a.m. Track and signals damaged.

13.4.1948
A steam locomotive fell down the wagon hoist at Waterloo on to the Waterloo & City Railway siding below. Line closed 11.40 a.m.–1.29 p.m. The locomotive was removed piecemeal.

20.4.1951
Three bicycles dropped on to the track between Leytonstone and Snaresbrook in the evening. Short-circuit delayed trains for half an hour. Signalling also damaged.

SECTION 3: MOVEMENT ACCIDENTS

JUNE 1892
* Passenger fell from a CSLR train in tunnel and was killed.

1.7.1907
* Guard killed at Euston, Hampstead line. He hit the tunnel portal whilst trying to change the destination indicator of the moving train.

26.8.1916
* Train at City Road started by guard before all passengers had alighted. One passenger killed.

19.11.1927
* A porter tried to close the gate of a moving train at Piccadilly Circus and was carried to the tunnel portal and fatally injured.

18.3.1948
* Guard fell from a westbound train between Liverpool Street and Bank and was killed.
21.9.1948
* A man trapped his arm in the doors of a train at Lancaster Gate after trying to force them open. He was dragged to the tunnel portal and killed.
10.10.1950
* A trespasser on the line between Highgate and East Finchley was killed by a train, the first car of which was derailed.
8.4.1958
* The guard of a westbound Piccadilly line train was killed after falling from it when leaving King's Cross.

SECTION 4: POWER FAILURES
(other than those caused by fires (see Section 5) or strikes (see Section 6))

13.3.1902
CSLR power failure, some passengers detrained.
3.10.1908
First power failure at Lots Road, 2.45 p.m. Short-circuit in a transformer, no current for two hours.
15.1.1911
Power failure at Lots Road 9.33–10.14 a.m. Burst in main steam pipe.
15.7.1912
Power failure at Lots Road at 8.15 p.m.
21.1.1919
Reduced power at Lots Road a.m. delayed trains.
28.8.1919
Power failure at Lots Road 4.30–5 p.m.
3.8.1920
Flooding caused by heavy rain reduced power at Lots Road 8 a.m.–noon.
JUNE–JULY 1926
Shortage of coal owing to coal strike reduced power and services by 25–30%. Some stations closed.
NOVEMBER–DECEMBER 1926
Power shortage at Lots Road caused by poor quality imported coal (British miners on strike). Reduction in train services.
7.7.1927
Fusing of h.t. cable between Lots Road and Wood Lane delayed trains on Piccadilly and Hampstead lines 3.21–3.52 p.m.
18.8.1927
Cable failure at Lots Road 8.52–9.26 a.m. All trains slowed, some stopped.
7.1.1928
Heavy rain flooded Lots Road, partial service only until midday.
26.7.1928
Power failure at Lots Road 11.33–11.48 a.m. owing to a switch defect.
30.8.1928
A fault at Tottenham Court Road caused a power failure on the Central line.

11.12.1929
Power failure at Lots Road owing to a mechanical breakdown. 7.12–7.27 a.m. Service not normal till 11 a.m.

14.7.1930
Generator failure at Lots Road 9.55–10.3 a.m. and a few minutes' interruption shortly afterwards.

22.1.1931
Two generators failed at Lots Road 6.41–7.18 p.m.

23.8.1937
Power failure at Stonebridge Park. No service north of Queen's Park 9.10 a.m.–midday.

15.3.1938
Power cable from Charing Cross substation earthed near Lambeth North. Bakerloo and Northern lines stopped 7.5–10.30 a.m. Normal service 2 p.m. Lambeth North station closed till evening.

24.10.1942
Power failure at Stonebridge Park a.m. affected service north of Queen's Park.

12–13.2.1947
National fuel shortage caused severe cuts in power and service. Minor reductions in service continued for some time.

8.2–21.3.1951
Power and services reduced to save coal.

15.5.1951
Failure at Mile End substation midday stopped all trains east of Liverpool Street.

18.12.1951
Defect on a train at Manor House set up a current surge which affected trains on the Piccadilly, Northern and Bakerloo lines 6–8.15 p.m.

7–8.12.1953
Power failure at Neasden 9.27 p.m. affected Stanmore branch and South Harrow–Uxbridge services. Power supplied from Lots Road from 10 p.m.

27.11.1955
Transformer fault, caused by a current surge, at Lots Road 4.5 p.m. Power cut, including all telephones, for almost an hour, but some trains moved after 20 minutes.

10.4.1956
Electrical fault at Lots Road. Power cut for 12 minutes.

SECTION 5: FIRES

16.1.1902
Locomotive fire between Elephant & Castle and Borough just after 7 a.m. Passengers detrained and service suspended for about an hour.

5.9.1902
Fire at Stockwell power station, cutting supply. All passengers detrained.

6.4.1905
Substation fire at Notting Hill Gate 5.45 p.m. Current off whole line for 1½ hours.

24.5.1908
Locomotive destroyed by internal fire at Clapham Common.

16.7.1908
Track caught fire at Moorgate Street crossover.

9.1.1917
Stonebridge Park station burned down.

7.11.1920
Serious arcing on a train between Piccadilly Circus and Oxford Circus 9.44 p.m. Consequent current surge caused arcing on a train at South Kensington one minute later.

10.4.1924
Sleeper fire between Chancery Lane and Post Office due to shoe arcing. Some stations closed.

19.1.1926
Cable fire affected signals between Camden Town and Golders Green. Service suspended during morning rush hour.

24.12.1944
Fire in escalator shaft at Paddington. Station closed until December 27, 1944, for outgoing passengers and until January 1, 1945, for incoming passengers.

SEPTEMBER 1945
Stonebridge Park station burned down.

24.8.1948
Eastbound Piccadilly line train fused and caught fire on South Harrow viaduct. Delay of one hour.

27.4.1951
Fusing on train near Liverpool Street. Eighteen passengers taken to hospital suffering from effect of smoke and fumes, three detained.

17.9.1955
Explosion and fire at Elephant & Castle substation.

21–22.5.1956
Serious fire in Goodge Street deep-level shelter (in use as army transit camp). Alarm given 9.45 p.m., May 21, 1956, fire not extinguished for twenty-four hours. Station closed, but trains continued to run through.

26.12.1956
Cable fire below signal box at Marble Arch about 11 p.m. Service suspended between Holborn and Queensway until 7.30 a.m., December 27, 1956.

28.7.1958
* Train fire (originating in cable receptacle box) 7.14 a.m., near Holland Park station, eastbound. One passenger died later from the effect of fumes, forty-eight passengers and three employees treated in hospital.

21.8.1958
Cable fire on platform at Notting Hill Gate (the cable was accidentally punctured by a workman's drill). Station closed for some hours and lifts and escalators at some other stations affected.

11.8.1960
Train fire (originating in cable receptacle box) 8.20 a.m., near Redbridge station, eastbound. Forty-one admitted to hospital. Normal service 12.30 p.m.
24.11.1960
Arcing caused by a fretted shoe lead on northbound Northern line train at Tottenham Court Road in the morning rush hour.
24.11.1960
Arcing caused by a loose side coupling chain on westbound Central line train touching conductor rail at Bank in the morning rush hour.

SECTION 6: STRIKES AND LABOUR TROUBLES

17–19.8.1911
Some reduction in service during the national railway strike.
19–22.8.1918
Partial strike of women, seeking war bonus paid to men. Normal service August 23, 1918.
24–27.8.1918
Strike for equal pay for women. Service partially maintained by closing some stations. Normal service August 28, 1918.
3–9.2.1919
Strike of drivers over the omission of meal times from working hours in the National Eight Hour Day Agreement. Lots Road men joined the strike from February 4th, no service on February 3rd except on the Hampstead line. Partial service on all lines by the evening of February 9th.
5–7.8.1919
CSLR men on strike in sympathy with Metropolitan Police strike. Service normal August 8, 1919.
27.9–6.10.1919
National Railway Strike. No service until September 30, 1919, when volunteers worked a partial service. Staff resumed work during October 6, 1919.
29.3.1924
Services disrupted by partial strike in sympathy with Tram strike.
5–12.6.1924
Unofficial strike at Lots Road from midnight June 4–5, 1924. On the morning of June 5th restricted services and many stations closed. Some uniformed staff joined the strike on June 6th and CLR was closed entirely on that day. Men returned to work from 11 p.m., June 12th and all closed stations reopened the following morning except South Kentish Town and Mornington Crescent. Services normal on June 14th but Mornington Crescent not reopened until July 2nd and South Kentish Town never reopened.
4–14.5.1926
General Strike. From May 4, 1926, a skeleton service was worked on the CLR with volunteers. Similar services on the other lines from May 6th and 7th. Finally, about one-third normal service ran. Normal working resumed May 15th.

4–10.2.1937
Drivers on Morden–Edgware and Bakerloo lines indulged in 'go-slow' tactics in protest against speed restriction indications.

1.3.1937
Drivers on the Morden–Edgware line indulged in 'go-slow' tactics in protest against reduced running times.

25.12.1944
Strike for two days' holiday in lieu of Christmas Day. Almost complete stoppage on all lines.

9.6.1958
Unofficial strike in support of bus strike. 10% cut in train service.

1.2.1960
Unofficial 24-hour strike. About 18% of normal service maintained by men remaining at work.

11.7.–20.7.1960
Unofficial strike of power station workers. Some disruption of rush-hour service on July 11, 1960, as emergency staff coped with mechanical troubles at Lots Road. Otherwise almost the whole normal service maintained on all lines.

29.1.1962
Unofficial 24-hour strike. Partial service maintained in central area by men remaining at work. Severe dislocation of road traffic.

3.10.1962
One day official strike by National Union of Railwaymen. Occasional trains ran on Northern and Central lines. Road congestion avoided as about 75% of normal rush-hour travellers stayed at home or slept in London overnight.

APPENDIX 3

Some Notes on Tube Tickets and Ticket Issuing Machines

written in collaboration with W. H. Bett

OF late there has grown up a practice of including in any railway history at least a few brief particulars regarding the tickets issued by the line or lines described. This development is to be commended, for apart from the specialized interest which attaches to the study of tickets in themselves, they form a valuable record of many varied aspects of the lines concerned. Railway history and geography, details of the services worked, and such features as the opening, closing and re-naming of the lines and stations, can all be found faithfully reflected in the tickets, as well as particulars more directly concerned with fares and charges. Each ticket, moreover, is a permanent record of the railway itself in action, for it represents a particular journey on a particular date, its very function being to set out the salient details in a handy, 'potted' form for checking at the time. If it survives, the record remains for all time, for the subsequent instruction of the railway student and historian.

In the case of tube railways there is perhaps some justification for treating this matter more fully than usual, for there were many specialized problems and practices arising in connection with tube fares and tickets which were closely bound up with the particular type of service the tubes were developed to give—intensive service over comparatively short distances. Fare and ticket practice thus differed in many interesting ways from the more standardized methods of the main-line railways, and broadly speaking we may recognize three successive periods:

(1) An *initial* period, from the first tube openings in the 'nineties up to the early years of the twentieth century, when fares and tickets tended to be of extreme simplicity. The tubes were then regarded as a separate, purely local facility, having little or nothing to do with the main lines, and usually copied American local transit practice in having a 'flat' fare irrespective of distance.

(2) A *middle* period of maximum complexity in which the multiplication of tube railways, introduction of through bookings between them and decay of the 'flat' fare led to a departure from the simplicity of the original tariffs and booking arrangements and the substitution of a medley of local and through fare rates, which was as yet unrelieved by mechanical aids and standardized devices. This phase covered the years from about 1906 to the World War I period and after.

(3) A *late* period during which matters were re-simplified, not by restricting the variety of bookings, but by the adoption of new and improved methods of handling them, notably the 'Scheme' ticket, and the

use of new rapid-acting machines and devices for ticket issue. Though the Bakerloo line had Scheme tickets as early as 1911, these developments really belong to the post-war period, mechanization proceeding rapidly from about 1921 onwards.

With this general picture in mind, we may proceed to consider, first the ticket histories of the tubes, line by line, as regards their early, individualistic days, and then the various specialized subjects which concerned all lines in later days when things had become more standardized.

CITY & SOUTH LONDON RAILWAY. This pioneer tube when first opened had a flat fare of 2d, collected at turnstiles without the use of tickets for single fares. However, in this era, if not at the very first, there were also return tickets of 4d for which a single-coupon Edmondson ticket was used, collected at the turnstile in lieu of cash on the return journey. The northward and southward extensions to Euston and Clapham Common respectively involved the abolition of the flat fare and thereafter Edmondson tickets were issued for single journeys. These were coloured in accordance with a code denoting destination station, to aid collectors. The code (which presents obvious signs of an increasing shortage of distinctive colours as the extensions proceeded) was as follows:

CLAPHAM COMMON—Pink.
CLAPHAM ROAD—Pink, diagonal black line.
STOCKWELL—Purple.
OVAL—Green.
KENNINGTON—Blue.
ELEPHANT & CASTLE—Primrose.
BOROUGH—Reddish-brown.
LONDON BRIDGE—Pale blue.
BANK—Buff, almost olive.
MOORGATE STREET—Orange.
OLD STREET—Lilac.
CITY ROAD—Red.
ANGEL—White.
KING'S CROSS & ST PANCRAS—White with red St Andrew's cross.
EUSTON—White with red 'U'.

Return tickets were also issued and were parti-coloured with the respective halves in the above colours according to destination.

While the CSLR was closed for reconstruction in the 'twenties, the tickets issued for the replacing bus service were in the same design as those of the London General Omnibus Co., values 1d–6d, but headed 'City and South London Railway, Auxiliary Omnibus Service', and there was also an Exchange Ticket, issued upon collection of through tickets from the other tubes to CSLR destinations.

WATERLOO & CITY RAILWAY. This line never had any intermediate stations, and the fare for many years was 2d, increased in recent times. Until reconstruction by the Southern Railway in 1940, fares were collected on the train and bell-punch tickets issued successively headed 'Waterloo & City Railway', 'London & South Western Railway' and 'Southern Railway'. A limited number of through bookings existed, to various northern main-line termini, and also to Finsbury Park. The latter tickets were also bell-punch, or in some cases Edmondsons carried in the conductor's pocket and clipped with a hand-punch on issue.

CENTRAL LONDON RAILWAY. This line had a flat 2d fare from opening until 1907. The tickets were white with a red safety background, torn from a roll and deposited in a box at the station of *origin*. They were overstamped with the initials of the issuing station. A blue 1d workmen's ticket also existed. These were almost exact replicas of New York Elevated tickets, and bore code numbers denoting the station of issue, for identification purposes, and for use in special traffic checks when passengers had to retain their tickets until reaching their destinations.

After the introduction of graduated fares (1d–3d), Edmondson card tickets were used. There was no destination colour code, but station initials or abbreviations (e.g. Bk.=Bank; T.C.R.=Tottenham Court Road) were printed in very large black type, with 'translations' in brackets beneath. The 3d all-the-way ticket was green; some 1d tickets were orange (there may have been other colours for 1d stages); the various 2d tickets were all brown, with different green overprints denoting, not exactly destinations, but the different 2d fare sections along the line, grouped according to station of origin in a rather curious manner, thus:

BANK or POST OFFICE to MARBLE ARCH—horizontal line.
CHANCERY LANE to QUEEN'S ROAD—circle.
BRITISH MUSEUM to NOTTING HILL GATE—vertical line.
TOTTENHAM COURT ROAD, OXFORD CIRCUS, BOND STREET or MARBLE ARCH to BANK or WOOD LANE—none eastbound, diagonal line westbound.
LANCASTER GATE or QUEEN'S ROAD to CHANCERY LANE or WOOD LANE—two vertical lines.
NOTTING HILL GATE to BRITISH MUSEUM—two horizontal lines.
HOLLAND PARK or SHEPHERDS BUSH or WOOD LANE to TOTTENHAM COURT ROAD—St Andrew's Cross.

The rather elaborate code applying to the 2d tickets *only*, suggests that the 'Twopenny Tube' still accorded a certain primacy to the 2d fare, and it is interesting to note that the brown colour of these 2d tickets later became the standard for all Central London tickets.

GREAT NORTHERN & CITY RAILWAY. This line had a 2d all-the-way fare with 1d stages from the start, but both the 2d ticket (which had a safety background, incorporating the fare, very like that of the CLR and Bakerloo) and the various 1d ones (which had safety backgrounds in various colours with a *motif* of oblique stripes, and fare in the text) were roll tickets of the general type used by the flat-fare lines. Return tickets were double-length roll tickets, a further American feature. By 1912 Edmondson tickets had come into use for local journeys in various (apparently arbitrary) colours, but with red overprinted numbers denoting destinations. The code was:

FINSBURY PARK—1.
DRAYTON PARK—2.
HIGHBURY—3.
ESSEX ROAD—4.
OLD STREET—5.
MOORGATE—6.

Colours included orange, pink, green and buff, but these had no discernible relationship to station of origin, station of destination or fare.

After the line was acquired by the Metropolitan Railway Co. in June 1913, that company introduced its own standard type of ticket.

BAKER STREET & WATERLOO RAILWAY. There was a flat-rate system similar to that of the CLR for a few months after opening (March 10–July 21, 1906) with similar roll tickets (blue background). From July 22, 1906, fares varying with distance between 1d and 3d by ½d stages were in force. The type of ticket cannot be stated with certainty, but a contemporary description likened them to 'cash register slips'! White uncoded Edmondsons were introduced on January 20, 1907, and were later replaced by similar tickets bearing numerical codes for the stations of both origin and destination (see below).

GREAT NORTHERN PICCADILLY & BROMPTON RAILWAY. This line had differential fares from the first, the original tickets having a destination colour code like that of the CSLR. This was (from examination of actual specimens):

FINSBURY PARK—Dark blue.
GILLESPIE ROAD—Yellow-buff.
HOLLOWAY ROAD—Green.
CALEDONIAN ROAD—Lilac.
YORK ROAD—Primrose.
KING'S CROSS—Medium blue.
RUSSELL SQUARE—Cerise (pink).
HOLBORN—Light green.
COVENT GARDEN—Orange.
LEICESTER SQUARE—White.
PICCADILLY CIRCUS—Slate.
DOVER STREET—Very pale brown.
DOWN STREET—Light blue. (The same sequence of colours was then repeated with the addition of a vertical red stripe.)
HYDE PARK CORNER—Dark blue with red stripe.
KNIGHTSBRIDGE—Yellow-buff with red stripe.
BROMPTON ROAD—Green with red stripe.
SOUTH KENSINGTON—Lilac with red stripe.
GLOUCESTER ROAD—Primrose with red stripe.
EARL'S COURT—Medium blue with red stripe.
BARONS COURT—Cerise (pink) with red stripe.
HAMMERSMITH—Light green, with red stripe.
(There were some slight variations in these colours.)

APPENDIX 3

Plain white tickets followed later with station code numbers like those of the Bakerloo and Hampstead (see below).

CHARING CROSS EUSTON & HAMPSTEAD RAILWAY. There were differential fares from the opening, and originally no coding; all tickets were buff. Later, numerical station codes were added as on the Piccadilly and Bakerloo. It is believed that these were on buff tickets at first, and on white tickets later. For the through bookings with the Golders Green–Hendon bus service, bell-punch type tickets, punched in a hand-punch, were used in the bus-to-train direction, and two-coupon Edmondsons from train to bus.

YERKES TUBES—STATION CODE NUMBERS. From autumn 1907 the three Yerkes tubes adopted a system of station code numbers for bookings local to the three railways and *from* the CLR. (From September 1, 1907, for CLR/CXEH bookings; other bookings from November 1907.) The origin and destination name and number appeared on each ticket, but with the number in larger type for easier identification by collectors. The Central London numbers were the same as those used earlier on the 2d flat-fare tickets, but for bookings *to* the CLR, initial letters were used for CLR stations instead of numbers, as for the local bookings on the CLR described above. Certain numbers were left blank, for probable future extensions. The system was abandoned when 'standard tickets' (see below) were adopted in 1909.

The code at its maximum extent was as follows:

CENTRAL LONDON
1. Shepherd's Bush.
2. Holland Park.
3. Notting Hill Gate.
4. Queen's Road.
5. Lancaster Gate.
6. Marble Arch.
7. Bond Street.
8. Oxford Circus.
9. Tottenham Court Road.
10. British Museum.
11. Chancery Lane.
12. Post Office.
13. Bank.
14. (Blank but presumably intended for Liverpool Street.)

BAKERLOO
15. Elephant & Castle.
16. Westminster Bridge Road.
17. Waterloo.
18. Embankment.
19. Trafalgar Square.
20. Piccadilly Circus.
21. Oxford Circus.
22. Regent's Park.
23. Baker Street.
24. Great Central.
25. Edgware Road.
26. (Blank but intended for Paddington.)

PICCADILLY
27. Hammersmith.
28. Barons Court.
29. Earl's Court.
30. Gloucester Road.
31. South Kensington.
32. Brompton Road.
33. Knightsbridge.
34. Hyde Park Corner.
35. Down Street.
36. Dover Street.
37. Piccadilly Circus.
38. Leicester Square.
39. Covent Garden.
40. Holborn.
41. Russell Square.
42. King's Cross.
43. York Road.
44. Caledonian Road.
45. Holloway Road.

46. Gillespie Road.
47. Finsbury Park.
48. Strand.
HAMPSTEAD
49. (Blank but presumably intended for Embankment.)
50. Charing Cross.
51. Leicester Square.
52. Oxford Street.
53. Tottenham Court Road.
54. Euston Road.
55. Euston.
56. Mornington Crescent.
57. Camden Town.
58. South Kentish Town.
59. Kentish Town.
60. Tufnell Park.
61. Highgate.
62. Chalk Farm.
63. Belsize Park.
64. Hampstead.
65. (Blank but intended for North End.)
66. Golders Green.

STANDARD TICKETS. The coding system mentioned above did not last long, and by summer 1909 a standard form of tube ticket was evolved, recognizable by two principal features: (1) the use of colours identifying the issuing line and (2) the appearance, above the title of the issuing company, of the word 'UndergrounD' in a dark 'cameo' block. These two features, however, did not always appear either simultaneously or in the same order, e.g. on the Hampstead, tickets are known to exist in the old colours of white or buff *with* the UndergrounD block, and in the new (blue) colour *without* it, and similarly for other lines. In fact, the use of standard colours without the cameo block was so widespread that it may be considered as a halfway stage in the process of changing to standard tickets. Probably the existence of old card stocks requiring using up, and the restricted availability of the 'UndergrounD' stereos in quantity governed the matter, but both features were general by the end of 1909.

On July 1, 1910, the London Electric Railway took over the Bakerloo, Hampstead and Piccadilly railways, and on the tickets the change was marked in two ways. The old company titles were replaced by 'London Electric Railway', and the ticket colours for the Bakerloo (originally lilac) and Hampstead (originally blue) were transposed.

From 1910, the colour code for the issuing line was as shown below. Subject to the exceptions noted, this prevailed on all ordinary tickets (including 'Schemes') until 1933.

CSLR—Orange, but lilac for Rapid Printer tickets, when later introduced.
CENTRAL LONDON—Brown.
BAKERLOO—Blue.
PICCADILLY—Pale yellow.
HAMPSTEAD—Lilac.

Third-class District Railway tickets were eventually pink, but the old green colour was very persistent, and appeared on some early Scheme tickets, and on certain non-Scheme tickets at quite a late date. First-class tickets (including some Schemes) were white.

The non-Scheme standard tickets, though eventually superseded by Scheme tickets for bookings within the Underground group and the Metropolitan Railway, remained in substantially the same form until 1933 (with the 'UndergrounD' block) for through bookings elsewhere.

APPENDIX 3

The somewhat complex development of the tickets on the Yerkes tubes from their opening until 1910 may be summarized as follows (all tickets Edmondson type unless otherwise stated):

	BAKERLOO	PICCADILLY	HAMPSTEAD
Stage 1.	(a) Blue roll, flat fare. (b) ? ? ?, varying fares.	Colour code for destination.	Buff uncoded
Stage 2.	White uncoded.	White uncoded.	White uncoded.*
Stage 3.	White coded.	White coded.	White coded.*
Stage 4.	Lilac uncoded.	Yellow uncoded.	Blue uncoded.†
Stage 5.	As stage 4, but with 'UndergrounD' cameo block added.†		
Stage 6.	Blue.	Yellow.	Lilac.

On all lines, 'London Electric Railway' in heading.

* Some or all of Hampstead tickets may have been buff.
† Some of Hampstead tickets were buff.

STRIP TICKETS. An early device to relieve pressure on booking offices was the 'Strip Ticket', i.e. strips of six single tickets, often sold at a small discount. The Central London, in flat-fare days, had a strip of 2d tickets for sale with no discount, in addition to the roll tickets which were sold individually, and later had books of 2d tickets bearing an array of all the 2d sections for punching, each section carrying the appropriate green overprint in miniature. From October 1, 1908, there was a general issue throughout the Underground group lines of tickets in sets of six, at reductions of up to $12\frac{1}{2}$% (see Chapter 6, page 134). These were on thin white card, each coupon of Edmondson size, with the UndergrounD block across the left-hand end, together with a very bold notation of the (ordinary single) fare, and worded 'Available for one journey on any day in either direction between ... and ... or intermediately. This ticket is one of a set of six, price ... Issued subject to conditions on back.' This ticket design was very like that of the first Scheme tickets introduced just over two years later, and was obviously its inspiration.

The Metropolitan Railway had similar strip tickets for some journeys, but in dark green, worded like ordinary singles, and somewhat narrower than an Edmondson. The GNCR also had them from January 1910. The UndergrounD strips were withdrawn from February 1, 1916.

TUBE THROUGH BOOKINGS. Through tickets between the various tubes were introduced progressively from 1906. Some very early ones had a separate coupon for each line, and for quite a time the titles of both lines were given in the heading. There were distinctive colour schemes for each combination of lines, which as far as is known were as follows, but it is possible that all the colour schemes quoted did not exist simultaneously. In most cases the colours applied in both directions, and known exceptions are noted.

CSLR to CLR—Yellow (reverse direction, upper half yellow, lower half red).
–Bakerloo—White with green horizontal band.
–Piccadilly—White with red horizontal band.

—Hampstead—White with yellow horizontal band (in reverse direction, tickets to Bank were dark orange with a large 'B' overprint; certain other bookings were buff with a purple band).
—GNCR—Green. (Note 1.)
—District—Primrose with blue vertical stripe via King's Cross and Piccadilly Line; with red vertical stripe via Elephant and Charing Cross, or Elephant, Piccadilly Circus and Earl's Court or South Kensington.
—Metropolitan—Green with two vertical stripes via Moorgate. (Note 6.)

CLR to Bakerloo—Grey, later green.
—Piccadilly—Blue. (Note 2.)
—Hampstead—Purple (in reverse direction, tickets to Bank were dark orange with a large 'B' overprint).
—GNCR—?
—District—Green, with red skeleton 'O' via Oxford Circus and Charing Cross; Green, with red skeleton 'N' via Notting Hill Gate.
—Metropolitan—? (reverse, 1st white, 3rd green, with red 'B' from stations north of Baker Street via Bakerloo).

BAKERLOO to Piccadilly—White, lilac vertical stripe.
—Hampstead—Right half white with purple upright bar, left half buff (reverse, right half buff, left half white).
—GNCR—White with thick red vertical line.
—District—Green, white vertical stripe. (Note 3.)
—Metropolitan—Buff, blue vertical stripe. (Note 4.)

PICCADILLY to Hampstead—Buff, red vertical stripe.
—GNCR—Red, white and buff.
—District—Green, three blue vertical lines, via normal routes. (Note 5.) Buff, with two blue vertical stripes, via King's Cross and Met.
—Metropolitan—Buff with one blue vertical stripe via Piccadilly Circus and Baker Street, two stripes via King's Cross, three via South Kensington. (Note 4.)

HAMPSTEAD to GNCR—Buff with blue band down centre.
—District—Green, with two vertical white stripes. (Note 3.)
—Metropolitan—Buff, blue vertical stripe. (Note 4.)

GNCR to District—Green, white horizontal band.
—Metropolitan—Green, red vertical stripe. (Note 6.)

NOTES

1. Some returns have purple horizontal band. Singles seen are somewhat faded, and may originally have had this.
2. Plain blue via Oxford Circus or Tottenham Court Road or British Museum/Holborn. Overprinted red skeleton 'N' added when via Notting Hill Gate and South Kensington or Gloucester Road. This 'N' appeared on other tickets valid via the Notting Hill Gate 'walkover'.

3. 3rd class on District Railway; for 1st, colours counterchanged, i.e. white with green vertical stripe.
4. 3rd class on Metropolitan Railway; for 1st, lilac with blue vertical stripe(s). These colours applied from all three Yerkes tubes *to* the Metropolitan. *From* the Metropolitan the colours were—1st, white; 3rd, green. First-class bookings from the Yerkes tubes were discontinued from November 1908, although they still applied from the Metropolitan.
5. 3rd class on District Railway; for 1st, white with similar blue lines.
6. *From* the Metropolitan: 1st, white; 3rd, green.

These special colours died hard. Standard LER tickets, with the 'UndergrounD' block, existed in *blue* for Piccadilly to Central London bookings, but for bookings between the three Yerkes tubes they were withdrawn when standard tickets were introduced. They were continued for other bookings, but obviously could not be used with the Scheme ticket, which appeared on the Bakerloo in 1911 and from the first included through bookings to all Underground lines, even the Metropolitan. They were abandoned generally about 1911, though there were certain exceptions which lasted much later. These were:

(i) The tubes and also the Metropolitan used a special dark-red colour for certain bookings where alternative routes were offered, e.g. round the Inner Circle from Baker Street to Charing Cross or cutting across it by the Bakerloo; to Hammersmith via the District or the Hammersmith and City route. Presumably this was in order that such tickets should be easily recognized for special scrutiny, careful punching and individual auditing after use. The corresponding first-class tickets were white with a vertical stripe of the same dark-red colour.

(ii) Return tickets to or from Ealing Broadway, available to return via either District or Ealing & Shepherd's Bush Railway (GWR-owned, CLR trains) had bright-red return halves, with outward half green on District issues and buff on GWR ones. This arrangement was a late introduction (as the E&SB line was not opened until 1920) and persisted into LPTB days. These were 3rd class only, as the CLR had no first.

SCHEME TICKETS. With the abandonment of flat fares, the introduction of through bookings and the extension of the tube system, a bewildering variety of tickets had to be stocked in the cramped tube booking offices. There was urgent need to simplify the ticket system, not only to relieve the pressure on storage space but also to speed ticket issue. The Underground group had achieved some slight reduction in the number of tickets by grouping destinations at the same fare, but real progress came with the introduction of the 'Scheme' ticket in 1911. The reduction in the variety of tickets was achieved by providing one ticket for each *fare* rather than for each *destination*; the fare was made prominent and the various extreme destinations which could be reached for this fare were shown in the form of a list (the 'scheme') for reference only, and not intended to be read by collectors. Bookings to intermediate stations to which the same fare applied were covered by the phrase 'or intermediately' in the availability clause. Scheme tickets were introduced for bookings from Bakerloo stations on January 1, 1911, from CSLR stations on September 1, 1914, and from the other tubes in the Underground group from 1922. Through bookings to other Underground lines, and to the Metropolitan, were included from the start. Other through bookings were included by agreement from time to time, but there were always exceptions—e.g. bookings to Ealing and Shepherd's Bush stations (except Ealing Broadway) were not included until about 1932, and bookings to the LMSR line north of Queen's Park always had separate tickets. Until 1933, Scheme tickets were in line-of-issue colours, and successive designs were as follows:

(1) (1911) Design generally similar to that of the Strip tickets, with UndergrounD block across left-hand end, but fare removed to right. 'Half-snip' system for half fares; serial number on right only; list of routes and destinations on reverse.
(2) (About 1924) Similar, but list of destinations divided into two columns headed 'To' and 'Change at', and certain conditions omitted from reverse of ticket.
(3) (Late 1925) Serial numbers at both ends, UndergrounD block deleted, and ticket now designed for halving for Child issues. Fare repeated (smaller) on bottom left-hand corner.
(4) (1926) Totally new design with Scheme of destinations vertically down centre of face, continued on reverse if necessary. Conditions and (abbreviated) title in two narrow margins. Fare very small, in lower margin at each end. Station of origin across left-hand end, repeated (smaller and in brackets) at right-hand end.
(5) (Soon after) Station of origin bolder, and same size both ends. Fare also bolder, and turned round to read 'vertically', with station of origin.
(6) (Soon after, again) Station of origin and fare *inverted* at right hand and hence legible whichever way up ticket presented.
(7) (About 1928) (Experimental only, and never completely replacing preceding.) Ticket wholly horizontal, and company title restored in full. Design similar to a main-line ticket, but with tightly packed 'scheme' replacing single destination.

Where the possible destinations for a given fare were reduced to one (e.g. from termini, with no junctions within range), this type of ticket became virtually identical with the 'standard' through ticket. The latter, however, had the UndergrounD block, and usually the fare at the right-hand end only (as separate Child tickets were supposed to be issued for through fares).

Thereafter, the London Passenger Transport Board on its take-over in July 1933 continued with tickets substantially similar in design to (6) (Design (7) was discarded), but in green for all lines, and with the new title in full across the end. There were certain subsequent modifications, chiefly typographical, and the conditions were concentrated into a single side margin. The London Transport Executive continued with similar Scheme tickets until, after limited experiments at six stations from May 31, 1948, extended to twelve more on January 1, 1950, the present 'Station of Origin' ticket (showing fare and station of origin, but no destinations at all) became universal for all local fares from July 29, 1950. It was extended to cover many bookings with British Railways from October 1, 1950, and further BR bookings have since been included.

TICKET MACHINES—BOOKING OFFICES. Concurrently with the later development of the Scheme ticket, the process of issuing tickets was also speeded by the introduction of machines.

In 1921 the 'Automaticket' machine was introduced at certain West End stations. This electrically powered machine (the type in general use for cinema admission tickets) carried pre-printed tickets in zigzag packs, and could issue any number of tickets from one to six (five in earlier models) in one operation. The first tickets were of normal cinema size (2 inches by 1 inch, much smaller than an Edmondson), but subsequently larger ones were used (finally, $2\frac{1}{2}$ inches by $1\frac{3}{16}$ inches, rather larger than an Edmondson). Feed-wheel holes between adjacent tickets left a semi-circular 'bite' at the end of each ticket. In 1932 thirty Automatickets were in use at Underground group stations, but these machines were all superseded by the Super-Printix and Rapid Printer types.

The manual 'Rolltic' machine was first introduced in 1922. Tickets were pre-printed, and similar to the largest size of the Automaticket, but were supplied in a roll. By turning a handle a ticket was dated, cut off and delivered to the counter; an issuing speed of up to 900 tickets an hour could be achieved by a nimble booking clerk. There were ninety-five Rolltics on the Underground group lines in 1932, but they have now been superseded by more modern types.

A radically different type of machine, the electrically powered 'Rapid Printer', was used from 1926. (Self-printing machines had been tried at Victoria, MDR, in 1914, and Piccadilly Circus in 1918.) This had been developed and was originally supplied by Allgemeine Elektricitäts-Gesellschaft (AEG) of Berlin, but from 1932 was made in Great Britain under licence by the Westinghouse Brake & Signal Co. (the ticket-machine activities are today conducted by a separate company—Westinghouse Garrard Ticket Machines Ltd). No pre-printed ticket stocks are used in these machines, as the ticket is printed from a roll of blank card, and automatically numbered, dated, and cut off. Issue is controlled by push button, flush with the counter, four tickets emerging each second while the button is pressed. The printing units, metal boxes measuring about $6\frac{1}{2}$ inches by $1\frac{1}{2}$ inches by $1\frac{1}{2}$ inches, incorporating the printing plates and the meter are interchangeable, so that a unit for one type or value of ticket can be replaced by another in a few seconds. The meter on each unit shows the number of tickets of that type issued, and is geared in with the numbering device.

Similar to the Rapid Printer is the smaller 'Mini-Printer' which ejects the ticket direct from the printing unit instead of by conveyor, and has (on London Transport) six units compared with the ten of most London Transport Rapid Printers. Today, 81% of all London Transport railway tickets are issued from self-printing machines.

A modern manual machine is the 'Ultimatic', which came into general use in 1955, after experiments from 1953. It was developed by the Bell

Punch Company from their 'Ultimate' machine, used on buses. Fanfolded pre-printed tickets are used, each ticket measuring $2\frac{1}{2}$ inches by $1\frac{3}{16}$ inches, with a complete feed-wheel hole in the centre and a half-hole at each end. The five compartments each hold two packs of 500 tickets; depression of one of the five levers dates and issues a ticket, but the booking clerk must tear it off against the serrated upper edge of the issuing aperture. London Transport use about 150 of these machines.

TICKET MACHINES—COIN-OPERATED. Pull-bar slot machines were used on the CLR from 1904, and later on other lines, with some grouping of different destinations at the same fare and the use of the 'or intermediately' device to reduce the number of stations listed. From 1908 onwards coin-operated electric machines were introduced on the three Yerkes tubes and the District Railway, and by 1913 there were 104 of these in use on the Underground. By 1928 there were no more than 44 pullbar and 121 electric machines, but by 1935 the number of slot machines had risen to 606. The older electric machines were converted to the self-printing method of operation.

Change-giving was a problem which impeded development. Some use was made of staff whose sole duty was to give change, and an elaborate passenger-operated change-giving machine was tried at various stations between 1932 and 1934. The solution lay in the combined ticket-and-change machine. This was first tried in November 1926, but the change-giving feature was not wholly satisfactory and was withdrawn. In 1930 came the 'bunch-hopper' machine which accepted combinations of copper coins all together, and printed the ticket from a plain roll on the same principle as a Rapid Printer. The earliest machines of this type were by AEG, but later examples were by Brecknell, Munro & Rodgers (today, Brecknell, Dolman & Rodgers). The familiar modern type of passenger-operated, ticket-and-change-giving machine with sloping illuminated information panel, supplied by the Brecknell concern, was introduced in June 1937 and is now standard, with about 750 in use.

The reduction in the number of ticket series achieved by the 'Scheme' and, later, the 'Station-of-Origin' tickets, the rapid issue made possible by modern machines and the use of auxiliary devices such as automatic change-giving machines have greatly speeded up the process of booking, and enabled it to keep pace with the intensive passenger handling which is the leading characteristic of the tube railways.

APPENDIX 4A

Traffic Results Compared with Estimates

BAKER STREET & WATERLOO RAILWAY

	Sellon Estimate	1906 (2nd ½ × 2)	1907	1908	1909	1910 (1st ½ × 2)
Passengers	35,000,000	13,599,790	20,599,871	26,277,927	28,245,086	29,615,472
Average Fare	1·86d	—	1·49/1·50	1·48	1·46/1·45	1·43
Passenger Earnings	£271,250	£88,528	£128,037	£162,362	£171,149	£176,766
Miscellaneous:						
Earnings }	10,500	{ 632	4,664	7,521	8,939	11,016
Receipts }		{ 6,016	2,732	2,629	3,779	4,306
From UERL	—	634	2,441	2,142	—	—
Gross Receipts	281,750	95,810	137,874	174,654	183,867	192,088
Working Expenses	135,281	66,654	86,053	90,116	84,868	84,158
Fixed Charges	35,600	29,156	34,150	36,225	35,481	35,022
Reserve	—	—	—	—	6,000	8,000
Gross Outgoings	170,881	95,810	120,203	126,341	126,349	127,180
Balance available for Dividend	110,869	Nil	17,671	48,313	57,518	64,908
plus brought forward	—	—	—	—	2,247	5,440
less carried forward	—	—	—	2,247	2,720	5,738
Used for Dividend	110,869	—	17,671	46,066	57,045	64,610

GREAT NORTHERN PICCADILLY & BROMPTON RAILWAY

	Sellon Estimate	1907	1908	1909	1910 (1st ½ × 2)
Passengers	60,000,000	25,868,538	34,436,978	37,494,725	39,946,098
Average Fare	1·86d	2·06/1·98	1·95/1·94	1·89	1·87
Passenger Earnings	£465,000	£217,844	£279,367	£295,608	£312,050
Miscellaneous Earnings }	10,000	{ 7,997	11,632	14,727	20,160
Miscellaneous Receipts }		{ 650	2,498	5,097	6,126
From UERL	—	5,270	—	—	—
Gross Receipts	475,000	231,761	293,497	315,432	338,336
Working Expenses	255,750	146,547	147,452	143,547	139,206
Fixed Charges	67,640	78,100	88,887	91,649	91,236
Stamp Duty	—	—	1,933	—	—
Reserve	—	—	—	9,000	12,000
Gross Outgoings	323,390	224,647	238,272	244,196	242,442
Balance available for Dividend	151,610	7,114	55,225	71,236	95,894
plus brought forward	—	—	14	5,860	12,048
less carried forward	—	14	5,860	6,024	17,944
Used for Dividend	151,610	7,100	49,379	71,072	89,998

CHARING CROSS EUSTON & HAMPSTEAD RAILWAY

	Sellon Estimate	1907 (2nd ½ × 2)	1908	1909	1910 (1st ½ × 2)
Passengers	50,000,000	19,762,704	25,238,002	29,387,162	30,528,710
Average Fare	1·86d	—	1·67/1·65	1·63/1·65	1·56
Passenger Earnings	£387,500	£140,544	£174,365	£200,530	£198,392
Miscellaneous Earnings	} 7,500	{ 8,112	8,585	9,791	12,862
Miscellaneous Receipts		{ 254	439	794	1,054
From UERL	—	11,254	—	—	—
Gross Receipts	395,000	160,164	183,389	211,115	212,308
Working Expenses	195,000	122,708	113,735	116,084	111,078
Fixed Charges	57,680	37,456	63,552	63,584	62,870
Reserve	—	—	—	4,500	6,000
Gross Outgoings	252,680	160,164	177,287	184,168	179,948
Balance available for Dividend	142,320	—	6,102	26,947	32,360
plus brought forward	—	—	—	6,102	1,206
less carried forward	—	—	6,102	603	1,120
Used for Dividend	142,320	—	—	32,446	32,446

APPENDIX 4B
Results Compared with Estimates

LONDON PASSENGER TRANSPORT BOARD

(a) *Nominal Amounts of Stock Issued and Outstanding*

	In Pro-forma Submitted 1931	At 30.6.38 and 30.6.39
4½% 'L.A.'	—	£9,835,036
4½% 'A'	£24,280,589	£23,843,249
5% 'A'	£12,403,950	£16,263,950
4½% 'T.F.A.'	£12,583,000	£12,583,000
5% 'B'	£23,557,818	£23,709,830
'C'	£26,752,423	£25,698,802

(b) *Net Revenue and Appropriation thereof*

(All in pounds)

	In Pro-forma	1937–1938	Difference from Pro-forma	1938–1939	Difference from Pro-forma
Net Revenue	5,661,991	5,265,034	Minus 396,957	4,769,322	Minus 892,669
Interest:					
Local Authorities	467,748	456,559	Minus 11,189	454,998	Minus 12,750
4½% 'A' Stock	1,092,627	1,072,946	Minus 19,681	1,072,946	Minus 19,681
5% 'A' Stock	620,197	813,197	Plus 193,000	813,197	Plus 193,000
4½% 'T.F.A.' Stock	566,235	566,235	Same	566,235	Same
5% 'B' Stock	1,177,891	1,185,492	Plus 7,601	1,185,492	Plus 7,601
Total Interest	3,924,698	4,094,429	Plus 169,731	4,092,868	Plus 168,170
Balance	1,737,293	1,170,605	—	676,454	—
Plus, from 'C' Stock Interest Fund	—	28,344	—	26,465	—
Balance available for 'C' Stock Interest, Reserve, etc.	1,737,293	1,198,949	Minus 538,344	702,919	Minus 1,034,374
'C' Stock Interest paid	(5½%) 1,471,438	(4%) 1,027,952	Minus 443,486	(1¾%) 385,482	Minus 1,085,956
Balance	265,855	170,997	Minus 94,858	317,437	Plus 51,582
Less—amounts payable to L.E.T. Finance Corp. and on CLR Stock	—	144,795	—	291,791	—
Balance	265,855	26,202	—	25,646	—
to 'C' Stock Interest Fund	—	26,202	—	25,646	—
to additional Interest on 'C' Stock	132,927	—	—	—	—
to Reserve	132,928	—	—	—	—

APPENDIX 5

North End Station and Hampstead Garden Suburb

THE Hampstead Railway had the doubtful honour of being the only tube to possess a station that was never opened. Powers were obtained in the company's 1903 Act to acquire the 2½ acres of land comprising Wyldes farmhouse, outbuildings and their immediate surroundings, with the intention of building a station (to be known as 'North End') between Hampstead and Golders Green.[1] The southern boundaries of this plot were conterminous with the Hampstead/Hendon boundary, thus obeying the letter (but certainly not the spirit) of the 1901 agreement with the Hampstead Borough Council to have no station within the borough north of Heath Street/High Street.

The Hampstead Heath Protection Society—ready, as ever, to defend the Heath from desecration—had approached the LCC by February 1903, asking them to buy the fields adjoining the station site to preserve them from building. No more was heard of this approach, but the cause was taken up by a newly formed Hampstead Heath Extension Council. This Council had been brought into being by the energy of Mrs Henrietta Octavia Barnett,[2] who, with her husband, had been deeply interested in social work in the East End for over thirty years. The object was to acquire from the Eton College Trustees the eighty acres forming the meadowland of Wyldes Farm, situated north of the proposed station, in order to retain them as open space. In an appeal issued in July 1903, Mrs Barnett welcomed the proposed tube station as allowing thousands of Londoners to enjoy fresh air and beauty for a 2d ride. Plans had already been prepared for laying out the meadows as 'small streets with small houses' (according to the appeal pamphlet). £39,000 was needed, but this was later increased to £44,000 because of road costs.

Enough money and firm promises were received for the land to be taken over in September 1904, and it remains today as the Hampstead Heath Extension, between Hampstead Way and Wildwood Road.

Whilst engaged in this campaign, Mrs Barnett conceived the idea of what became Hampstead Garden Suburb, for people of all classes to live together, in planned, open conditions. The Hampstead Garden Suburb trust was registered in March 1906, and within a year had acquired 243 acres from the Eton College Trustees. The first sod was cut on May 2, 1907, and the first two cottages opened on October 9, 1907. Development proceeded steadily, and subsequent purchases of land increased the total area to nearly 800 acres.

[1] Today the initial section of Hampstead Way cuts across the plot. Wyldes farmhouse still exists; for many years it was the home of Sir Raymond Unwin.

[2] 1851–1936. Made a Dame of the British Empire Order in 1924.

APPENDIX 5

Meanwhile the construction of the Hampstead line had been progressing, and the station tunnels, platforms and stairs to the intended lower lift landing were built at North End. The southern ends of the station tunnels are on the centre line of Hampstead Way; the surface station was intended to be on the north side of the road, opposite Wyldes farmhouse. In March 1906, North End was included in the Hampstead lift contract with three lifts, but in the following month there was the first sign of lost enthusiasm, when the UERL Works Committee resolved to have only one 23-foot shaft (for two lifts) and a stair-shaft. Soon afterwards the proposal for a station was abandoned, before any vertical shafts were driven or surface works started. The tunnels have since been used for storing various items, including permanent-way supplies, sound-insulation materials and documents. The platforms have been replaced by short wooden stages. In 1927 the surface plot was sold, and a house now known as 'No. 1 Hampstead Way' erected on part of it.

The station site came to be known as 'Bull and Bush', and is thus called in the LTE railway traffic circulars, but this is a misleading title as the public house bearing the same name is about a quarter of a mile distant.

Various reasons have been propounded for the abandonment of the project, including opposition by local residents and the anticipated traffic not being enough to justify lowering the average speed for the whole line. However, the crux of the matter was that the Underground group had early acquired a keen nose for potential housing estates. Much of the area round the station was already permanently devoted to public open space (Hampstead Heath and Golders Hill Park); the Hampstead Heath Extension dealt the *coup de grâce* to the chances of developing sufficient traffic to justify the capital, maintenance and operating costs of a station at North End. The group drew the correct inference from the facts, and cut its losses. But there was one consolation. Had there been no plans for North End station, there would probably have been no Hampstead Garden Suburb.

Index

N.B.—There may be more than one reference on the page given

Aberconway, Lord, 150, 239
Abercrombie, Sir Leslie Patrick, 309, 356–7
Accidents, 156, 175, 215, 271, 305, 338, 365–77
Acton Town (Mill Hill Park), 102, 118, 201, 210, 212, 213, 214, 217, 332, 336
Acton Vale, 90
Acton Works, 189–90, 269, 340
Acworth, Sir William, 45
Adams, Holden & Pearson, 183, 236, 258, 359
Advertisements, 21, 133
Air raids and damage: precautions 1914–1919, 155–6, 309; 1915–1918, 156; precautions 1931–1940, 294–8; 1940–1945, 290, 296, 298–306, 308, 340
Aldenham depot, 250–1, 262, 263, 328
Aldwych (Strand), 43, 81, 108, 109 (plan), 142, 301, 313, 367
Aldwych branch and service, 94, 108, 109 (plan), 135, 163, 228, 301, 311, 313, 332, 341
Alexandra Palace, 98, 200, 245, 247, 255, 256, 257, 258–9, 260–3, 261 (map), 310, 327–9
Alperton, 102, 211, 214, 215, 334
American Car & Foundry Co., 118, 119
American influence, 70–4, 102, 121, 123, 126, 128, 168, 170, 227, 375
American investment, 70–4, 79, 152
Angel (Islington), 34, 35, 48, 49, 50, 51, 68, 69, 77, 176–7
Architecture 359–60. *See also:* Easton, Green, Heaps, Holden, Stations and station design
Archway (Highgate), 22, 66, 75, 108, 117, 120, 153, 154, 163, 176, 185, 193, 206, 224, 254, 255, 257, 337
Arnos Grove, 199, 203, 205, 207, 209
Arsenal (Gillespie Road), 108, 115, 136, 137, 223, 296

Ashbury Railway Carriage & Iron Co. Ltd, 50, 56, 61
Ashfield, Lord: birth and early career, 126–7; comes to London and UERL, 127; director, UERL, 128; organizes publicity 1907–1909, 132–3, 361; traffic promotion 1908–1909, 133; tries to erase 'tube', 133; managing director, 140, 167; and LGOC merger, 149, 152; knighted, 152; Ministry of Munitions, 157; Board of Trade, 157, 167; name queried, 157; and Common Fund, 157–8; Baron, 167; Development Scheme 1921, 167; on workmen's fares, 192; godfather to tube baby, 21; on pirate buses, 193–4; on future (1925), 193–4; on growth, 353; chairman LPTB, 242; valediction UERL, 243; Standing Joint Committee with main lines, 244; on LPTB finances, 292; member, REC, 297; member BTC, 314; death, 314
Associated Electrical Industries Ltd, 344, 345
Associated Equipment Co. Ltd, 193, 243

Babies, born in tube stations, 300
Baby, born on tube train, 21
Baker, Sir Benjamin, 30, 37, 40, 55
BAKERLOO LINE: proposed, 34, 35; begun, 37, 39; London & Globe regime, 36–9; bought by Yerkes, 72–3; extension of time, 77; amalgamation proposal, 87–8; Bills 1903–1904, 88–9; debentures, 91; work resumed, 94; construction methods, 96–7; cars delivered, 106–7; test runs, 107; opens, 107–8; track, 112; description, 111 *et seq.*; 'Bakerloo' adopted, 110; light loads, 124; publicity, 131; services and train formation 1909, 135; and NWLR, 138–9; part of LER, 140;

BAKERLOO LINE: (cont.)
  Paddington extension, 37, 138–9, 144–5; Queen's Park extension, 146–7, 158–9; connection to LNWR, 146–7, 158–60; to Willesden and Watford, 158, 160–1, 186, 229, 266, 269, 312; trains 1917, 163; proposed junction with CSLR/Hampstead at Waterloo, 178–9; rolling stock unsuitable, 187; Camberwell extension, see Camberwell; to Finchley Road and Stanmore, 246, 247, 263–8; lengthening stations, 247, 269; blue stripe trains, 280; additional trains and seven-car trains, 308; tunnel rebuilt, 309; improvement scheme 1949, 329
Baker Street, 35, 36, 94, 108, 143, 162, 238–9, 264–8, 335
Baker Street & Waterloo Railway, see Bakerloo
Balfour, Beatty & Co. Ltd, 212, 334
Balham, 183, 184, 185, 293, 296, 303–4
Bank, 34, 35, 47–8, 51, 52, 55, 56, 153, 154, 164, 176, 177, 185, 192, 287–91, 296, 304, 310, 311, 332, 334, 365, 368, 371. See also Lothbury
Bank–Monument subway, 201, 225, 274
Barkingside, 238, 322
Barlow, Peter William, 27, 28, 29
Barnato Bros, 166
Barnett, Mrs Henrietta Octavia, 388–9
Barons Court, 108, 114, 117, 120, 212, 214, 220, 336
Barry, Sir John Wolfe, 59, 85
Battersea Park, 29
Beach, Alfred E., 28
Becontree, 194
Belsize Park, 108, 302, 303
Bethnal Green, 301, 305, 316–17, 326
Betjeman, John, 142
Beyer-Peacock Ltd, 49
Birmingham Railway Carriage & Wagon Co. Ltd, 60, 188, 189, 230, 283, 341, 344, 345
Blackhorse Road, 331
Blake Hall, 327
Blumenfeld, R. D., 103
Board of Trade, see Trade
Bombs, see Air Raids
Bond Street, 56, 58, 59, 192, 296, 350
Bone, Stephen, 219
'Boodle', 63, 64

Booking clerks, 122, 383
Booking offices, 122, 154, 383–4
Borough (Great Dover Street), 29, 30, 46–7, 155, 164, 174, 175, 176, 177, 301, 367, 369
Boston Elevated Rly, 102, 116
Boston Manor, 193, 201, 212, 213, 216, 360
Bott & Stennett, 98
Bounds Green, 202, 203, 204, 205, 206, 209, 304, 356, 360
Bowes Park, 356
Brand, Charles & Son Ltd, 169, 183, 202, 254, 265, 273, 274
Brakes and brake blocks, 119, 171, 189, 231, 249, 280, 282, 285, 346
Brent, 170–1, 233, 253, 306
Bristol Carriage & Wagon Co. Ltd, 51
British Electric Traction Co. Ltd, 138, 149–50, 151
British Museum, 56, 57, 59, 153, 222–3, 301
British Thomson-Houston, 50, 58, 60, 94, 97, 100, 101, 104, 118, 161, 186, 207, 214, 231, 282, 283, 285, 326
British Transport Commission, 314–15
Broad Street, 142–3, 144, 160, 322
Brockley Hill, 262, 263, 327, 329
Brompton & Piccadilly Circus Railway, 41–2, 68, 69, 71–2, 77–8, 81, 82, 83, 86, 87. *Then see* Piccadilly line
Brompton Road, 41, 82, 108, 137, 220
Brondesbury, 77, 138
Brown, Billy, 306–7
Brown, H. G., 116
Brunel, Sir Marc Isambard, 27
Brush Electrical Engineering Co. Ltd, 51, 56, 100, 118, 138, 145, 153
Buckhurst Hill, 324
Burnt Oak, 171, 172, 253, 355
Buses and bus routes: good routes, 41; competition with tubes, 45, 110, 125, 128, 130, 134; feeders to tubes, 172, 186, 200, 209, 321, 355–6; competition between bus operators threatens tubes, 193–4, 198, 240; conditions at Finsbury Park, 197–8; buses temporarily replace tubes, 174–5, 255, 256, 257, 306, 320, 374; passengers lost to Underground, 346–7; capacity c.p. tubes, 358. *See also* Hendon

Bushey, 146
Bushey Heath, 169, 248, 250-1, 255, 257, 260-3, 261 (map), 310, 327-9, 357

Cable traction, 29, 30, 35, 47, 65
Caffin & Co. Ltd, 323
Caledonian Road, 108, 136, 137, 365
Camberwell, 237-8, 251, 269-70, 310, 329-30, 340
Camden Town, 34, 36, 108, 112, 117, 120, 172-4, 173 (plan), 176, 224, 248, 302, 303, 304, 328, 337, 370
Cammell-Laird & Co. Ltd, 162, 187, 188, 189
Canonbury & Essex Rd, *see* Essex Road
Canons Park, 239-40, 264, 266, 268
Capacity of tube trains, 321-2, 358
Carpenders Park, 161, 335
Cars, tube, features of: centre doors, 145; air doors, 187, 225, 226, 228-9; heaters, 230; wartime lighting, 297-8, 307; window net, 306; straps and grips, 307; door fault detection lights, 342, 344; larger windows, 342; door space changes, 343; fluorescent lights, 343, 344; rubber suspension, 343. *See also:* Brakes, Passenger Door Control, PCM, Weak Field, Wedglock, Ventilation
Cars, tube, shortage of, 163, 165
Cars, tube, types of: CSLR, 50-1, 61, 176; W&C, 53, 288, 291; CLR, 56-7, 59, 61, 226, 273; GNCR, 100, 151, 260; Bakerloo original, 118, 228; Piccadilly original, 118, 160-1, 187, 228; Hampstead original, 119, 160, 171, 228; LER 1914, 145-6, 160, 228; CLR Brush 1914, 153, 161, 168; Converted trailers 1914, 160, 228; LER/LNWR Joint, 160, 186, 229-30; Cammell-Laird 1920, 187, 228, 286; Experimental 1923, 188, 278-9; Standard stock 1923-1925, 188-9, 280; CLR modernized 1925-1928, 226, 227; Standard stock 1926-1934, 226-31, 286, 322, 340, 341-2, 343-4, 346; UCC Experimental 1930, 230; Nine-car trains, 253-4, 263, 341; 1936 stock, 280-2, 341; 1938 stock, 282-6, 308, 341, 342; 1949 stock, 340-1; 1956 stock, 344-5; 1959 stock, 344-5; Cravens 1960 stock, 345-6; Rebuilt trailers, 346; 1962 stock, 344-5
CENTRAL LINE: beginnings, 33, 34, 35; and CWE, 41; finance of, 44; opens, and description, 55-61; success of, 67, 74; loop proposals, 68-9, 78, 81, 88, 90; fires, 92, 369; fire precautions, 93; co-ordination with UERL, 129-33; extension to Wood Lane, 137; extension to Liverpool Street, 34, 35, 143-4; extension to Ealing, 148, 167-9; taken over by UERL, 151-2; Underground group influences, 152; Thames Valley extensions, 153; Gunnersbury extension, 153, 167; and Common Fund, 158, 235; trains in 1917, 163; junction with LSWR at Shepherd's Bush, 167; right-hand running, 168; to LPTB, 242-3; 'Central Line', 254; eastern extensions, 238, 245, 247, 248, 273-5, 301, 316-17, 319-27, 357; western extensions, 246, 247, 248, 273-5, 315, 317-19, 323-4, 325-7, 357; stations lengthened, 247, 272-3; converted to LT tube standards, 272-3; overcrowding, 321-2; rolling stock difficulties, 322, 340, 344; train lengths, 340, 341-2. *See also:* Cars, Ealing & Shepherd's Bush Railway
Central London Railway, *see* Central Line.
Chalk Farm, 108, 146, 147
Chambers Committee, 315
Chancery Lane, 55, 56, 201, 225, 296, 302, 304, 311, 332, 367, 370
Chapman, James R., 82, 102, 104, 112, 113, 123
Charing Cross (Embankment), 40, 59, 75, 77, 81, 82, 95, 108, 114, 115, 141-2, 163, 164, 171, 172, 180 (plan), 181, 183, 185, 191, 192, 224, 248, 277, 295, 296, 309, 333, 340, 365, 366, 369; link line to Kennington 177, 178-83, 180 (map); loop terminus at, 141-2; 180 (plan), 296, 365
Charing Cross (Strand), *see* Strand
Charing Cross and Waterloo Electric Rly, 30-1
Charing Cross, Euston & Hampstead Rly, *see* Hampstead line

INDEX

Charing Cross, Hammersmith & District Rly, 68, 69, 78, 80
Chicago, Yerkes' activities in, 62–4
Chigwell, 325
Chingford, 79, 85, 310, 330, 331
Chiswick, 90
Chiswick Park, 212, 214
Church, R. F., 54
City & Brixton Rly, 68, 89
City & Crystal Palace Rly, 77
City & North East Suburban Electric Rly, 68, 69, 79, 80, 81, 85, 88
CITY & SOUTH LONDON RAILWAY: beginnings and opening, 29–32; results, 33, 44, 51–2; Islington extension, 34, 35, 48; description, 46–52; Euston extension, 48, 89; fires, 92, 369; co-ordination with UERL lines, 130–3; taken over by UERL, 151–2; enlargement of tunnels and junction with Hampstead line, 152, 153, 165, 167, 172–8; tunnels searched, 155; and Common Fund, 158, 235; power supply rearranged, 162; Sutton extension proposed, 178–9; Morden extension, 178–86, 232, 293, 354, 355; connection to Hampstead line at Charing Cross, 177, 178–83, 180 (plan); nomenclature after 1926, 185, 192; to LPTB, 242–3 (*See* Northern line after 1926)
City & West End Rly, 41–2, 68
City of London & Southwark Subway, 29–32
City Railway, 176, 185
City Road, 48, 176, 301, 367
Clapham Common, 31, 48, 49, 50, 153, 174, 175, 176–7, 179, 183, 184, 193, 293, 296, 302, 337, 370
Clapham Junction, 79, 80, 81, 89
Clapham Junction & Marble Arch Rly, 89
Clapham North (Clapham Road), 48, 49, 177, 302, 303
Clapham Road, *see* Clapham North
Clapham South, 183, 302, 303
Clay, London, 21, 22, 27, 348
Cliff, John, 196, 242
Closures of stations, 23, 47, 175, 176, 220, 221, 223, 259, 290, 295–6, 301, 309, 319, 368, 371. *See also* Strikes
Coats of arms, 57
Cochrane, John & Sons Ltd, 222, 273

Cockfosters station and depot, 199, 200, 201, 202, 203, 206, 207, 208, 209, 282
Colindale, 171, 172, 249, 253, 306, 367
Colliers Wood, 183, 184
Committees, Parliamentary, *see* Parliamentary
Common Fund, 157–8, 193, 235
Control, Traffic, 295, 338
Cooper, Patrick Ashley, 242, 244
Co-ordination between London operators: of fares 1907–1909, 129–31; of publicity 1907–1909, 131–3; of facilities, 141–3; mergers 1912–1913, 149, 151–2, 154; Common Fund 1915, 157–8; Government moves 1919–1920, 165–6; pirate bus problem, 193–4; towards full co-ordination 1924–1933, 240–2; between LPTB and main lines, 244, 297, 357; between LTE and BR, 315–16. *See also* Through Bookings
County of London Plan, 309
Covent Garden, 21, 108, 109 (plan), 117, 137, 233, 361
Cranley Gardens, 262, 329
Cravens Ltd, 345
Cricklewood, 77, 138
Crompton & Co. and Crompton Parkinson & Co. Ltd, 50, 176, 282, 283
Cross, George, 352–3
Crouch End, 262, 327
Crouch Hill, 259, 262
Croxley Green, 146, 160, 230, 269
Crystal Palace, 29, 30, 77, 131, 311
Cuningham, Granville, 21, 58
Curves, 47, 54, 82, 111, 123, 203, 350
Cut-and-cover construction, 26, 27, 274–5

Daily Express, 59, 103
Daily Mail, 55, 124, 306
Dawkins, Clinton, 80
Debden, 325, 326
Decapod locomotive, 80
Deep shelters, Government, 302–3, 370
Denham, 248, 323, 327, 357
Depths, 46, 51, 52, 56, 66, 67, 111–12, 115, 141–2, 143, 144, 159, 184, 257, 267, 303
Derby, BR works, 345

Design, Underground, 358-60
Development (Loan Guarantees & Grants) Act, 199, 221, 239
Development Schemes: 1921, 167; 1922-1926, 167, 185; 1930-1932, 199, 223, 237, 330; 1935-1940, 245-8, 250-1, 254-63, 264-86, 316-29
Dick, Kerr & Co. Ltd, 54, 100
Dimsdale, Sir Joseph, 41, 42
District Railway, *see* Metropolitan District Rly
Dollis Hill, 265, 266, 267
Dover Street, *see* Green Park
Down Street, 41, 108, 137, 201, 220-1, 301
Drayton Park, 98, 99, 100, 101, 137, 150, 151, 260, 263, 327-9
Dreiser, Theodore, 62
'Drico', 338, 342
Drivers, 121, 122, 249, 366
Drummond, Sir Hugh, 178
Durnsford Road Power Station, 53, 54

Ealing & Shepherd's Bush Rly, 90, 148, 153, 167-9, 247, 275, 314, 315, 318, 336, 382
Ealing & South Harrow Rly, 102, 116, 117, 215
Ealing Broadway, 90, 148, 167, 193, 210, 297, 315, 323, 334, 336, 365
Ealing Common, 216
Earl's Court, 40, 42, 72, 77, 78, 89, 104, 108, 114, 142, 189, 220, 247, 275, 277
East Acton, 168
Eastcote, 270, 366
East Finchley, 245, 247, 254, 255, 256, 257, 258, 259, 263, 315, 337, 368
Easton, J. Murray, 360
Edgware, 75, 89, 167, 169-72, 176, 200, 233, 239-40, 245, 247, 250-1, 253, 254, 255, 256, 257, 259, 262, 263-4, 297, 310, 327-9, 352, 366. *See also:* Edgware & Hampstead Rly, Watford & Edgware Rly
Edgware & Hampstead Rly, 75, 76, 88, 89, 169
Edgware Road, 38, 59, 89, 108, 117, 138-9, 144, 145, 201, 223, 238-9, 247
Electric traction, beginnings of, 30, 35
Electric Traction Co. Ltd, 44, 55

Elephant & Castle, 29, 38, 46, 47, 49, 50, 108, 115, 117, 154, 158, 159, 160, 162, 164, 174, 176, 177, 180 (plan), 237, 247, 269, 295, 311, 330, 367, 369, 370
Elliot, Sir John Blumenfeld, 22, 315, 347, 362
Elstree South, 262, 263, 329
Embankment, *see* Charing Cross
Emergency Procedures, 337-8, 342, 366. *See also* Telephones
Employees, *see* Staff
Enfield, 197, 199, 209, 310, 330
Enfield West, *see* Oakwood
English Electric Co. Ltd, 259, 288
Epping, 317, 325, 327
Epsom, 186, 252
Epstein, Jacob, 236
Escalators, 53, 142-3, 144, 153, 159, 162, 176-7, 181, 184, 191-2, 206, 218, 219, 220, 221, 222-3, 224, 225, 247, 248, 257-8, 265, 266-7, 276, 277, 304, 307, 308, 313, 317, 318, 320, 322, 333-5; speedray control of, 277
Essex Road, 99, 100, 137, 332
Euston, 29, 34-5, 36, 37, 48, 49, 50, 59, 68, 89, 115, 146-7, 152, 153, 154, 155, 160, 162, 163, 172, 173 (plan), 174, 176, 177, 185, 192, 248, 331, 333-4, 337, 366, 367
Euston Road, *see* Warren Street
Evacuation of schoolchildren, etc., 297
Evening News, 110, 132
Express tubes, 252, 302, 310

Fairlop, 322, 332
Fares: CSLR, 51, 374; W&C, 54, 374; CLR, 55, 129-30, 152, 375; London Suburban Rly, absurd, 80; GNCR, 101, 375; Piccadilly, Hampstead and Bakerloo, 122, 130-1, 376-9; co-ordination, 129-31; reduced, 1909, 134; E&SB, 168; fluctuations 1917-1925, 165, 192-3; Cockfosters line, 209; Stanmore line, 240, 263-4, 266; Northern line extensions, 256; Edgware and Stanmore lines discrepancies, 263-4, 293; Bakerloo extensions, 266; LPTB increases, 292-3, 308; superstandard and substandard, 293; main line reduced to LT, 256, 317, 318. *See also* Through Bookings

396 INDEX

Farrer, Rt Hon. Lord, 73
Feltham works, 227
Field, Marshall, 65
Fifty-five Broadway, see St James's Park
Figgis, Samuel, 66-7, 76
Figgis, T. P., 47
Finance and financial results: comments on, 21, 22,; CSLR, 33, 52; Bakerloo original, 36-9; difficulties of attracting, 44-5, 198; CLR, 67; Yerkes, original, 71-4; UERL 1903-1904, 90-2; GNCR, 99, 101; UERL 1905-1906, 125-6; UERL 1906-1908; 127-9; LER 1910, 140; GWR contribution, 144; LNWR contribution, 147; UERL 1912, 149; take-overs 1913-1914, 152; American investment fades, 152; Common Fund, 157-8, 193, 235; cost of tube railways, 165, 252, 273; LER and CSLR loans 1919, 165; difficulties 1919, 166; CSLR modernization, 172; no subsidy 1926, 185; dividends and finances 1918-1925, 193; 1930 programme, 200-1; finances 1926-1932, 234-6; winding up UERL, 243; 1933-1939, 244, 291-3; New Works programme 1935, 246; 1939-1948, 297, 308; Victoria line, 330-1
Finchley Central (Church End), 254, 256, 257, 258, 259, 262, 263, 310
Finchley Road, 238-9, 246, 247, 264-8
Finsbury Park, 34, 43, 72, 76, 94, 98-9, 108, 111, 114, 117, 153, 193, 195-203, 223, 233, 247, 256, 257, 258, 260, 263, 330, 331
Fire and fires, 51, 92-3, 303, 305, 342, 343, 369-70
First class on tubes, 151, 380-1
Floodgates, 295-6
Fog, 22, 101
Forbes, James Staats, 39, 42, 70
Ford, H. W., 142
Foundation Co. Ltd, 170, 183
Fowler, Sir Henry, 65
Fowler, Sir John, 55
Free rides, 111, 155, 185, 203, 213
Fulham, 68, 77, 78, 80, 81, 82

Gabbutt, Edmund, 30
Galbraith, W. R., 35, 37, 54, 96, 110
Gants Hill, 273-4, 320-1, 353

Ganz electrification system, 71
Gatemen, 121, 228
Gauge, 51
Geddes, Sir Auckland, 157
Geddes, Sir Eric, 166
General Electric Co. of America, 56, 58, 79, 94
General Electric Co. Ltd, 214, 231, 282, 283, 326, 344, 345
Ghost, a tube, 21
'Ghost' station, see North End
Gibb, Sir George Stegmann, 85, 105-6, 111, 129-31, 140
Gill, Arthur Eric Rowton, A.R.A., 236
Gillespie Road, see Arsenal
Gloucester Railway Carriage & Wagon Co. Ltd, 145, 188, 230, 344
Gloucester Road, 40, 77, 108, 136, 247, 248
Golders Green, 65, 66, 75, 76, 95, 107, 108, 114, 117, 119, 120, 136, 153, 163, 167, 169-72, 185, 189, 190, 192, 209, 225, 233, 249, 250, 253, 306, 337, 351-2, 370
Goodge Street (Tottenham Court Road), 59, 108, 275, 302, 303, 370
Gradients, 54, 66, 112, 159, 174, 184, 189, 203-4, 257-8, 260, 272-3, 320
Grange Hill, 319, 325, 326
Great Central, see Marylebone
Great Dover Street, see Borough
Great Eastern Railway, 80, 85, 143, 144
Greater London Plan, 309, 356-7
Greathead, James Henry, 27-8, 29, 30, 31-2, 34-5, 36, 47, 52, 54, 55, 57
GREAT NORTHERN & CITY RAILWAY: beginnings, 34, 35, 42; and Finsbury Park, 72, 98; Lothbury extension, 77, 99, 150, 238; construction, opening and description, 98-101; co-ordination with UERL, 130-3; Sunday service, 137; non-stop trains, 137; proposed extensions and junctions, 150; taken over by Metropolitan, 150-1; improvements, 151; powers to connect with LNER and BR(E), 196, 247; through running to Alexandra Palace, etc. 1935 scheme, 245, 247. See Northern City line for subsequent references

Great Northern & Strand Railway, 43, 72, 76, 81, 82, 94, 108. *See* Piccadilly line for subsequent references
Great Northern, Piccadilly & Brompton Railway, *see* Piccadilly line
Great Northern Railway, 34, 42–3, 72, 98–9, 150, 196
Great Western Railway, 37–8, 90, 139, 145, 148, 210, 211, 244, 246–8, 301, 315, 318
Green, Leslie W., 114, 359
Green Belt, the, 250, 323, 327–8, 353, 356–7
Greenford, 273, 315, 318, 323, 324, 326, 360
Green Park (Dover Street), 41, 108, 164, 201, 221, 277, 296, 301, 304, 331, 334
Greenwich, 75
Greenwich Power Station, 294, 326, 339
Guards, 121, 228, 343, 367, 368
Gunnersbury, 153, 167
'Gutter title', 110

Hainault, 273, 275, 319, 320, 321, 322, 323, 325, 326
Hamilton, Lord Claude, 144
Hamilton, Lord George, 140
Hammersmith, 78, 89, 94, 108, 110, 114, 117, 149, 153, 192, 201, 209, 210, 211, 212, 214, 217, 315, 366;
—City of London tube schemes, 74–5, 77–85, 89, 90, 310
Hammersmith & City Railway, 37, 210, 212, 284
Hammersmith, City & N.E. London Rly, 85, 89, 90, 310
Hampstead, 22, 34, 65–7, 75, 88, 108, 111, 112, 117, 120, 163, 233 249, 335, 388–9
Hampstead Garden Suburb, 233, 388–9
Hampstead Heath, threat to, 66–7, 76, 388–9
HAMPSTEAD LINE: beginnings, 34–5; progress, 36; Yerkes and, 64–6; and the Heath, 66–7; extensions proposed, 68–9; 1902 Bills, 75; 1903 Bill, 87–8; 1904 Bill, 89; 1905 Bill, 90; Finance, 91; construction, 95, 98; cars delivered, 107; opens, 108, 111; description, 111 *et seq.*; publicity, 131–2; services, 134–6: part of LER, 140–1; proposed connection with LNWR, 147; trains in 1917, 163; connection with CSLR, *see* CSLR; Edgware extension, 167, 169–72, 233, 355; link to Kennington, 178–83; nomenclature after 1926, 185, 192. *See* Northern line for subsequent references
Hampstead, St Pancras & Charing Cross Rly, 34, 35
Hanger Lane, 317, 318
Hare, T. J., 38, 110
Harrow & Wealdstone, 160, 312
Hay, Sir Harley Hugh Dalrymple, 65, 96, 219, 222
Heaps, S. A., 170, 267, 359
Hendon, 75, 170–2, 175, 208, 253, 306, 355; railway bus to Golders Green, 120, 136, 149, 209, 377
High Barnet, 200, 247, 255, 256, 257, 258, 259, 263, 315
Highbury, 99, 100, 248, 331, 335
Highgate, 247, 254, 255, 256, 257–8, 259, 301, 368. *See also*: Archway (the original 1907 station), Wellington Sidings
Hillingdon, 366
Hoe Street, 331
Holborn, 29, 43, 72, 77, 81, 82, 90, 94, 108, 109 (plan), 114–15, 117, 135, 153, 164, 201, 222–3, 301, 332
Holden, Dr Charles H., 204, 215, 219, 236, 359–60
Holland Park, 56, 335, 342, 343, 370
Holloway Road, 108, 115, 136, 208
Hopkinson, Dr Edward, 49
Hornsey, 43
Hounslow Central, 210, 216
Hounslow East, 210
Hounslow West, 75, 105, 210, 213, 216
Housing development and tube railways, 65, 136, 168, 169, 171, 211, 233, 248, 250–2, 270, 351–5, 357
Hungarian Railway Carriage & Wagon Works, 118
Hurst Nelson & Co. Ltd, 51, 97
'Hustlers', 190–1
Hyde Park Corner, 41, 108, 117, 136, 201, 220, 221, 296, 301

Ice and snow, troubles and precautions, 231–2, 338–9
Ickenham, 248, 332

Ilford, 194, 238, 245, 320, 353, 355, 357
Ilford & District Railway Users' Association, 238, 245, 321
Inner Circle, the, 26, 34, 71

Joel, S. B., 166
Johnston, Edward, 359
Johnston type, 359
Jones, W. Kennedy, 166
Junctions, confluent, to be avoided, 68-9

Kennedy, Professor A. H. W., 54
Kennington (New Street), 46, 47, 77, 164, 174, 176, 177, 178-83, 180 (map), 182 (plan), 192, 248, 252, 253, 295, 296, 310, 311, 337
Kennington Road, *see* Lambeth North
Kensal Green, 160
Kentish Town, 36, 88, 108, 201, 224
Kilburn, 77, 266, 267
Kilburn Park, 158-9, 170, 277
Kingsbury, 239, 266
King's Cross, 34, 35, 43, 48, 108, 115, 154, 174, 176, 177, 201, 223, 247, 276, 296, 307, 331, 332, 335, 337, 368; connecting link, 189-90
Kings Road Railway, 68, 69, 77
King William Street, 31, 46, 47, 50, 51, 75, 155, 301
Knapp, Zac Ellis, 123
Knightsbridge, 41, 81, 89, 90, 94, 108, 163, 201, 220, 274, 296, 301

Lambeth North (Kennington Road: Westminster Bridge Road), 88, 89, 108, 117, 183, 269, 295, 304, 369
Lancaster Gate, 56, 335, 368
Land values, increased by tube railways, 86, 251, 351-3
Late night services and last trains, 135, 145, 157, 208, 313, 332
Latham, Lord, 314, 315, 327, 329
Leeds Forge Co. Ltd, 56, 145, 188
Leicester Square, 42, 90, 108, 115, 192, 221-2, 248, 253, 307, 334, 351, 360; control room, 295; regulating room, 337
Level crossings, tube trains over, 248
Leyton, 79, 85, 245, 273-4, 315, 316-17, 326

Leytonstone, 317, 326, 367
Lifts, 47, 48, 57, 99-100, 114, 115, 143, 152, 163, 176-7, 192, 220, 221, 275, 335
Light railway working, 76, 169
Lillie Bridge depot, 95, 107, 117, 119-20, 176, 189, 212-13, 214-15, 277, 279
Liveries, *see* Cars, types of
Liverpool Overhead Railway, 92
Liverpool Street, 34, 35, 142, 143-4, 156, 162, 249, 273-4, 277, 301, 316, 317, 321, 322, 326, 336, 368, 370
Lloyd George, of Dwyfor, Earl, 110, 111, 166
Loch, Lord, 36-7, 38
Locomotives: CSLR, 49-50, 51, 176, 369, 370; W&C, 54; CLR, 56, 59, 61, 273; GER Decapod, 80; LER, 97; GNCR, 100-1; Sleet locos, 338-9
London & Globe Finance Corporation, 36-7, 38-9, 73
London & Home Counties Traffic Advisory Committee, 196-7, 240-2, 252
London & North Eastern Railway, 170, 196, 200-1, 238, 244, 245, 246, 247, 254, 315, 321, 328, 356; tube cars owned, 283; tube scheme, 238
London & North Western Railway, 36, 38, 106, 146-7, 160-1; suburban electric lines, 159, 160; tube scheme, 146-7; tube tunnels, 147
London & South Western Railway, 34, 44, 52, 89, 153, 210-11
London & Suburban Traction Co. Ltd, 151, 243
London Bridge, 29, 30, 47, 49, 52, 164, 176, 177, 296, 337, 365
London, Brighton & South Coast Railway, 131
London Central Electric Railway, 31
London County Council, 43, 45, 59, 85, 86, 87, 94, 158, 192, 241, 355, 356
London County Council Tramways, 150, 153-4, 166, 175, 197, 211
LONDON ELECTRIC RAILWAY: formation, 140-1; bus powers, 148; and Common Fund, 158, 235; takes over E&H, 169; dividends, 193, 235; to LPTB, 242-3

London Electric Transport Finance Corp. Ltd, 246
London General Omnibus Co. Ltd, co-ordination with UERL, 130; taken over by UERL, 148–9; and Common Fund, 158, 193, 235; competition from other bus operators, 167, 193; acquires competitors, 240; to LPTB, 242–3. *See also*: Buses, Co-ordination, Through Bookings
London, Midland & Scottish Railway, 214, 230, 244, 256, 269, 328, 356
London Passenger Traffic Conference, 130
LONDON PASSENGER TRANSPORT BOARD: formation, 241–4; Joint Committee with main lines, 241, 244; Passenger Pooling Scheme with main lines, 244, 297; Frank Pick on, 243–4; financial results and difficulties, 291–3, 387; economics of depend on London growth, 353; Government control, 296–7; 'a vigorous if sometimes uncertain role', 357; to LTE, 314
London Plan Working Party, 310, 327
London Road (Southwark) depot and substation, 38, 94, 106, 119, 162, 180 (plan), 189, 236, 269
London Suburban Railway, 80, 81, 85
London Traffic Act, 193, 196, 234, 240
London Traffic Advisory Committee, 166
London Traffic Authority (proposed), 166
London Traffic Board (proposed), 86, 87, 348
London Traffic Combine, 132
London Transport Board, 315
LONDON TRANSPORT EXECUTIVE: formation, 314–15; Inquiry, 315; co-ordination with BR, 315–16; retrenchment, 332–33; public relations, 333; to LTB, 315
London United Electric Railways, 78, 80, 82–3, 84
London United Tramways, 39, 78, 83, 84, 122, 130, 151, 153, 242–3
London, Walthamstow & Epping Forest Rly, 147
Lothbury, 77, 99, 150, 238
Lots Road Power House Joint Committee, 148, 243
Lots Road Power Station, 41, 71, 75, 91, 102–5, 159, 160, 162, 167, 172, 183, 190, 214, 233–4, 236, 278, 368–9, 371, 372
Loughton, 275, 323, 324–5, 326, 360
Lovatt, Henry & Co. Ltd, 167
Low, Professor A. M., 188

McAlpine, Sir Robert & Sons Ltd, 202, 212
McIver, Sir Lewis, 83, 84, 87
McKenzie, Holland & Westinghouse, 57, 159–60, 168
Maida Vale, 37, 159, 161, 238, 277, 296, 365
Main Line Railways: traffic to tubes at termini, 34, 139, 307; tubes running over, 147, 357; pooling and co-operation with LPTB, 241, 244, 297; tubes across London for, 311; abstraction of traffic from, 355–6; tubes' relationships with, 34, 356, 357; suburban services of, c.p. tubes, 321–2, 325, 349, 365. *See also* names of the companies.
Mandelick, W. E., 123
Manor House, 197, 198, 199, 200, 202, 203, 204, 205, 206, 208, 209, 277, 316, 369
Mansion House, 40, 80, 89
Maps (in text): whole system, 24; Holborn, Aldwych, Covent Garden, 109; Camden Town, Euston, 173; Charing Cross, Waterloo, Kennington, 180; Kennington (tracks), 182; Edgware, Bushey Heath, Barnet, Cockfosters, Stanmore, Finsbury Park, etc., 261
Maps, UndergrounD, 132, 316
Marble Arch, 56, 57, 58, 77, 81, 86, 89, 138, 139, 201, 225, 277, 296, 370
Marylebone (Great Central), 36, 37, 38, 108, 117, 163, 248, 308, 311
Mather & Platt Ltd, 31, 49
Matthews, Sir Ronald, 255, 321
Maybury, Sir Henry, 197, 242, 244
Metropolitan-Cammell Carriage & Wagon Co. Ltd (and predecessors), 60, 100, 118, 186, 188, 227, 230, 231, 281, 283, 344–5. *See also* Cammell-Laird
Metropolitan District Deep Level Railway, 40–1, 42, 75, 77–8, 82, 89, 301

Metropolitan District Electric Traction Co. Ltd, 71, 72–4
Metropolitan District Railway, 26, 28, 39–42, 70–1, 75, 89, 102, 105, 130–3, 158, 210–12, 213, 235, 236, 242–3
Metropolitan Electric Tramways, 145, 149–50, 151, 170, 197, 242, 243
Metropolitan Railway, 26, 34, 35, 37–8, 60, 65, 71, 85, 89, 104, 129–33, 144, 150, 161, 210, 213, 239–40, 241, 242–3, 263–4, 270; tube scheme, 238–9
Metropolitan line track rearrangement, 264–8
Metropolitan-Vickers & Co. Ltd, 282
Midland Railway, 210
Mid-London Electric Railway, 31
Mid-Metropolitan Railway, 29
Mile End, 273–4, 316–17, 369
Mill Hill East, 255, 257, 258, 259, 262, 315, 327–9
Mill Hill Park, see Acton Town
Mill Hill, The Hale, 262, 327–9
Milnes, G. F. & Co. Ltd, 50–1
Moore, Henry, 236
Moorgate, 47, 48, 49, 50, 51, 164, 172, 174, 175, 176, 177, 192, 247, 276, 277, 296, 337, 370
Moorgate (GNCR), 98, 99, 260, 263, 276, 329
Morden, 178, 179, 183, 184, 185–6, 192, 193, 252, 293, 311, 337, 355, 358; depot, 183, 184, 185; Station Garage, 185–6
Morden–Edgware line, see Northern line
Morgan, John Pierpont, 79, 80, 83, 84
Mornington Crescent, 36, 89, 108, 117, 173 (plan), 332, 337, 366, 371
Morrison, of Lambeth, Lord Herbert Stanley, 241, 246
Morrison, Robert, Lord, 196
Moscow Metro, 222, 321
Mosquitoes, in tubes, 300
Mott, Basil, 55
Mott, Charles Grey, 31, 41
Mott, Hay & Anderson, 238
Mowlem, John & Co. Ltd, 47, 52, 98, 141, 143, 144, 147, 162, 172, 181, 191, 219, 222, 273
MP, trapped in lift, 335
Multiple-Unit cars, 93–4, 100
Muswell Hill, 259, 262, 327, 329

Neasden, 266, 268, 311
Neasden Power Station, 103, 151, 172 214, 240, 269, 339, 369
Neasden Works and Depot, 265, 268–9
Newbury Park, 247, 274, 315, 319, 320, 321, 322, 326
New Works Programmes, see Development Schemes
New York, 28, 375
Noise reduction, 188, 278–80
Nokes, G. A., 110, 124, 155
Non-stop trains, 135–7, 161, 220, 233
Nord de la France, les ateliers du construction du, 118
North Acton, 168, 273, 275, 317, 318, 319
North Cheam, 311, 353, 355
North Ealing, 214, 216, 332
North East London Railway, 68–9, 79, 80, 85, 90
North End (Hampstead), 88, 110, 388–9
NORTHERN CITY LINE: so named, 255; connection to LNER and extension to Alexandra Palace, 245, 247, 255, 256–9, 260, 261 (map), 310, 327–9; conversion to LT tube standards, 259–60; transfer of stock to Acton, 260, 329; 1948 extension proposals, 311. See also Great Northern & City Railway
NORTHERN LINE: 'Morden–Edgware line', 192; increasing traffic and problems of, 233, 248–54, 263–4, 293, 355; extension to East Finchley and running over Barnet & Edgware lines: 245, 247, 254–9, 261 (map), 310, 327–9, 357; extension to Bushey Heath, 248, 250–1, 255, 257, 261 (map), 260–3, 310, 327–9, 357; increasing the capacity of, 252–4; nine-car trains, 253–4; fares and traffic, 293; 'Northern line', 254; automatic signalling, 336–7. See also: City & South London Rly, Hampstead line
Northfields, 193, 199, 209, 212, 213, 214, 216; depot, 212, 214
North London Railway, 38, 42, 130, 144, 150–1, 356
North Metropolitan Electric Power Supply Co. Ltd, 207, 243
Northolt, 323, 324, 326

North Weald, 325, 327
North West London Railway, 77, 86, 89, 138–9
Notting Hill Gate, 56, 58, 92, 248, 334, 369, 370
Nuttall, Edmund, Sons & Co. Ltd, 274

Oakley, Sir Henry, 42–3, 59
Oakwood (Enfield West), 202, 203, 204, 205, 206, 207, 208, 209, 297, 360
Office construction, 357, 358
Oldbury Carriage & Wagon Co. Ltd, 51
Old Colony Trust Company, 74, 91
Old Street, 48, 49, 162, 177, 296
Ongar, 247, 317, 320, 324, 325, 326–7, 328, 357
Opening dates, 363–4
Osterley, 216, 360
Otis Elevator Co. Ltd, 115, 123, 290
Oval (Kennington Oval), 46, 164, 175, 177, 252, 296, 302
Overcrowding, 22, 137–8, 163, 165, 251–2, 282, 317, 321–2
Oxford Circus, 35, 39, 56, 108, 115, 143, 154, 163, 164, 191, 225, 278, 296, 310, 331, 335, 370
Oxford Street, see Tottenham Court Road
Ozone, 59

'Padded Cells', 50, 176
Paddington, 37, 38, 77, 138–9, 144–5, 153, 162, 269, 277, 301, 305, 370
Palmers Green, 79, 81, 85, 88, 89, 90
Pardoe, J. W., 196, 203
Paris Métro, 92, 349
Park Royal, 102, 215; power station, 169
Parliamentary Committees, 1863, 26; 1864, 26; 1892, 34–5, 349; 1901, 67–9, 74; 1902, 75–82; 1919, 165–6
'Passenger-open' door controls, 283–4, 342, 345
'Passimeters', 170, 177
P.C.M. control, 282, 284, 345
Pearson, Charles, 26
Pearson, S. & Son Ltd, 99, 101
Peckham, 68, 75, 77
Perivale, 317, 318, 360
Perks, R. W., 64–5, 70, 72, 84, 124

Perry & Co., 37, 39, 52, 97
Piccadilly & City Railway, 68, 69, 79, 80
Piccadilly Circus, 21, 35, 41, 72, 77, 81, 82, 89, 108, 114, 115, 117, 163, 164, 208, 217–19, 236, 269, 307, 313, 367, 370
Piccadilly, City & North East London Rly, 80, 82, 83
PICCADILLY LINE: beginnings, 40, 72, 76, 77, 82, 87; extensions proposed, 89, 90; finance of, 91; construction, 94, 98; cars delivered, 107; opens, 108, 110; description, 111 et seq.; services, 134–5, 136–7; nonstops and skip-stops, 135–7; to LER, 140–1; Richmond extension, 153, 211, 212; trains in 1917, 163; six-car trains, 187; connection to CSLR, 189–90; Cockfosters extension, 195–209, 261 (map); 356; western extensions, 199, 209–17; central area rebuilding, 217–23; 1949 improvement scheme, 332; speed control signalling, 336
Pick, Frank: joins Underground, 358–9; commercial manager, 359; encourages artists, 359; and new type face, 359; and station architecture, 359–60; creates London Transport 'image', 358–60; assistant managing director, UERL, 359; on Piccadilly extensions, 197–201; on south-east London tubes, 237; on common management, 241; managing director, 242; vice chairman and chief executive officer LPTB, 242; on the LPTB, 243–4; member SJC, 244; letters to *The Times*, 1937, 250; on patience of strahangers, 357; on substandard fares, 293; on London population and new works, 353–4; plans evacuation, 297; member REC, 296–7; retires, 297; date of death, 358
Plessey Tunnel Factory, 301, 320
Pneumatic Despatch Co. Tube, 155
Pneumatic propulsion, 28–9
Poole Street Power Station, 99, 100, 151
Post Office, see St Paul's
Post Office Pneumatic tube, 29, 155
Post Office Tube Railway, 162

Power supply and plant: CSLR, 49, 162; W&C, 53, 289; CLR, 58, 190; GNCR, 99, 100, 151; Yerkes tubes, 102–5; E&SH, 102; LNWR, 160–1; E&SB, 169; Edgware extension, 172; Morden extension, 184; Piccadilly extensions, 207, 214; power from alternative outside sources, 190, 248, 294, 339; Northern line extensions, 259; Central line extensions, 326–7; failures, 368–9. *See also:* Greenwich, Lots Road, Neasden, Stonebridge Park, Substations
Press, the, and tubes, 55, 107, 110, 124, 127, 232, 333, 361–2
Price, John & Co., 55. *See also* Rotary Excavators
Price & Reeves, 96, 98
Protective clauses, 44–5
Publicity, 131–3, 153, 185, 256, 266, 316, 333, 347, 359 361

Queensbury, 240, 266
Queen's Park, 146, 158–9, 160, 163, 269, 336, 366
Queens Road, *see* Queensway
Queensway, 56, 57, 335

Railgrinders, 279
Railway Executive Committee, 296–7, 301
Railway (London Plan) Committee, 309–10
Rammell, T. W., 28–9
Ravenscourt Park, 210–12, 214
Rayleigh, Lord, 59, 74
Rayners Lane, 213, 270–1, 332, 358, 360, 366, 367
Redbridge, 274, 320, 326, 343, 371
Regent's Park, 35, 37, 89, 108, 114, 161
Reno spiral conveyor, 115
Ribblesdale, Lord, 75, 348
Ribblesdale Committee, 75–7
Richmond, 153, 167, 210, 211, 212
Rickmansworth, 146, 160, 230
Rigby W., & Co, 48
Road-rail co-ordination, *see* Co-ordination, Through Bookings
Robinson, Sir Clifton, 122
Roding Valley, 320, 325, 326
Rolling Stock, *see* Cars, Locomotives
Rotary Excavators, 55, 96–7
Routes 'A', 'B', 'F', 'G', 'J', 'K' and 'L', 311

Routes 'C', 'D', 'E' and 'H', 310
Royal Commissions: on Metropolitan Railway Termini, 25; on London Traffic, 58, 70, 73, 85–7, 88, 151, 348; on Distribution of Industrial Population, 354, 356–7; on Local Government, Greater London, 358
Ruislip, 270; depot, 273, 275, 324
Ruislip Gardens, 323, 324, 326
Ruislip Manor, 270, 332
Rush Hours, 314, 347, 358. *See also:* Overcrowding, Staggering
Russell Square, 43, 108, 137, 201, 223, 233

St James's Park offices, 236
St John's Wood, 266, 267
St Paul's (Post Office), 56, 58, 223, 246, 272, 276, 302, 304, 370
Safety precautions, 116. *See also* Emergency Procedures
Scott, Walter & Co. Ltd (Scott, Walter & Middleton & Co. Ltd), 30, 48, 55, 98, 147, 190
Seats: air doors but fewer seats, 226, 230; tube and steam train comparison, 321–2
'Sekon', G. A., *see* Nokes
Selfridges, 350–1
Sellon, Stephen, 92, 124–5, 385–6
Sentinel steam wagons, 202
Service and speeds: CSLR, 51; W&C, 54, 290, 291; CLR, 58, 137; Yerkes lines, 120; 1907–1913, 134–8; 44 trains an hour, 136; GNCR, 151; to Queen's Park, 159; to Watford, 160–1; 1917, 163; to Ealing, 168; Edgware extension, 171–2; during CSLR reconstruction, 174–5; CSLR 1924, 176; during Charing Cross reconstruction, 181; Morden extension, 184; Cockfosters extension, 208–9; Piccadilly western extensions, 213, 232; 1919–1933, 232–3; Stanmore 1932, 239; Northern 1937–1939, 249–54; Northern extensions, 256, 257; proposed Bushey Heath, etc., 263; Bakerloo extensions, 266; Uxbridge 1938, 271; 1950–1952, 313–14, 332–3; short trains in slack hours resumed, 313–14; Central extensions, 317–18, 321, 322–3, 324, 325; 1955–1962, 347

Seven Sisters, 331
Sheltering in tubes, 156, 256, 294, 295, 298–305
Shepherd's Bush, 22, 33, 55, 57, 59, 78, 80, 81, 89, 90, 148, 153, 167, 168, 192, 193, 210, 273, 365
Shops and tube railways, 218, 350–1
Siemens, Werner von, 30
Signalling: CSLR, 49, 117, 177; W&C, 55, 117, 289; CLR, 57, 117, 152; GNCR, 101; E&SH, 102, 116, 117; Yerkes tubes, 116–18, 207, 336; Queen's Park extension, 159–60; E&SB, 168; Edgware extension, 171–2; Camden Town junctions, 174; rebuilt CSLR, 177; Kennington 1926, 183; Morden extension, 184; Piccadilly extensions, 207, 213–14; train describers, 207; Stanmore branch 1932, 240; speed control, 257, 269, 322, 336; Northern extensions, 257, 258–9; Northern City, 260; Bakerloo extension, 268; Bakerloo resignalled, 269; Uxbridge branch, 271, 332; Central extensions, 319, 325–6, 327; semaphores go, 336; route setting and remote control interlocking, 336–7; automatic programme machines, 336–7
Sleet locomotives and tenders, 338–9
Smell, of CLR, 58–9
Smoking, 229
Snaresbrook, 320, 367
South Acton, 213, 217, 332
South Ealing, 213, 216–17, 267, 366
South Eastern Railway, 35, 52, 95
Southern Railway, 178–9, 211, 237, 244, 287–9, 315, 349
Southgate, 79, 80–1, 197, 198, 199, 202, 203, 204, 205, 206, 208, 352, 355, 360
South Harefield, 323
South Harrow, 102, 209, 210, 212, 213, 215, 360, 370
South Kensington, 40, 41, 42, 77, 81, 89, 94, 108, 115, 153, 247, 301, 350, 370
South Kensington and Knightsbridge and Marble Arch Subways, 29
South Kentish Town, 108, 136, 301, 371
South Kenton, 269
South Metropolitan Electric Tramways and Lighting Co. Ltd, 151, 243

South Ruislip, 323, 324
South Wimbledon 183, 310
South Woodford, 320, 326
Spagnoletti, C. E., 49
Speyer Bros, 73, 74, 83, 84, 91
Speyer, Edgar: early career, 73; director UERL, 91; chairman UERL, 105–6; speeches on opening of tubes, 110, 111; approach to LCC, 127; sombre mood, 128; on 1912–1913 mergers, 151; 'a de facto Traffic Board', 151; attacked 1915, 156–7; Asquith on loyalty of, 157; resigns, 156–7; leaves England, 157; death, 157; shares held, 166
Speyer, James, 91, 128
Sprague, Frank Julian, 30, 57, 93–4, 335
Staff: working conditions, 121–2; war bonus, 157–8; women, 159, 308; staff magazine, 153; effect of air doors, 228; ticket checking, 315; shortage, 333; West Indians, 333. *See also:* Strikes, Uniforms
Staggering of working hours, 252, 314, 316
Stamford Brook, 210–12
Stanley, Albert H., see Ashfield
Stanmore, 239–40, 246, 263–8, 293, 329, 339
Starting trains, 121, 190–1
Stations and station design: CSLR, 46–7, 48; W&C, 52; CLR, 57, 152, 272–3; GNCR, 99; Yerkes tubes, 111, 112, 113–16, 359; second-best sites, 361; posters cover names, 133; bull's-eye name device, 133, 152, 359; Charing Cross 1914, 142; misleading names, 142, 168, 183, 206, 221; Queen's Park extension, 159; barriers and queues, 163–4, 290; E&SB, 168; Edgware extension, 170–2, 359; CSLR reconstructed, 176; Charing Cross link, 181; Morden extension, 183–4, 359; too close to each other, 183; floodlit, 184; searchlights, 184, 203; rebuilding and improvement after 1920, 190–2, 217–25, 275–7, 333–6; busiest, 142, 191; destination indicators, 192, 207; Piccadilly extensions, 204–6, 215–16, 360; lighting, 205, 277, 307, 313; light direction signs, 224; Stanmore branch, 239; Northern

Stations and station design: (cont.) extensions, 257–8; Bakerloo extensions, 266–8; Uxbridge line rebuilt, 270–1; gap lights, 277; platform numbers, 277; loudspeakers, 308, 333; cleaned up, 313; Central line extensions, 316–25, 360; reconstruction 1945–62: 333–6; attractiveness and convenience, 356, 360; general architectural comment, 359–60. *See also:* Air Raids, Closures, Escalators, Lifts, Sheltering, Suicides

'Stay-in' gestures by passengers, 249, 333

Steam suburban lines compared with tube, 321–2, 349, 356

'Stentorphone', 191

'Stock Exchange Ramp, a', 84

Stockwell, 31, 32, 46, 47, 49, 50, 51, 153, 163, 174, 177, 189, 293, 302, 303, 369

Stonebridge Park, 161, 335, 370; power station, 160, 161, 369

'Stop and Proceed' Rule, 365, 366

Strand (Aldwych), *see* Aldwych

Strand (Charing Cross), 75, 88, 90, 95, 108, 114, 117, 120, 141–2, 275, 295, 296, 337, 365

Straphanger, patience of, 357

Straphangers' Revolt, 249, 333

Stratford, 273–4, 316–17, 338, 366

Streamlined trains, 281–2

Strikes, 249, 371–2

Substations, 49, 58, 104, 151, 159, 162, 169, 172, 183, 184, 207–8, 214, 221, 240, 259, 262, 326, 337–8, 339–40, 369, 370

Suburban growth and tube railways, 351–5, 357. *See also* Housing

Suburban tubes fed on abstracted traffic, 355–6

Suburban tubes filled up with terminal traffic, 355

Suburban tubes to support the rest, 354

Sudbury Hill, 214, 215, 360

Sudbury Town, 215, 359

Suicides and anti-suicide pits, 204

Sunday services, 51, 135, 137, 157, 290, 332

Sundays, stations closed on, 157, 290, 332

Sutton, 167, 178–9, 186

Swiss Cottage, 265, 266, 267

Talbot, George, 55

Telephones, guard to driver, 228, 289, 338, 342; tunnel, 119, 289, 337–8

Temple, 76

Thames Ironworks, 50

Theatre trains and traffic, 135, 136, 137, 185, 233, 351

Theydon Bois, 325

Thomas, John Pattinson, 203, 253, 254, 300–1

Through bookings and facilities, 101, 122, 130–1, 134, 144, 145, 153–4, 158, 159, 192–3, 209, 308, 375, 379–81. *See also* T.O.T.

Ticket machines, 219, 224, 383–4

Tickets: general survey, 373–83; seasons, 133–4, 145, 159, 192–3; strip tickets, 134, 379; books of, 134; collected on trains, 145; tube available on buses, *see* Through Bookings; shopping, 351; checking at inward barriers, 315; standard tickets, 378; scheme tickets, 381–2, 384; station-of-origin tickets, 383, 384

Times, The, 66, 245, 352

Tooting Bec (Trinity Road), 183, 184, 296

Tooting Broadway, 178, 183, 184, 249, 337, 367

'T.O.T.', 153–4, 192–3

Tottenham, 68, 79, 84, 209, 330–1

Tottenham Court Road (Oxford Street), 56, 89, 108, 114, 115, 138, 154, 163, 190–1, 192, 206, 224, 248, 253, 254, 267, 276, 278, 296, 304, 367, 369, 371

Tottenham Hale, 331

Totteridge & Whetstone, 256, 259

Tower Subway, 27–8, 97

Track, 51, 53, 60, 61, 100, 112–13, 144, 184, 204, 259, 272–3, 279–80, 288–9, 326; railgrinding, 279; welding, 279–80, 326

Trade, Board of, 59, 68–9, 71, 92–3, 157–8; London Traffic Branch, 87

Trade Facilities Acts, 166–7, 172, 179

Trafalgar Square, 38, 108, 114, 115, 191, 278, 295, 304, 350

INDEX 405

Traffic and results: CSLR, 33, 44, 51; CLR, 67, 74, 152; UERL estimates, 91–2, 385–6; GNCR, 101, 151; Bakerloo, 124, 125, 137, 143, 385; Piccadilly, 125, 385; Hampstead, 125, 351, 386; wartime increases, 155, 165, 307; increases 1905–1919, 166; increases 1919–1932, 236; rebuilt CSLR increases, 178; reduced by competing buses, 198; abstracted from main lines, 355–6; vigorously pursued (Piccadilly extensions), 208,–9, 355; encouraged by new stations, 223, 277; Edgware extension growth, 233; LPTB results, 291–3, 387; rush-hour increases, 290; 1946–1962 pattern, 314, 346
Traffic guides, 191
Train describers, 207, 217, 266
Trams: competition with tubes, 45, 101, 125; Yerkes and American, 63–4; conditions at Finsbury Park, 197; tram loading islands at tube stations, 206; trams as tube feeders, 170, 209, 225; capacities compared, 358. *See also*: London County Council, London United, Metropolitan Electric, South Metropolitan, Through Bookings
Transport, Ministry of, 166
Travolators, 290–1
Trinity Road, *see* Tooting Bec
'Tube', use of the word, 133, 360–1
'Tube Chaos', useful headline, 361
Tube 'Mania', the, 67–9, 74–90
Tube railways: evolution of, 27–9; conception of function, 147, 312, 351; vast plans for new lines 1948, 309–12; increasing importance of, to London, 346–7, 362; lack of planning, 348–9, 354; lack of lines in east, north-east and south London, 349; following streets, 111, 349–50; relief to streets, 358; and urban traffic movement, 358; convenience and speed, 361; democratic nature of, 361; popular with visitors, 361
Tube Refreshments Special, 300, 340
Tube tunnels: diameters, 28, 29, 35, 40, 46, 48, 52, 58, 98, 112, 141, 143, 144, 147, 153, 159, 162, 171, 172, 174, 181, 184, 203–4, 262, 265, 274, 302; construction, 27, 28, 37, 55, 96–7, 99, 110, 171, 174, 175, 181, 183, 202, 254, 258, 265, 273, 274–5, 302, 312, 331; lighting of, 100, 112, 289; air-pressure relief devices 203, 258; entrance protection, 214, 258, 265, 326; CLR realigned, 272–3; Joosten soil consolidation, 274, 333; concrete segments, 274, 331; new methods, 27, 331; anti-noise measures, 279–80, 317
Tufnell Park, 108
Turnham Green, 89, 210–12, 315
Turnpike Lane, 198, 201, 202, 203, 204, 205, 206, 209, 316
'Twopenny Tube', 55

'UndergrounD', 114, 132–3, 347, 378
Underground Consolidated Electric Railways, 87
UNDERGROUND ELECTRIC RAILWAYS COMPANY OF LONDON LTD: formation, 73; directors and chairman, 74; LCC suspicious of, 86; leasing of tubes to, 86, 87; amalgamation of tube companies, 87–8; funding of, 90–2; 'contractors' for construction of tubes, 97; continued existence necessary, 105; finances 1905–1908, 125–9; voluntary liquidation, 128–9; finances 1908–1912, 129; Lots Road sale, 148; LGOC bought, 148–9; and Metropolitan Rly, 150; and LCC Tramways, 150, 166; interest in tram companies, 151; CLR and CSLR bought, 151–2; finances to 1919, 166; finances and dividends 1926–1932, 234–6; London must grow to suit it, 353; moves to common management with LCC, 241; winding up, 243
Underground group, 132
Uniforms, 122, 159, 308
Union Construction & Finance Company Ltd (Union Construction Co. Ltd), 227–8, 230
Uxbridge, 105, 122, 213, 246, 247, 270–1, 293, 332

Valentine, Alexander Balmain Bruce, 315
Ventilation, 58–9, 116, 121, 208, 248, 278, 342
Vibration, 59–61, 74

Victoria, 66, 68, 69, 75, 77, 86, 89, 138, 139, 190, 310, 330, 331
Victoria line ('Route C'), 97, 310, 315, 330–1

Walker, Sir Herbert, 178, 244
Walker, Price & Reeves Ltd, 98
Waltham Abbey, 68, 79, 85, 88, 90
Walthamstow, 68, 69, 79, 85, 90, 147, 209, 310, 330–1
Walworth, 237
Wanstead, 274, 301, 320
Warren Street (Euston Road), 108, 201, 220, 224, 331, 360
Wartime, tubes in, 155–64, 190, 290, 294–309, 340
Warwick Avenue, 159, 277, 365
Waterloo, 34, 35, 37, 38, 39, 52–5, 90, 94, 108, 114, 178, 179, 180 (plan), 181, 192, 277, 289, 295, 296, 307, 309, 311, 333, 334, 366, 367
WATERLOO & CITY RAILWAY: beginnings, 34, 35; construction, 37, 52; finance, 44; opening and description, 52–5; GNCR connection proposed, 150; ownership, 44, 242; deterioration and unpopularity, 286–7; modernization, 287–91; war damage, 290
Waterloo & Whitehall Railway, 29
Watford & Edgware Railway, 89, 169, 251, 262
Watford High Street, 89, 146, 160, 169, 312
Watford Junction, 146, 160–2, 163, 269, 312
Wayleaves, 35, 350
Weak field control, 229, 249, 280, 345
Wedglock couplers, 281, 284
Wellington car depot and sidings (Highgate), 250, 259, 280, 329
Wembley Central (Sudbury & Wembley), 146
Wembley Park, 238–40, 264–6, 268 335–6
West Acton, 168
West & South London Junction Rly, 68, 77
West Finchley, 256
West Hampstead, 265, 266, 267
Westinghouse Brake & Saxby Signal Co. Ltd, 116, 171, 240, 282, 285, 289, 383

West Kensington, 78, 89, 94, 117, 120, 210, 214. *See also* Lillie Bridge
Westminster Bridge Road, *see* Lambeth North
Weston Street signal box and shaft, 48
West Ruislip, 246, 247, 248, 318, 323, 324, 326
Whistler, James Abbott McNeill, 103
Whitechapel & Bow Rly, 26, 317
White City, 318–19
Widener & Elkins, 63, 64, 65
Willesden Green, 238, 266, 268
Willesden Junction, 146, 160, 189
Wimbledon, 75, 178
Wimbledon & Sutton Railway, 167, 178–9
Wimpey, Geo., & Co. Ltd, 318
Windsor, Lord, 67, 75
Windsor Committee, 77–82
Woodford, 319, 320, 321, 322, 323, 324, 325, 326
Wood Green, 43, 72, 76, 197, 198, 199, 200, 201, 202, 203, 204, 205, 206, 207, 208, 209, 360
Wood Lane, 137, 148, 153, 168, 248, 318–19; depot and power station, 57, 58, 189, 272–3, 274, 319, 324
Woodside Park, 256, 259
Woolwich, North and South, tube schemes, 28, 90
Workmen's fares, 35, 43, 45, 69, 141, 192
Wright, Whitaker, 36, 38–9

Yerkes, Charles Tyson: early career, 62; imprisoned, 62; activities in Chicago, 63–4; interest in Hampstead tube, 64–6; outing to Golders Green, 65–6; organizes District Railway takeover, 71; forms Traction Company, 71; acquires more tube companies, 72–3; alliance with Speyer and formation of UERL, 73–4; Edgware ambitions, 76; why he came to London, 81–2; promises to buy British, 82; triumph over Pierpont Morgan, 83–4; Clapham Junction and Marble Arch, 89; experiments with fireproofing, 93; experimental line 102; death, 105; payments outstanding, estate sold up, 105
York Road, 43, 108, 117, 136, 223, 279
Yorke, Lt.-Col. H. A., 68